Porosity in Carbons

Porosity in Carbons: Characterization and Applications

Edited by

John W. Patrick

Carbon Research Group, Department of Chemical Engineering
University of Technology, Loughborough,
Leicestershire
UK

Halsted Press
An imprint of John Wiley & Sons, Inc.
New York Toronto

First published in Great Britain 1995

Library of Congress Cataloging-in-Publication Data

Available upon request

ISBN 0 470 23454 7

Printed and bound in Great Britain.

Preface

The idea for this book arose from the realization that porosity in carbons meant different things to different people depending largely on the different applications of the various carbon materials. On the one hand, those involved in gas purification or respiratory protection are concerned primarily with the *micro*porosity, and at the other extreme, the user of carbon in the form of metallurgical coke is concerned with the *macro*porosity because of its influence on the mechanical properties of the coke. Between these extremes there is a range of applications which rely on different aspects of the nature of the porous structure and the characterization required reflects the particular application in mind. Clearly the characterization of the wide diversity of porous structures presents some problems. However, recent developments have produced some solutions, e.g. computerized image analysis has facilitated the measurement of pore shape and size especially at the macropore end of the size range, and highlighted others such as the difficulty of determining an accurate characterization of the pore shape and size in two-dimensional carbon–carbon composites.

The eleven chapters of this book present an analysis of the current methods of characterization and the role of various aspects of the porosity in some representative and diverse applications of carbons. The book has taken some time to pull together and the contributing authors are thanked for their forbearance and understanding and also for their wholehearted endeavour which has made this book possible.

John W. Patrick
Professor of Carbon Science
Director, Carbon Research Group
Department of Chemical Engineering
Loughborough University of Technology

List of Contributors

Professor Alan Bailey

Royal Military College of Science
Cranfield University
Shrivenham
Swindon
Wiltshire SN6 8LA
United Kingdom

Dr Jane F. Byrne

American Norit
P.O. Box 790
Marshall
Texas 75671
USA

Professor Frank J. Derbyshire

Center for Applied Energy Research
University of Kentucky
3572 Iron Works Pike
Lexington
Kentucky 40511-8433
USA

Dr Marit Jagtoyen

Center for Applied Energy Research
University of Kentucky
3572 Iron Works Pike
Lexington
Kentucky 40511-8433
USA

Brian T. Kelly (deceased)

Reactor Services
AEA Technology
NRL-S
Springfields
Nr. Preston
Lancashire
United Kingdom

Professor Harry Marsh

Department of Engineering and Energy Processes
Southern Illinois University
Carbondale
Illinois 62901
USA

Dr Timothy J. Mays

School of Materials Science
University of Bath
Claverton Down
Bath
Avon BA2 7AY
United Kingdom

Professor Brian McEnaney

School of Materials Science
University of Bath
Claverton Down
Bath
Avon BA2 7AY
United Kingdom

Dr Heinz-Jürgen Mühlen

DMT–Gesellschaft für Forschung und Prüfung mbH
DMT–Institut für Kokserzeugung und Kohlechemie
Franz-Fischer-Weg 61
D-4300 Essen 13
Germany

Dr Norman D. Parkyns

British Gas plc
Research and Technology Division
Gas Research Centre
Ashby Road
Loughborough
Leicestershire LE11 3QU
United Kingdom

Professor John W. Patrick

Carbon Research Group
Department of Chemical Engineering
University of Technology
Loughborough
Leicestershire LE11 3TU
United Kingdom

Dr David F. Quinn

Royal Military College of Canada
Department of Chemistry and Chemical Engineering
Kingston
Ontario K7K 5LO
Canada

Professor Francisco Rodríguez-Reinoso

Department de Química Inorgánica
Universidad de Alicante
Alicante
Spain

Professor Kenneth S. W. Sing

Department of Chemistry
University of Exeter
Exeter EX4 4QD
United Kingdom

Professor H. Fritz Stoeckli

Institut de Chimie
University of Neuchâtel
Switzerland

Dr Michael Thwaites

Center for Applied Energy Research
University of Kentucky
3572 Iron Works Pike
Lexington
Kentucky 40511-8433
USA

Professor Karl H. van Heek

DMT–Gesellschaft für Forschung und Prüfung mbH
DMT–Institut für Kokserzeugung und Brennstofftechnik
Franz-Fischer-Weg 61
D-45307 Essen
Germany

Dr Alan Walker

Carbon Research Group
Department of Chemical Engineering
Loughborough University of Technology
Loughborough
Leicestershire LE11 3TU
United Kingdom

Contents

1 Introductory Overview

Jane F. Byrne and Harry Marsh
Department of Engineering and Energy Processes,
Southern Illinois University,
Carbondale, Illinois 62901, USA.

1.1 Origins and structure of porosity

Introduction

Porous carbons, also called activated carbons, have been in use for thousands of years. Their application in water purification can be dated back to 2000 BC when the ancient Egyptians used charcoal to purify water for medicinal purposes.[1] The major development of activated carbon began during World War I, when hard granular activated carbon (GAC) was manufactured for use in gas masks. In the last 50 years the technology involved in active carbon manufacture has advanced. Powdered active carbon (PAC) is extensively used in water purification processes together with granulated active carbon. PAC is associated with larger pore diameters than GAC.

Activated carbons are used for gas- and liquid-phase adsorption processes. Nearly 80% of about 300000 tons per year is used for liquid-phase applications and approximately 60000 tons per year for gas-phase applications; these figures are based on worldwide usage.[2] Granulated active carbons are mainly employed for gas adsorptions and separations.

The large internal surface areas of activated carbons are the result of the activation process. Two routes are used to increase porosity: chemical and 'physical' activation. Chemical activation involves the addition of compounds such as $ZnCl_2$ and H_3PO_4 to the parent feedstock prior to carbonization. The mechanism by which the porosity increases is poorly understood: chemical activation is not now favoured by industry because of environmental considerations.

'Physical' activation refers to gasification of the carbon by gases such as carbon dioxide and steam in the range 700–1100 °C, rather than physical manipulation in the usual sense of the word. This is the most commonly used method in the production of activated carbons and possible models of the mechanism involved to create and open up porosity, in particular microporosity, are discussed below.

Activated carbons are manufactured from a wide variety of materials but those most frequently used on a commercial scale are peat, coal, lignite, wood and coconut shell.[3] However, one of the problems of using active carbons derived from agricultural by-products is the variability in feedstock resulting in limited control over extent and size distribution of resultant porosites. Therefore, in addition to carbons from natural products, other precursors are required from which to produce carbons with reproducible pore size distributions to meet specifications. Such precursors are the synthetic resins and other polymeric material such as polyvinylidene chloride, $(CH_2Cl_2)_n$. The most desirable method of producing activated carbon is by adjusting processing conditions to control precisely the properties of the end-product. This is more cost-effective than changing the precursor to suit changing requirements, but necessitates a consistent material. Variations occur in precursors in terms of structure and mineral matter content, associated with locality of growth, climate and cropping. Mineral matter is often responsible for preferential gasification during the activation process resulting in meso- and macropore channelling and pitting, and not the preferred microporosity.

Figures 1.1 and 1.2 are scanning electron micrographs of surfaces of activated carbons. Figure 1.1, at low magnification, shows that the macroporosity of the

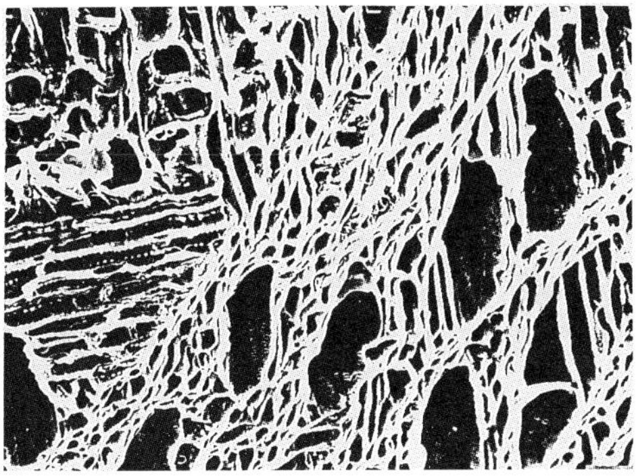

Fig. 1.1 Scanning electron micrograph of surface of wood-based activated carbon, showing macroporosity

parent wood is retained during carbonization. Figure 1.2, at a higher magnification, of a different carbon, shows small sizes of macroporosity, ~100 nm in diameter, created by gasification. Microporosity, dominantly responsible for adsorption capacity, cannot be resolved by scanning microscopy. The macroporosity of these photographs constitutes the transport facility within the carbons.

Carbon materials

Porous carbons are a non-graphitic form of carbon, characterized by internal surface areas ranging from 500 to ~3000 $m^2\ g^{-1}$. X-ray analyses show that the structure of activated carbons is composed of carbon atoms in arrangements which approximate roughly to a short-range, layered (lamellar) structure. Some of these layers are stacked only approximately parallel to each other over distances of up to a few nanometres.[4] Consequently, X-ray diffraction shows line broadening of only the principal graphite diffraction bands. This broadening is usually interpreted in terms of dimensions of a 'hypothetical' crystallite. Although the crystallite concept has uses when comparing structures in carbons, it has to be stressed that the crystallite does not exist as such within these structures. In fact, to call these non-graphitic activated carbons microcrystalline

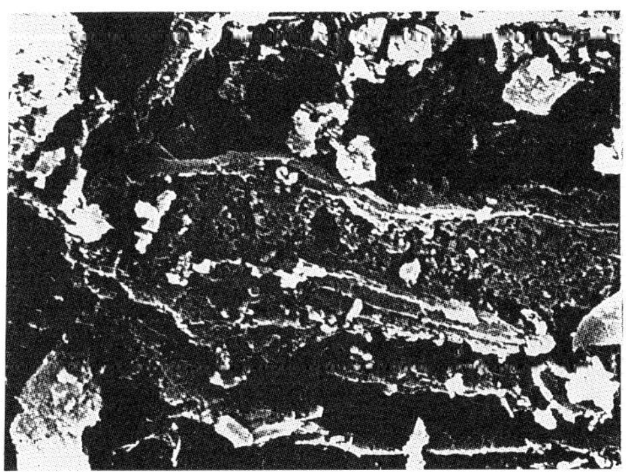

Fig. 1.2 Scanning electron micrograph of surface of commercial activated carbon showing small size of macropores, ~100 nm diameter, created by gasification during activation

introduces confusion and misconceptions. There is the need to adopt structural models which must explain the processes of pyrolysis, carbonization, activation and adsorption all of which are structure-related. Further, there is the need to have a practical understanding of the concept of surface area in porous solids. It is the space both within and between these roughly parallel layers of carbon atoms [also called lamellar constituent molecules (LCM)], and other carbon atoms not within the layers, that constitutes the microporosity which is responsible for the large internal surface areas possessed by these carbons. It is this ability to adsorb relatively large quantities of a diverse range of molecules (metal ions, gases, organic compounds, etc.) which makes porous, activated carbons so useful in purification and separation processes.

This book is concerned with porous activated carbons and the process of adsorption within the porosity of such carbons. As the understanding of structure in carbons and the relationship of structure to porosity is not straightforward, it is considered useful to set out a few definitions and terms used to describe carbon materials, and carefully evaluated by the International Union of Pure and Applied Chemistry (IUPAC).[5]

The atomic number of carbon is six and its atomic weight is 12.010 amu. Its electron configuration in the ground state is $1s^2 2s^2 2p^2$, and it has valencies of 2, 3 or 4. Carbon has the ability to bond to itself (catenation) via sp^3 (diamond-like) and sp^2 (graphite-like) hybridization. The sp^2 hybrid has three outer electrons in equal bonding orbitals which are directed in a plane at 120° to each other. The fourth electron in the third p-orbital is capable of forming π-bonds with neighbouring atoms.

Elemental carbon exists in the two crystallographic forms. There is diamond with its tetragonal bonding imparting the hardness to the crystal, and there is graphite which has a lamellar structure in which each lamella is composed of carbon atoms arranged in six-membered ring systems. Within the rings, the adjacent carbon–carbon distance is 245.6 pm. In the hexagonal form of graphite the lamellae are arranged relative to each other with an ABABAB stacking sequence in which the atoms of lamella A all lie above/below each other, and likewise with lamella B. The interlamellar disance is 335.4 pm. In the less stable rhombohedral form, the lamellae are arranged relative to each other in an ABCABCABC stacking sequence (Figure 1.3). The recently-discovered fullerenes[6] possess spherical or cylindrical structures based on the graphite lattice. By introducing five-membered ring systems into the six-membered

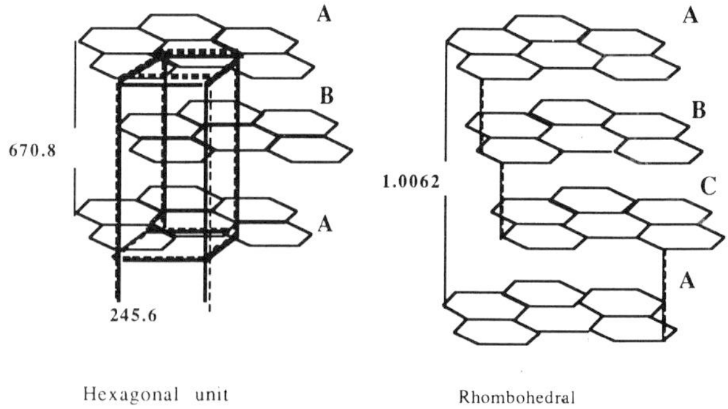

Fig. 1.3 Model of hexagonal and rhombohedral structures of graphite

systems the lamellae adopt spherical shapes with about 70 carbon atoms per sphere. Cylindrical fullerene-type structures also exist. As the base structural unit is the graphitic lamella the fullerenes are essentially a derivative of graphite.

Owing to the weak bonding between layers, graphite behaves as a lubricant. It is also able to conduct electrons (which are delocalized) within planes via the σ-bonds. However, electrons cannot pass easily between layers and hence graphite exhibits a strong degree of anisotropy.

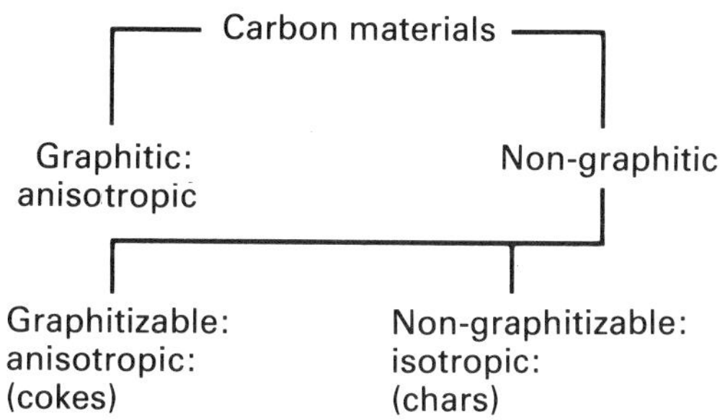

Diamond has no role in the development of activated carbons and attention here is directed to relationships between the structure of the graphite crystal and structure in carbons, i.e. the many forms of carbons which are manufactured industrially. Perfect graphite is a rare form of carbon; the majority of carbon is found in less ordered structures which are chemically heterogeneous, such as coals, cokes and chars. These types of carbon are described as *graphitic* or *non-graphitic* depending on the degree of crystallographic ordering and represent an intermediate between the living organic precursor and single crystal graphite. Graphitic carbons possess three-dimensional symmetry and non-graphitic carbons do not.

The non-graphitic carbons are divided into two further categories, the *graphitizable* and the *non-graphitizable* carbons[5,7] (Figures 1.4 and 1.5).

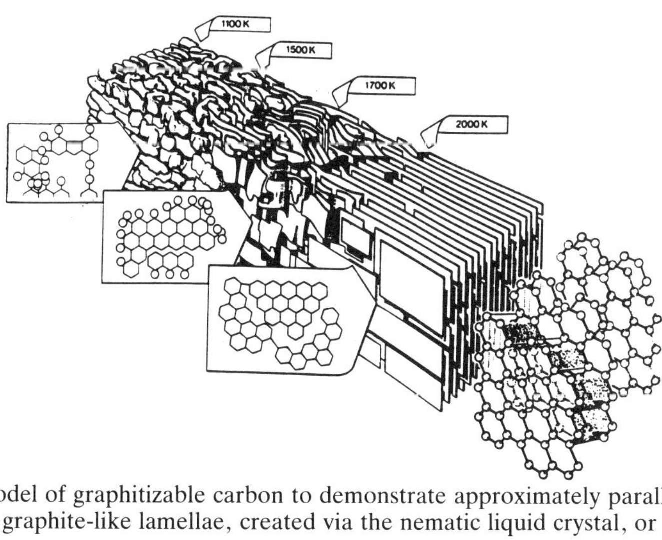

Fig. 1.4 Model of graphitizable carbon to demonstrate approximately parallel stacking of imperfect graphite-like lamellae, created via the nematic liquid crystal, or mesophase (Marsh[8])

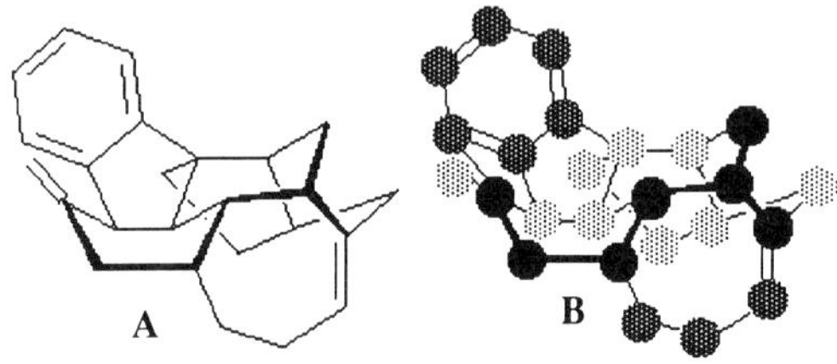

Fig. 1.5 Model of non-graphitizable carbon to show (A) the crumpled, defective nature of LCM and (B) the location of carbon atoms in three dimensions

Graphitizable carbons are non-graphitic carbons which can be converted into graphitic carbon by heat treatment.[8] The degree of graphitization depends on the heat treatment temperature (HTT) and the time allowed at temperature to anneal the structure. A non-graphitizable carbon is a non-graphitic carbon which cannot be transformed into a graphitic carbon solely by heat treatment (under inert conditions) up to temperatures of 3500 K under atmospheric or lower pressures. The graphitizable carbons, called *cokes*, are formed commercially by the pyrolysis of aromatic petroleum and coal-tar pitches. In addition, model polyaromatic compounds (Figure 1.6) and polyvinyl chloride ($C_2H_2Cl_2$) also carbonize to graphitizable, anisotropic cokes.

The crystallinity of the *coke* is established via the association, or layering together, of large lamellar polyaromatic hydrocarbon-type molecules (~1000 amu) formed during the carbonization process. This association of molecules is a new phase termed aromatic, nematic, discotic liquid crystals (or mesophase) formed from within the fluid phase of carbonization systems (Figure 1.4). The non-graphitizable carbons, called *chars*, constitute carbonization products from wood, activated carbons, brown coals, lignites and highly volatile bituminous coals and others. Such parent materials do not pass through a fluid phase during pyrolysis and carbonization. Chars have a short-range structural order, <10 nm, roughly layered, and are isotropic in all properties (Figure 1.5). *Activated carbons* are porous carbon materials, derived from chars, which have been

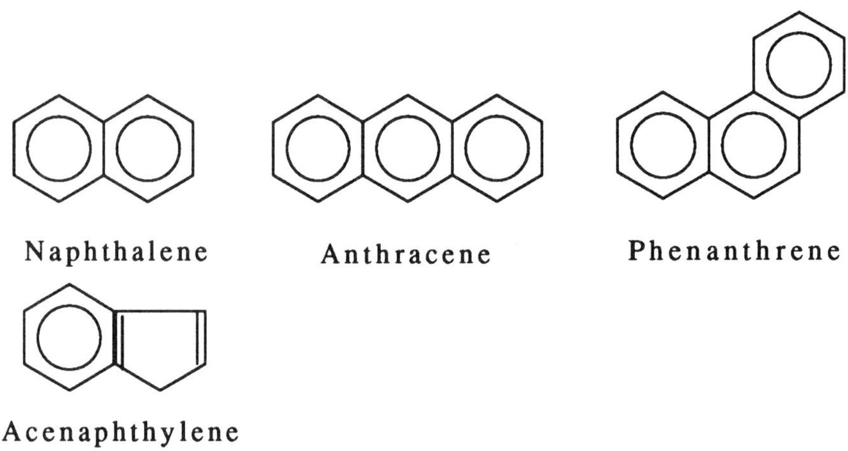

Fig. 1.6 Model aromatic compounds which carbonize to graphitizable, anisotropic cokes

partially oxidized or gasified with steam, carbon dioxide and air during or after carbonization in order to increase their porosity. Porous carbons are selected for specific properties. *Molecular sieve carbons* are created for use to separate components of mixtures, both liquids and gases. *Catalyst support carbons* are created to support metal catalyst particles, e.g. platinum, within their porosities, such systems having applications in gaseous hydrogenation reactions.

Structure in carbons

There exists today a wide variety of commercial carbons including delayed cokes from delayed coking operations using aromatic feedstocks, metallurgical cokes, activated carbons, carbon blacks, carbon fibres and polycrystalline graphites in many formats, etc. In terms of structure they are all composed of imperfect sections of graphitic lamellae. It is possible to imagine taking these perfect lamellae and 'cutting' them into sizes varying from about 1.0 nm to about 500 μm. The smaller sizes, 1.0–10 nm, can be crumpled and defects of structure can be introduced. They can be 'bonded' together to create a three-dimensional network. Such a structure, isotropic in physical properties, approximates to that of the porous, activated carbons. Because of the crumpled, defective nature of the constituent lamellar-based molecules (LCM), their packing density is low, $\sim 1.0\ g\ cm^{-3}$, and it is the spaces or voids between the LCM which constitute porosity. The graphitizable cokes are made up of LCM which are larger, less defective and less crumpled than those in the porous carbons. Accordingly, they stack parallel to each other over relatively larger dimensions of micrometres. The densities are higher, approaching $2.0\ g\ cm^{-3}$ (the density of pure graphite is $2.267\ g\ cm^{-3}$), and they have little or no porosity with sizes comparable to those of molecules, i.e. nanometres (Figure 1.4). Hence cokes are not used as parent materials for activated carbons.

These differences in packing arrangements of the LCM impart widely different properties to the carbons. In the carbon blacks the LCM are arranged parallel to the surfaces of the carbon blacks. This imparts a low porosity and graphitizability. The glassy carbons are similar to the activated carbons but have a denser packing of the LCM to create a harder carbon with a narrow size range of porosity. The pyrolytic carbons, formed from deposition from the vapour phase, have extensively developed LCM, with high density packing and hence are extremely graphitizable. This makes the point that although carbons exist with a wide range of structural, physical and chemical properties, their existence depends on the extreme flexibility of the structural associations possessed by these LCM. Further, within one class of these carbons, i.e. the activated carbons, wide variations in properties are caused by differences in the detail of size, shape, deformation and 'bonding' together of the LCM generated from the individual parent feedstocks during pyrolysis and carbonization. Extents of bonding, or cross-linkages, within the carbon are critical as they control the hardness or abradability of the carbon, particularly after the activation process when internal carbon material is removed by gasification.

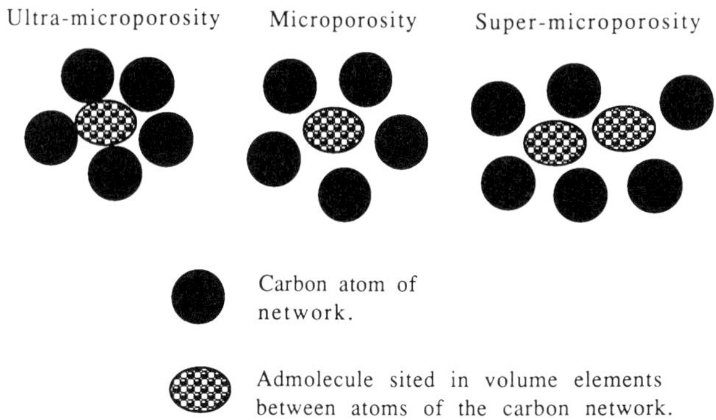

Fig. 1.7 Models of sizes of microporosity showing an admolecule of constant size in porosities (volume elements) of increasing size (over-simplified)

Porosity: definitions

Based on the experiences of adsorption chemistry, total porosity is classified into three groups. The origins and structure of porosities are elaborated upon below. IUPAC[9] classifies porosities as follows:

- Micropores width less than 2 nm
- Mesopores width between 2 and 50 nm
- Macropores width greater than 50 nm

It has also been useful to classify micropores further into ultra- (<0.5 nm width) and super- (1.0–2.0 nm) micropores, these definitions being relevant when considering adsorption behaviour (Figure 1.7). Micropores are considered as being about the size of adsorbate molecules and accommodate one, two or perhaps three molecules. Mesoporosity is wider and is characterized by hysteresis loops during adsorption and desorption at high relative pressures of adsorption. Macroporosity has little interest for the surface chemist. They are transport pores to the interior of particles and are considered as external surface. These definitions are not based on direct measurements, nor should they imply rigid size characterization. Rather, it is more meaningful to describe adsorption processes as being microporous and mesoporous in nature. Microporous behaviour occurs in terms of the ratio of admolecule size to pore size rather than on the absolute size of either. Some activated carbons contain all of these sizes of porosity associated with the botanical composition of the material[10] (Figure 1.8).

It is the space within and between the LCM, otherwise called porosity, which is the subject of this book. The word pore is derived from the Greek 'πορoσ', meaning passage, and distinguishes it from an isolated void.[11] Porosity in carbons has disadvantages in addition to advantages. In nuclear graphites, for example, porosity is undesirable as it allows for gasification within the carbon structure, resulting in corrosion and loss of mechanical strength, whereas its presence in activated carbons accounts for their usefulness.

The dominant purpose in studying physical adsorption in porous solids is to characterize the 'nature' of the porosity. The term 'nature' includes the following structural aspects of the porosity:

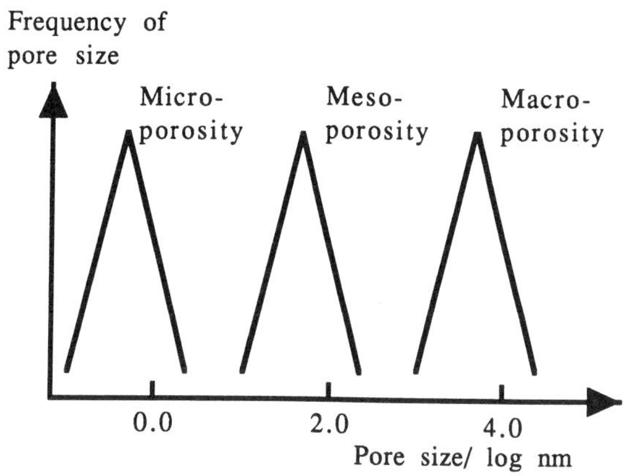

Fig. 1.8 The trimodal distribution of porosity found in carbons from natural precursors

- pore-size distribution, or potential energy distribution;
- average pore size;
- the shapes of the pores;
- the relative positions of porosities of different sizes;
- the surface chemistry of the pore;
- diffusion paths controlling rates of adsorption.

Experimental adsorption data, used to evaluate structural aspects of adsorbents, are set out as the *adsorption isotherm*, i.e. the variation of extent of adsorption (mmol g^{-1}) with equilibrium partial pressure, p/p^0 (p^0 = saturation vapour pressure), at constant temperature. The *isobar* describes the relationship between amounts adsorbed and temperature, at constant pressure, and the *isostere* describes the relationship between temperature and pressure of adsorption, at constant amounts of adsorbate on the surface (Figure 1.9).

Microporous, isotropic carbons contain porosity (volume elements) with dimensions approximating those of molecules. The IUPAC definition of microporosity restricts the term to volume elements of size <2.0 nm. Within this

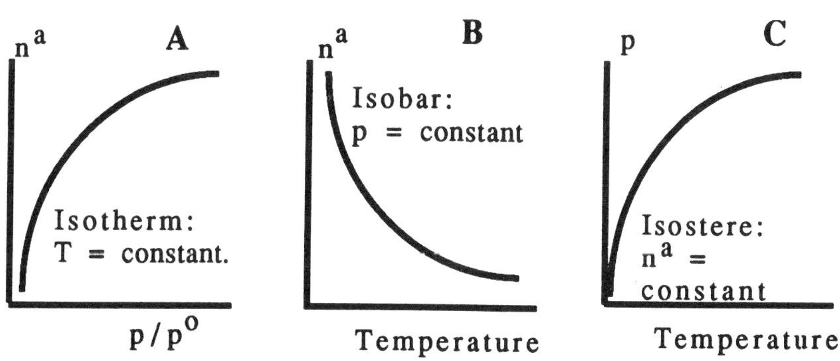

Fig. 1.9 Adsorption data presented as (A) an isotherm, (B) an isobar and (C) an isostere

defined range of porous carbons, i.e. the microporous carbons, there exists an infinite number of such carbons each with its own specific microporosity. No other solid material, apart from carbons, exhibits such diversity of properties and facility to generate microporosities. However, such diversity does itself create problems, namely, if a continuous range of properties exists, the procedural question arises as how best to create an activated carbon with the desired properties. To create a specification for the desired properties of an activated carbon is easy. How to treat and manipulate a source material to give an activated carbon with those specifications is an exercise in trial and error associated with the experience of developers.

Carbonization: creation of porosity

A detailed look at the process of pyrolysis and carbonization of a parent feedstock and the processes of generation of an isotropic porous carbon and its activation illustrates the complexity of the reactions. The parent materials themselves, wood, coal or resin, have diverse structures. Conditions of carbonization are variable, including rates of heating, final heat treatment temperature, soak time and ambient gases. The conditions of activation to develop further microporosity within carbons constitute a family of variables, including reaction temperature, reacting gas and duration and extent of activation. Each of these parameters influences the detail of the conversion of the parent polymer material to a solid carbonaceous product. It is necessary, in order to understand the limits to the characterization of microporosity, to consider how microporosity is generated during a carbonization and an activation process, and how structure in activated carbons relates to the wider family of carbon materials.

A microporous material can be generated from almost all naturally occurring organic materials, e.g. beet sugar sludges, blood, bones, cereals, coals, coconut shells, distillery waste, fish, fruit stones, kelp and seaweed, leather waste, lignite, nut shells, peat, pulp-mill waste, rice hulks, rubber waste, sawdust and wood. As organic, agricultural waste is usually available in large quantities, there are no shortages of starting materials, but the suitability of the starting material is another matter. There are the costs of purchase, collection and transportation, the proximity of manufacturing industry to the commercial buyer and, of critical importance, the suitability of the parent material to produce, consistently, an activated carbon which meets the predetermined specifications. Choice of a parent material is determined by factors other than ease of availability. There is the problem of variability within the parent material, on a day-by-day basis, or on a seasonal basis. There are no unequivocal 'ground rules' or procedures of control over these precursors whereby any required, specified carbon can easily be made from any organic precursor. In fact, the very specific needs required of activated carbons preclude the great majority of activated carbons that can be made from readily available sources. Hence the search is an extensive one. Parent materials of commercial importance represent only a small fraction of possible organic precursor materials.

These naturally occurring materials, ranging from woods, banana skins and fruit stones to coals, are three-dimensional polymeric or macromolecular networks, composed of cellulose and lignin associated with their botanical morphologies. Coals have their own aromatic, macromolecular structural net-

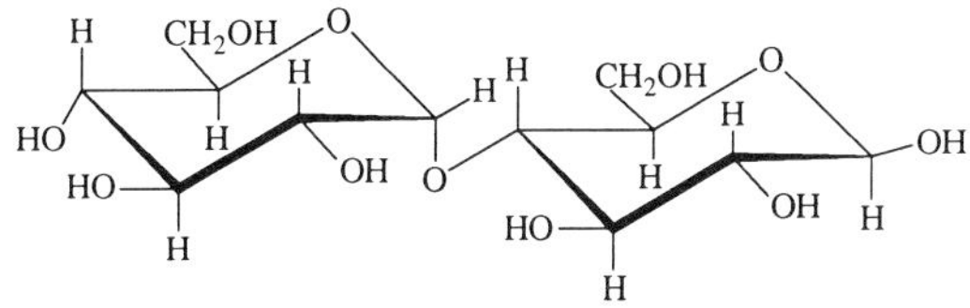

A cellulosic-type precursor (A).

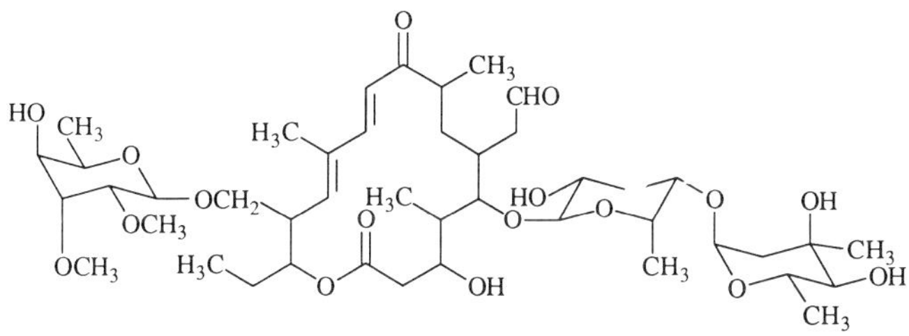

A possible structure intermediate between cellulose and an aromatic carbon (B).

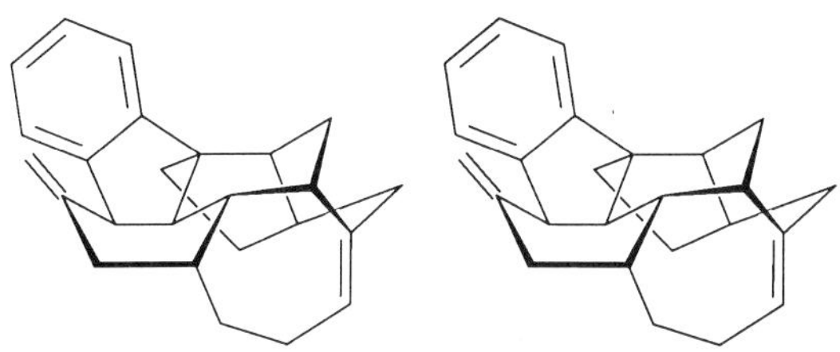

Possible, highly defective carbonaceous structures which, when interconnected, create microporosity (C).

Fig. 1.10 Structure of a cellulosic-type precursor (A) and of guideline intermediate (B) leading to an aromatic carbon material (C)

works, coal-rank related, which are a function of coalification processes. As such, they exhibit low internal surface area values and have no commercial value as adsorbents. During pyrolysis (heat treatment to <700 °C when major changes occur in structure) and carbonization (heat treatment to >700 °C when the development of aromatic carbon occurs), a process of disruption or decomposition of the parent polymeric network is initiated (Figure 1.10). Free radicals are generated and considerable strain energy is introduced into the lattice. The cellulosic structure, on pyrolysis, loses small molecules as volatiles such as water and carbon dioxide together with a complexity of aliphatic acids, carbonyls, alcohols, etc. These evolutions do not occur at a single decomposition temperature, but over a range of temperatures. As small molecules are removed from the original macromolecular network, the resultant chemically reactive lattice immediately 'zips-up' around the vacancies created by loss of the volatiles. As a result, there is no melting. Consequentially, a new lattice is continuously created with a composition of higher C/H and C/O ratios due to preferential loss of hydrogen and oxygen (Figure 1.10).

The newly-created lattice, richer in carbon than the original cellulosic lattice, possesses considerable strain energy and is not in thermodynamic equilibrium

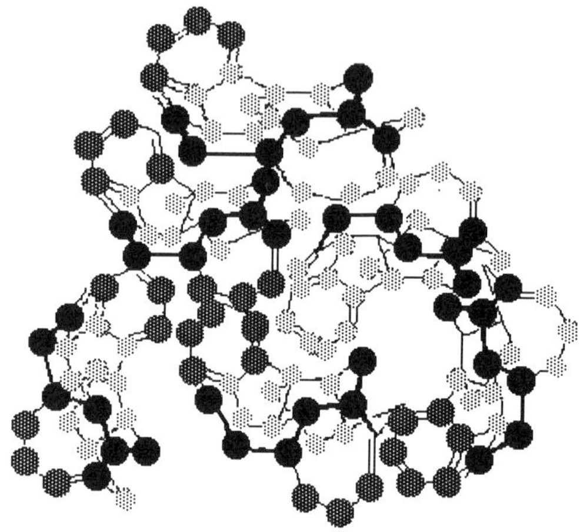

Fig. 1.11 Model of defective LCMs positioned relative to each other in three dimensions to create structure in isotropic, porous carbon

because the more stable state is the graphite-like lamellar constituent molecules (LCM). Hence, with increasing heat treatment temperature (HTT) into the carbonization range, the unstable network becomes more carbonaceous and more aromatic as the carbon atoms re-adjust their positions and approximate to the six-membered ring systems which are the building blocks of the LCM. With increasing HTT there is created a macromolecular network made up of small clusters of the six-membered ring systems in defective non-lamellar configuration, crudely approximating to the lamellae of graphitic carbon, i.e. the LCM are bonded together (Figure 1.11). Further increases in HTT cause further removal of hydrogen, oxygen, nitrogen and sulphur, causing the network to become more carbonaceous and aromatic. The defective nature of the clusters of ring systems and the random bonding together of these clusters or LCM ensures a density of the carbon considerably less than that of single-crystal graphite ($2.267\ \mathrm{g\,cm^{-3}}$), i.e. about $1.0\ \mathrm{g\,cm^{-3}}$. It is the spaces or volume elements in between these bonded LCM which constitute the microporosity of the carbon.

The chemistry of the initial macromolecular network is dependent on the parent material, e.g. different woods have different macromolecular networks. Consequently, each of the pyrolysis reactions, involving released volatiles and the establishment of a new carbon lattice, creates different carbon lattices. Accordingly, the detail of (LCM) size, the number and strength of bonding positions of the LCM and the space (porosity) between the LCM, differ from one lattice to another. Microporous, activated carbons differ in their adsorbing ability. However, to relate differences in adsorption behaviour precisely to structural parameters is a task not yet accomplished. In addition, meso- and macroporosities are created by shrinkages during carbonization.[12]

Models of porosity

Because of differences in size of these LCM and of differences in their modes of attachment to each other, the sizes and shapes of the spaces trapped between the LCM also differ. In fact, it is extremely unlikely that any two such shapes approximate closely to each other. It is possible to estimate the number of micropores in a carbon. For an activated carbon with an effective surface area of $\sim 1000\ m^2\ g^{-1}$, the monolayer coverage, $n^a{}_m$, is about $10\ mmol\ g^{-1}$. Using the Avogadro number of 6×10^{23}, and allowing for some volume filling, the estimate is that about 10^{21} micropores are capable of adsorbing an adsorbate, and all are distinguishable.

It is a misconception that porosity in carbons is a series of interconnecting tubes or slits. Nor is the tree-branch model helpful. A more realistic concept considers that the carbon atoms of solid porous carbon form a covalently bonded three-dimensional network within which imperfect lamellar-type arrangements can be recognized. This network is of low density ($\sim 1.0\ g\ cm^{-3}$), and it is the volume elements (of zero electron density) between the carbon atoms which determines microporosity. As these volume elements exist between the carbon atoms they have the sizes of atoms/molecules. Therefore, the 'structure' of the carbon atom network and the 'structure' of the microporosity are intimately related. Interconnections between the microporosity are of various sizes and, as significant, the interconnections are themselves part of the microporous system. This means that all sizes of porosities form part of both the adsorption system and of the transportation system (Figures 1.12 and 1.13). Or, a micropore could be both an adsorption site and part of a passage, 'ποροσ'.[11]

Such a model, as opposed to the tube model, is forced upon us not only by the logical analysis of mechanisms of formation of microporous carbons but also by consideration of the molecular sieve properties of carbons. Microporous carbons adsorb molecules, not just of one size but with a range of sizes. As the size of the adsorbate molecules increases, so the extents of adsorption decrease as porosity of smaller size is excluded from the process. Comparison of Figures 1.12 and 1.13 illustrates this exclusion. Data on molecular sieve effects yield a pore-size distribution. The range of this pore-size distribution is a function of the source and mode of preparation of the activated carbon. The tube model has to be

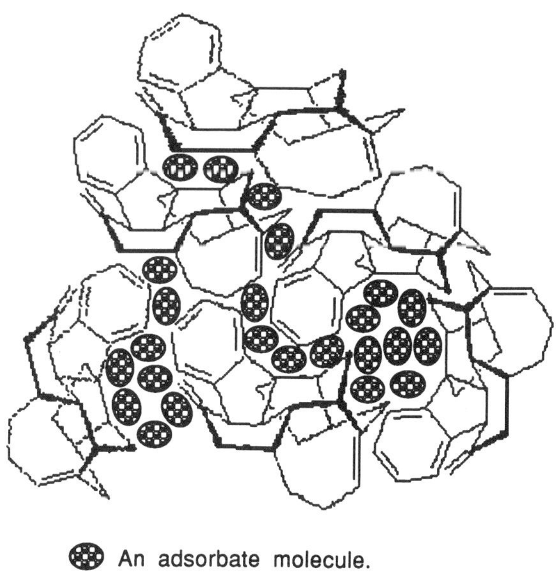

Fig. 1.12 Model of defective LCMs positioned relative to each in three dimensions to create porosity in an isotropic, porous carbon. Some micropores contain adsorbate molecules

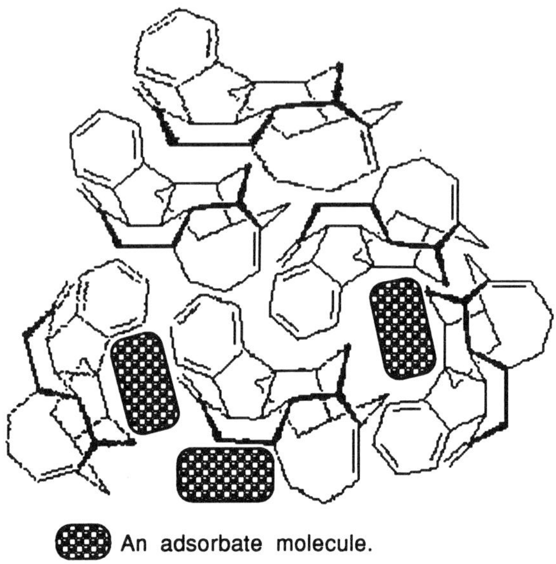

Fig. 1.13 Model showing adsorption of large admolecules in larger sized porosity (molecular sieve effects) leaving smaller sizes of porosity unoccupied. Compare with Figure 1.12

abandoned because no mechanism exists for the creation of a tubular system during pyrolysis and carbonization (Figure 1.14). An intertwined (but not interconnected) network of tubes, each tube being of constant but different diameter, would explain both the molecular sieve effects and adsorption capacity. The evidence of structure available to us, of which scanning electron microscopy and high-resolution phase contrast transmission electron microscopy are the most powerful, reveals no such tubular structures.[13] An interconnected network of tubes of various diameters allows into the microporosity only molecules which could pass through the smallest 'orifices'. Hence, molecular sieving is not acceptable to the model. A satisfactory model of the adsorption process is the filling of individual adsorption sites, volume elements or what you will, where each adsorption site has as nearest neighbours about six other sites of different size and shape. Perhaps the best way to envisage microporosity is a series of interconnecting volume elements, somewhat akin to structures in zeolites but with each cage (volume element) being of a different size and shape, and connected in three dimensions in a random way, without pattern (Figure 1.13).

Microporosity within a carbon changes with heat treatment temperature (HTT). The covalent carbon atom network is not thermally stable. With increasing HTT >700 °C, carbon atom displacement and relocation within the network occur to create the more stable graphite-like LCM. As the micropore is a description of 'where carbon atoms are not' then movement of carbon atoms changes the microporosity. In the range 600–800 °C the adsorption capacity changes little for these carbons. For carbons from polyfurfuryl alcohol, the adsorption capacity is reduced to zero at 1000 °C. Cellulose carbons maintain capacity to 1200 °C and polyvinylidene chloride to 1600 °C. These differences reflect variations in cross-link densities within the carbon atom networks.[14] The detail of these structural changes was reported by Oberlin.[15]

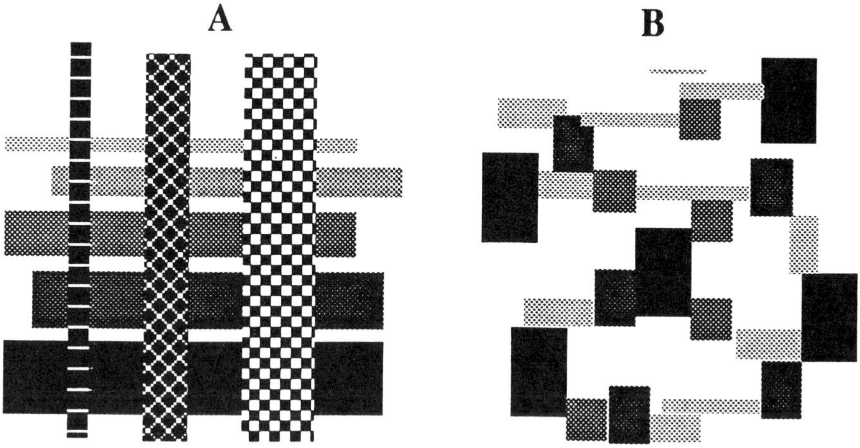

Fig. 1.14 Tube models of porosity: (A) intertwined; (B) intertwined and interconnected

The activation process

Carbons produced from parent materials such as wood, nut shells and coal are microporous. However, if the carbonization process is influenced chemically, or the resultant carbons are internally gasified by steam, carbon dioxide or air, then the microporous content is enhanced considerably.

Chemical activation is usually limited to woody precursors. This process involves carbonizing the parent material after impregnation with, e.g., phosphoric acid, zinc chloride, sulphuric acid, potassium sulphide, carbonates of alkali metals and metal chlorides.[2,16,17] The common feature of these compounds is their ability to act as dehydrating agents which influence the pyrolytic decomposition and inhibit tar formation. The temperature range for chemical activation is 400–800 °C. However, chemical activation poses problems such as lack of adequate control over porosity development. The activation agents are considered hazardous chemicals and because of pollution regulations are best avoided where possible.

Physical activation is the development of porosity by gasification with an oxidizing gas at 700–1100 °C. Commonly used gases are carbon dioxide, steam and air. They can be used singly or in combination. Several types of oven can be used in the manufacture of activated carbons, i.e. rotary kilns, multiple hearth furnaces and fluidized beds. The type of oven used determines the properties of the activated carbon produced. In general, a decreasing mesoporosity is observed in the order: fluidized bed, multiple hearth furnace, rotary kiln.[18]

During chemical activation the chemistry of pyrolysis is changed and a different, macromolecular network system is established with enhanced porosity. The chemical detail of this remains unexplored (or unexplorable). To understand the process of 'physical' activation it is helpful to use the model of porosity outlined above. This model states that carbons consist of networks of bonded carbon atoms, with a limited extent of ordered structural arrangements, and that porosity is the continuous, interconnected assembly of volume elements between the carbon atoms. It would appear that the ratio of volume element

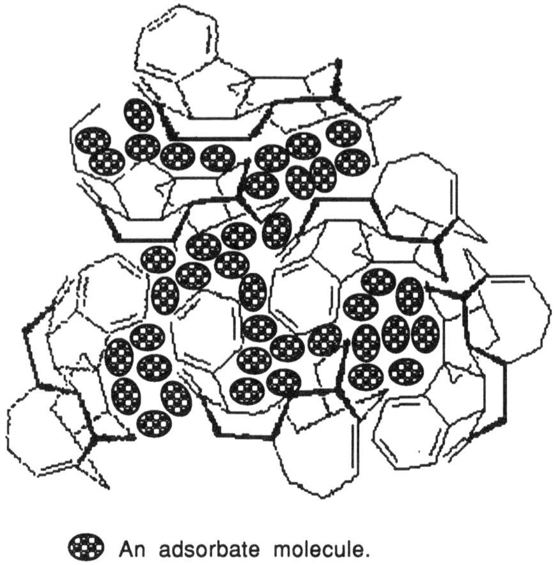

Fig. 1.15 Model illustrating the process of physical activation by selective removal of carbon atoms to enhance adsorption capacity. Compare with Figure 1.12

space to carbon atom space is not optimized during normal carbonization procedures. Hence, if carbon atoms can be removed, extremely selectively, from within the carbon particles, then the above ratio will be enhanced (Figure 1.15). Because of structural defects and strain energy differences within the defective carbon atom networks, not all carbon atoms have the same reactivity to an oxidizing gas molecule, such as water, carbon dioxide or oxygen.

When gasifying conditions are optimized in terms of oxidizing gas, temperature, flow rate and pressure, an acceptable compromise exists between 'selective' burn-off (oxidation) of carbon atoms and 'external' burn-off, which is simply removal of carbon material without any increase in porosity. This happens with fast burn-offs, e.g. as with oxygen. Thus, the extent of development of porosity and the characteristics of this porosity are a function of:

- the structure of the initial carbon;
- the presence of catalytic inorganic impurities in the carbon;
- the oxidizing gas;
- temperature of gasification;
- pressure of the gas;
- duration of gasification;
- particle size of the carbon.

Generally, the slower the reaction the more extensive is the development of porosity within the particle. The gas molecule, e.g. carbon dioxide, must penetrate into the interior of the particle to remove a carbon atom from that position. Thus, many carbon atoms within the porous carbon are bypassed during penetration without undergoing reaction or gasification. There are at least two reasons for this: the sensitivity of carbon atom 'reactivity' to relative position within the carbon network is critical; also of importance are stereo-effects, i.e. the steam, carbon dioxide or oxygen molecule needs appropriate space to orientate itself for the reaction which leads to gasification. During this process an oxygen atom has to be transferred and bonded to the carbon atom of the network, and the resultant carbon monoxide has subsequently to be 'desorbed' from the network. These are extremely complicated processes and may never be understood other than in general terms. In principle, however, it is easy to appreciate how different activation conditions influence the mechanistic detail of carbon atom removal, so creating a wider range of porosities in carbons. This is where activation becomes a matter of experience as well as understanding.

The reactions of carbon with steam and carbon dioxide to remove carbon atoms from the network within solid carbon are as follows:

$$\begin{array}{lll} C+H_2O = CO+H_2 & & \Delta H = +117\ \text{kJ mol}^{-1} \\ C+CO_2 = 2CO & & \Delta H = +159\ \text{kJ mol}^{-1} \\ C+CO_2 \leftrightarrow C\cdot(O)+CO & & \\ C\cdot(O) = C\cdot + CO & & \end{array} \qquad (1.1)$$

where () represents the chemisorbed state.

Steam is the preferred activation gas because the water molecule has smaller dimensions than the carbon dioxide molecule, and consequently the use of steam leads to:

- faster diffusion into the porous network;
- easier access into the micropores;
- a faster reaction rate, i.e. approximately three times faster than the carbon–carbon dioxide reaction at a temperature of 800 °C and a pressure of 10 kPa.[19]

The steam reaction is product-inhibited; hydrogen is strongly adsorbed on the active centres of the carbon surface and this reduces the reaction rate. In some way, these inhibition reactions also influence the selection of carbon atom removal.

$$\begin{array}{lll} C+H_2O & \leftrightarrow & C\cdot(H_2O) \\ C\cdot(H_2O) & \leftrightarrow & C\cdot(O)+H_2\uparrow \\ C+H_2 & \rightarrow & C\cdot(H_2) \\ C+C\cdot(O) & \rightarrow & C\cdot(CO) \\ C\cdot(CO) & \rightarrow & CO\uparrow \end{array} \qquad (1.2)$$

The reaction of carbon dioxide with solid carbon is known as the Boudouard reaction. The mechanism and kinetics of this reaction have been extensively studied.[20] Activation with carbon dioxide is inhibited by hydrogen and carbon monoxide. The carbon monoxide acts by increasing the rate of the back reaction $[C\cdot(O)+CO = C\cdot + CO_2]$. Rand and Marsh[21] showed that the addition of carbon monoxide to the reacting gas resulted in the development of a better microporous structure in addition to decreasing the rate of gasification. Carbon dioxide activation requires higher temperatures than steam activation and is generally used as a mixture of flue gas plus steam.

Although pore development is dependent on temperature and concentration of reactant gas, the carbon precursor also determines the characteristics of the activated carbon produced. The origin of the carbon precursor and the nature of the porosity evolved during activation has been examined by Rodríguez-Reinoso *et al.*,[22] who activated plum and peach stones under identical conditions. They observed that peach stones yielded active carbons with molecular sieving behaviour towards alcohols and a very narrow microporosity (ultramicroporosity). The microporosity and macroporosity were much more developed in carbons prepared from plum stones. Iley *et al.*[23] have also shown that active carbons from almond shells are good for the selective adsorption of gases whereas olive stones are best for adsorption from solution.

The mineral matter in parent materials used for the manufacture of activated carbons must not be assumed to be inert during the activation (physical process). Some are gasification catalysts, i.e. they promote rates of gasification of the carbon in immediate contact with the catalyst. Consequently, preferential reaction occurs at this location. The effect is to produce channelling, pitting and surface erosion as modelled in Figure 1.16. It is likely that the porosity seen in the carbon surfaces of Figure 1.2 are gasification channels or pits associated with mineral occlusions. The size of the channel or pit is a function of the size of the mineral matter particle and hence these porosities are in all sizes. The porosity of an impure activated carbon differs from that using a demineralized precursor (by treatment with hydrofluoric acid).

The adsorption process

Porosity in solids has been studied extensively by the *adsorption process*.[24] When a solid surface, whether or not it has porosity (it is not relevant), is exposed to a gas or vapour, then the concentration of the gas/vapour at the solid surface (i.e.

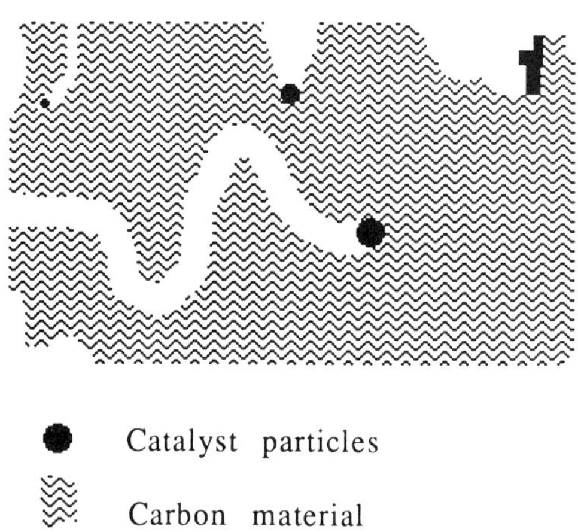

Fig. 1.16 Model of channels, pits and surface erosion created during gasification of a carbon containing catalytic mineral particles

within a molecular or atomic distance) is greater than in the gas/vapour phase. That is, an excess of molecules exists at the interface. The phenomenon which causes this excess is called 'adsorption', not to be confused with 'absorption' which involves bulk penetration into the solid.

Physical adsorption must be distinguished from chemisorption. The physical adsorption (*physisorption*) process is very general and is comparable to condensation (liquefaction). Physisorption occurs to a greater or lesser extent for all gases/vapours with all solid surfaces, increasing with increasing pressure and decreasing temperature. There are no chemical bonds formed. On the other hand, *chemisorption* involves the formation of chemical bonds. Major differences between chemisorption and physisorption are the following:

- Heats of physisorption are of the same order of magnitude as heats of liquefaction of the adsorbate; heats of chemisorption are of the same order as those of bulk-phase chemical reactions.
- Physisorption occurs in any gas–solid system provided that the temperature and pressure are suitable. Chemisorption only occurs if the gas is capable of forming a chemical bond with the solid.
- Physisorbed layers are desorbed by reducing pressure or by increasing temperature moderately. Chemisorption requires higher temperatures to break the chemical bonds.
- Physisorbed layers can be many molecules thick dependent on conditions, whereas chemisorption only occurs as a monolayer.

The following nomenclature (Figure 1.17) describes adsorption systems:

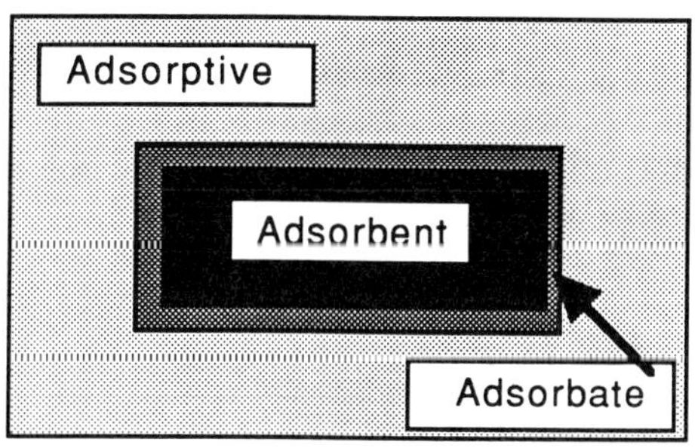

Fig. 1.17 Nomenclature for an adsorbing system

Adsorptive	the gas or vapour prior to being adsorbed.
Adsorbent	the solid material on which adsorption occurs.
Adsorbate	the gas or vapour when adsorbed. The terms admolecule and adatom are also used in this context.
Absorption	penetration by the adsorbate into the 'bulk' of the adsorbent.
Adsorption	the attachment of the adsorbate to the surface (internal and external) of the adsorbent.
Sorption	the term given when absorption and adsorption occur simultaneously, or cannot be distinguished.

Adsorption is always accompanied by a decrease in entropy and free energy of the system. Therefore, from

$$\Delta G = \Delta H - T\Delta S \tag{1.3}$$

it follows that ΔH_{ads} is negative, i.e. adsorption is always an exothermic process and that the extent of gas adsorption increases with decreasing temperature. The forces responsible for adsorption are van der Waals or London dispersion forces. These arise from induced dipole–induced dipole interactions. Dispersion forces result from rapid fluctuations in the electron density of one atom, inducing an electrical moment in a second atom.[24] However, if the adsorbate has a permanent dipole or multipole then additional interactions occur because of charge distributions induced in the adsorbent and interactions of these moments with any permanent field of the solid.

Pore-filling mechanisms

Porosity in solids is characterized from information contained within the adsorption isotherm. Unfortunately, this information is obtuse and heavily 'coded' and this renders difficult (or impossible) a single unequivocal interpretation of adsorption behaviour and modelling of structure of porosity. This accounts, in part, for the continuing debate to be found in the literature over the interpretation of adsorption phenomena. Here, it is considered helpful if a model of the pore-filling process is described as precisely as can be imagined. This could prevent serious misconceptions and incorrect modelling.

The use of the term 'surface' can itself be misleading and it is useful to say a few words about this aspect. Common phraseology is, 'adsorption took place on surfaces of the porosity.' This choice of words generates in the mind an image of a flat plane, much larger than the admolecule, on to which the admolecule is attached, either localized or with mobility. In fact, there is no surface as such in the vicinity of the admolecule when it is adsorbed within the adsorbent. In another sense, the admolecule never 'sees' a surface of the adsorbent. What the admolecule 'sees' are atoms of adsorbent separated from other adsorbent atoms, and sometimes there is just enough space (a volume element) for the admolecule to locate itself between several of these adsorbent atoms (Figures 1.12 and 1.13). For activated carbons, the containing atoms are carbon, oxygen or hydrogen or whatever is in the carbon material.

During an adsorption process, adsorbent molecules diffuse through this network of interconnected micropores. When working with 1 g of a porous adsorbent (surface area ~1000 $m^2\,g^{-1}$), with its 10^{21} micropores, the number of diffusion paths to an individual micropore must be in the region of 10^{20}. Thus, an adsorbate molecule passes from one micropore to the next, its residence time being a function of the size of the micropore ranging from about 10^{-12} s to perhaps years. The process of physical adsorption is dynamic. When a state of equilibrium is established between the adsorbed phase and the adsorbent, there is a continuous process of replacement of the adsorbed molecules within a pore. The frequency of change is of the order of 10^3–$10^{10}\,s^{-1}$. There is always diffusion into and out of the porosity.

Initially, during adsorption at low relative pressures, e.g. $1.0 \times 10^{-10}\,p/p^0$, an 'instant photograph' would reveal that the admolecules are spending most of their time in the smallest (or most energetic) volume elements of the porosity. All the other volume elements are acting as transport pores. With these very short relative residence times, amounts adsorbed in this transport porosity are negligible. The admolecule initially prefers the smallest volume element, or the volume element constructed, e.g. with oxygen atoms which offer, by reasons of polarity, an energetically more favourable site with a longer residence time. Thus, with progressive increases in relative pressure, adsorption takes place in volume elements of progressively increasing size (or decreasing energy). Some volume elements are capable of holding two or three admolecules and so establish the process of cooperative adsorption. This is an interesting phenomenon whereby the presence of one admolecule in a pore capable of holding two causes the volume element to behave as if it was a much smaller volume element giving preferred adsorption to the second admolecule. Figure 1.18 models are progressive filling of microporosity during an adsorption process to establish an adsorption isotherm.

The concept of surface area

The concept of 'surface area' of porous solids presents difficulties and is a topic for certain discussion and debate, not to mention direct argument. It may therefore be useful to set out certain guidelines as to what may be assumed to be

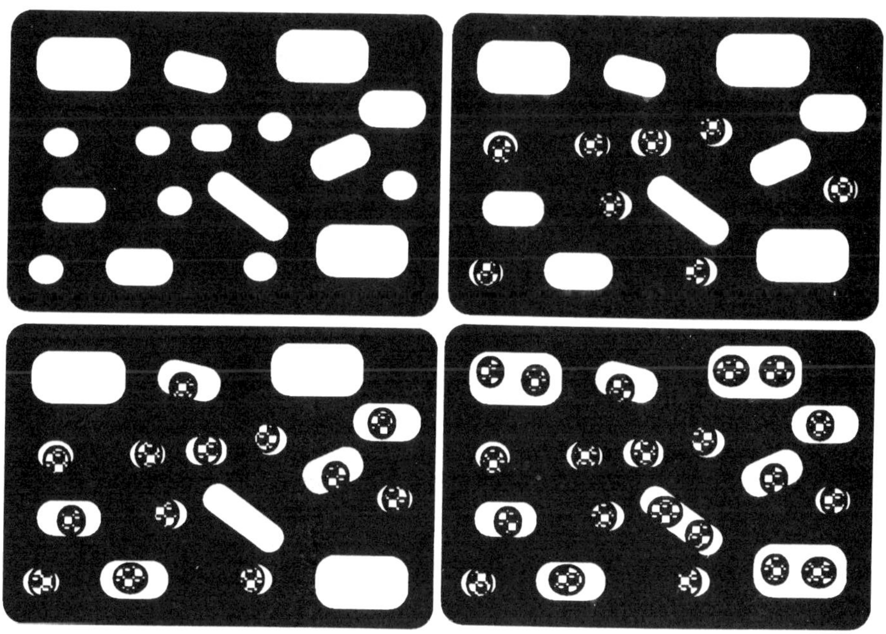

Admolecules sited in volume elements of carbon atom networks.

Fig. 1.18 Model of progressive filling of microporosity during an adsorption process to establish an adsorption isotherm

acceptable and what is not. When such properties of a solid, e.g. weight or temperature, are measured by several workers and the values disagree with each other, it is safe to assume that either equipment or operation is at fault. Values can agree to within four significant figures. However, for a porous solid, the concept of a single surface area value, easily and reproducibly determined, does not exist. There is no single experimental method that is a 'standard method' or an approved method to be used for the determination of a surface area. A porous solid will present to the experimentalist a surface area value which is a function of the methodology of the experiment. This does not imply that we have a hopeless situation on our hands, meaning that we have no means of truly characterizing these porous solids. No, the situation is in one sense very simple. The fact has to be accepted that the surface area of a porous solid is a variable parameter (Figure 1.19).

Perhaps it is best, at this stage, to have a closer look at the concept of surface area. The concept of porous solids being composed of three-dimensionally interconnected micropores (volume elements), with sizes about those of molecules, and not ordered in any way such that the approaches of crystallography are inapplicable, raises the question of where and what is the surface. The mind automatically thinks in terms of a surface as being two-dimensional and more or less flat. This is the macro-world and is not applicable to microporosity where the solid material, i.e. the carbon, has a structure of carbon atoms, arranged with such disorder that porosity is created because of imperfect structural alignments. When modelled, this type of structure has nothing within it which can be labelled as 'surface'. Thus, we go back to the description of adsorption as a

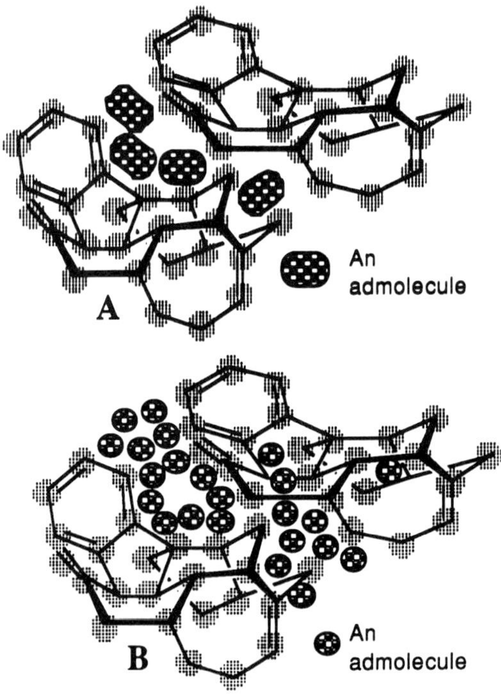

Fig. 1.19 Model of filling of volume element in a porous carbon to illustrate the problem of defining a surface area. Model A adsorbs few larger molecules. Model B adsorbs more smaller molecules

process of progressive filling of individual sites or micropores. The concept of progressive coverage of a surface is therefore inaccurate and misleading. Different adsorbates respond to the adsorption potentials of the interconnected micropores very specifically and the resultant isotherm is a description of that response. As a result, the subject of adsorption chemistry is dominated by discussion of possible interpretations of informations contained within the isotherms, interpretations of modes of pore filling by the adsorbate and the structure of the porosities. In a particular experiment or situation, the surface area of the porous solid is the surface area which is being used, or is effective, in the experiment or situation being studied.

The only methodology available to the surface chemist to measure surface area with any accuracy is in terms of the adsorption of adsorbates into the porosity of the adsorbent. However, the surface chemist, in reality, is not measuring an area, he is estimating monolayer coverage, in units of millimoles of adsorbate per gram of adsorbent. He then has to convert this number into a surface area. In one sense there is no fundamental need to do this. The internal surface area of a porous solid is a curiosity function. All that is required of a porous solid is a knowledge of how much it will adsorb of a given adsorbate under defined conditions of concentration and temperature. However, from the point of view of understanding the processes of adsorption, and of comparing carbon with carbon, the concept of surface area is established in the literature and will remain. However, it is important to recognize that it is a variable parameter of convenience, that it cannot be measured absolutely and that its numerical value is determined by what are in some cases arbitrary conversions of amounts of adsorbate into an assumed monolayer coverage.

The adsorption of nitrogen at 77 K, the determination of the isotherm and its interpretation by the BET equation to give monolayer coverage is often assumed and quoted as being the preferred method to determine the surface area of a solid. A BET surface area is considered to be correct and beyond dispute. In fact, for non-porous solids such as carbon blacks, particulate silica or titania this assumption is valid. However, for a porous solid, the nitrogen isotherm is a description of the response of the porous solid to this particular adsorbate and is no better or no worse than any other adsorbate. To understand something of how the porosity or volume elements are arranged relative to each other in a given carbon requires a knowledge of several isotherms because each adsorbate is giving its own, individual information concerning the adsorbent. Thus, there can be a surface area value derived from the adsorption of water, benzene, chloroform, carbon dioxide, neon, argon, nitrogen, etc., of which nitrogen has no special merit. The reason why the nitrogen isotherm figures so dominantly in discussions and publications is simply the ease with which the complete isotherm can be determined.

Instead of having a single surface area with which to characterize a porous solid, it is useful to distinguish between three definitions of surface area. The reasons for this are associated with the practicalities of usage of activated carbons. First, the *geometric surface area* (GSA) of a particle is the geometric shape area and does not include meso- and microporosity. Second, there is the generally used *total surface area* (TSA) concept. This term does not have a precise definition. The TSA is considered to be the maximum surface area which a porous solid can exhibit, without complications of multilayer adsorption or activated diffusion effects. In one sense, it is the envisaged internal surface area, i.e. envisaged as the area of projection of a *monolayer* of adsorbate situated in the porosity. The concept of the TSA as being the summation of the geometric areas of the micropores or volume elements does not apply. For a porous carbon made up of 1 nm wide isometric micropores, the geometric area is in the region of $6 \times 10^{-18} \times 10^{21}$ m^2 g^{-1}, i.e. about 6000 m^2 g^{-1}. This is an unacceptably high number. The TSA can also be considered as that area which is available to a reacting gas molecule, e.g. carbon dioxide, steam or oxygen, for reaction at temperatures, 300–1000 °C, to remove a carbon atom within the porosity. The TSA is then used as a normalizing function to compare the reactivities of different carbon materials to oxidizing gases, the reactivity having units of g m^{-2} (TSA) s^{-1}. The experimental determination of TSA is still a matter for discussion. A practical answer to the problem is to have carbon dioxide as an adsorbate at operating temperatures of 273–293 K, and using the Dubinin–Radushkevich (DR) equation (page 33) to interpret the isotherm. This approach provides a value for micropore volume that converts to a surface area which, in terms of internal consistency between data and general acceptability, appears to be a measure of TSA as defined above. This approach is used currently until something better can be identified.

The third definition of a surface area is the *effective surface area* (ESA), envisaged as that area which the material appears to possess, as measured from adsorption isotherms using a defined adsorbate. As examples, for a given carbon, the TSA can be envisaged as being 1000 m^2 g^{-1}. The nitrogen isotherm predicts a surface area of 5000 m^2 g^{-1}. It is not relevant to debate the structural validity which facilitates a value of 5000 m^2 g^{-1}. There is no method of 'correcting' the isotherm data or its processing to obtain a more acceptable value. It has to be accepted, without debate, that the material is behaving as though it was an adsorbent with this value of surface area, i.e. its effective surface area is 5000 m^2 g^{-1}. The debate has to centre around the problem of explaining how the material behaves in this way.

Rates of adsorption

Studies of the chemistry of adsorption are dominated by discussions of the adsorption isotherm and its relationship to structures containing porosity. The adsorption isotherm is a description of equilibrium states of an adsorbate within an adsorbent. Rarely, in the literature, are there discussions of rates at which equilibria are established. In industrial usage of adsorbents the following aspects are crucial:

- extents of adsorption of an adsorbate usually at low levels of relative concentration;
- the retentivity of the adsorbate, i.e. the ability of the adsorbent to 'hold on to' the adsorbate when the relative concentration of the latter is significantly and rapidly lowered;
- the *rate* at which the adsorbent adsorbs the adsorbate.

In molecular sieve carbons, as used, e.g., with oxygen–nitrogen separations, rates of adsorption–desorption are critical to the acceptability of a carbon. Rates of adsorption–desorption cannot be predicted from the adsorption isotherm and

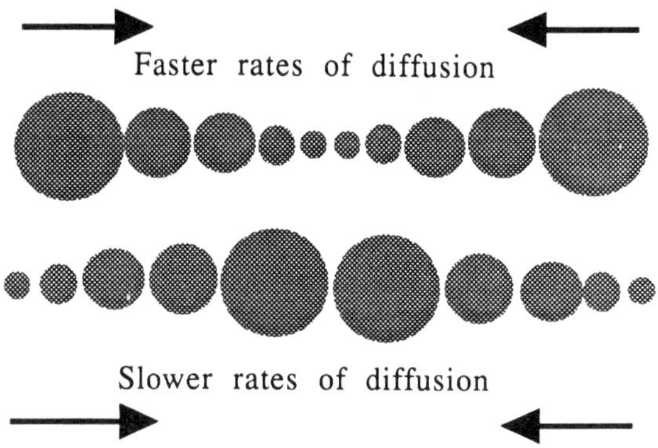

Fig. 1.20 Models to illustrate how different relative positions of volume elements of porosity influence rates of adsorption

must be determined directly. The adsorption isotherm is a description of the distribution function of adsorption energies (approximating to a pore-size distribution). The isotherm has nothing to say about positions relative to each other of the different volume elements of the porosities. It is the way in which the different volume elements are connected relative to each other which controls the rate of diffusion of an adsorbate into the adsorbent. This is a property which is extremely difficult to control and design. Figure 1.20 models, in a simple way, this aspect of rates of adsorption versus structure of porosity. In the upper series of pores, adsorption processes occur via larger pores into smaller pores. In the lower series, adsorption occurs via smaller pores into larger pores. Although this model is too simple, it makes the point that the relative positions of pore sizes influence rates of adsorption (diffusion). Further, the relative positions of the porosity (volume elements) are not randomly arranged, one relative to the other. Otherwise, although the isotherms of such materials would be identical, so also would be diffusion properties of adsorbate into the adsorbent. The differences in the way the volume elements are located relative to each other within the carbon must therefore be a function of the properties of the parent material and subsequent treatments. Hence, we can begin to appreciate the fact that adequate activated carbons are not generated, regardless, from any feedstock. Byrne[25] studied a series of microporous carbons of varied origins. Using equilibrium adsorption data and rate data from flow microcalorimetry, no correlations were found between size-related parameters and diffusion data.

Problems and possibilities

The subject area of the analysis and use of porous solids falls into two categories. First, there are the fundamentals of adsorption science which search for an understanding of what happens to an adsorbate molecule when it is taken up by an adsorbent, and seek to establish what the effects of changing the molecular size, shape and polarity of the adsorbate and also the size, structure and surface chemistry of the adsorbent are. Second, there are the applications of porous carbons to industrial processes of purification and separation. A feature of this subject area is to relate the first category to the second.

A problem area centres around this concept of surface area and its usefulness to characterize carbons for industrial use. It is inappropriate to ask for the correct surface area. It is inappropriate to ask for the best method of analysis of porosity. The ethos of this subject area has to centre around the attitude that no direct, unequivocal method of characterization exists. The adsorption isotherm, in its own right, offers much information about the structures of porosity and states of adsorbate within this porosity. Much of the information is 'veiled' and reliance is placed on adsorption equations (models) to pull the veil aside. The models are simply 'best guesses' of what is happening in the system. When admolecules are used as probes, the system responds uniquely to each probe and the answers differ accordingly. This is not to state that one method is wrong and another is right, or that one method is fundamentally more correct than another. It is that each approach gives partial information or characterization of porosity. The solution to the problem is to integrate the answers provided from as many experimental techniques as are available and to create a model which accommodates everything, consistently. This is expensive and takes time.

Relating theoretical studies to the industrial scene, where an activated carbon has to work for its living, is another problem area. As indicated above, a tailor-made carbon to meet tight specifications has to be searched for and involves scanning through a large number of samples. Short-circuits are necessary to limit possible experimentations or tests. One approach is to test the sample only in terms of what it is intended to do. If it is required to adsorb trinitrophenol from an alcohol solution, then this becomes the test. To adsorb nitrogen at 77 K from the vapour phase, to characterize porosites fully, is not helpful. The isotherm describes only how nitrogen behaves in porosity at 77 K, and this may have limited relevance to adsorption of phenols from the liquid phase.

It was noted by Freeman *et al.*[26] and Marsh[27] that an experimental adsorption isotherm is a very specific curve. It is capable of being linearized by the three principal adsorption equations (see below). However, if a random curve is drawn by hand, and it looks like an isotherm, it will not linearize using Langmuir coordinates, etc. (Figure 1.21). Curve B is, e.g., an experimental isotherm. It is linearized by the Langmuir equation. Curves A and C are, e.g., hand-drawn isotherms and do not linearize. The Langmuir equation is therefore something more than a linearizing device. So, what is so specific about the experimental isotherm? It is that the isotherm is telling us how many pores it is filling at each level of site energy or adsorption potential. The total isotherm is a description of

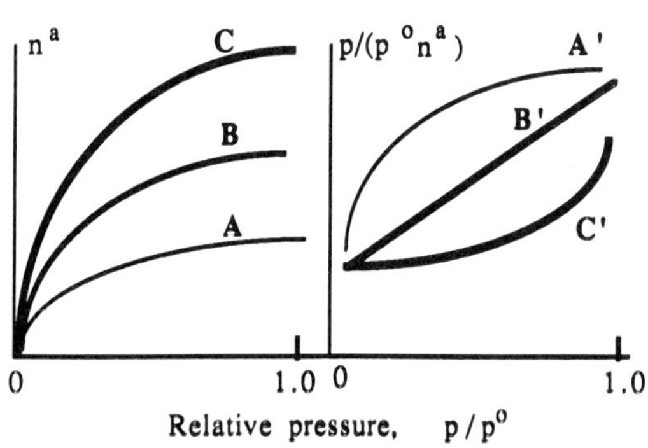

Fig. 1.21 Experimental and hand-drawn isotherms to illustrate the specific information contained within an isotherm

the distribution of adsorption potential within the carbon. The experimental isotherm, when linearized, is stating that the distribution of adsorption potential, of about 10^{21} sites per gram, is describable in terms of a mathematical equation of statistical distributions, e.g. a Rayleigh or Gaussian distribution.[28,29] Should the distribution become distorted or skewed during processes of manufacture and cease to follow recognized distribution functions, then linearization does not occur. This is simply demonstrated by measuring the adsorption isotherm not of a single carbon, but of a mixture of carbons, say 25 wt % of each of four different carbons. This isotherm defies analysis. The main causes of distortion or skewing are as follows:

- *Activated diffusion effects*, where rates of adsorption into the smallest sizes of porosity are so slow that equilibrium is not established within the time scale of the experiment.
- *Molecular sieve effects*, where adsorption into the smaller sizes of porosity is excluded because the adsorbate molecule is too large to enter into these smaller volume elements.
- *Pore-filling effects*, where the size of a pore (volume element) is such that not only is there adsorption on the 'walls' of the pore but sufficient space remains for further adsorption within the volume of the pore. Thus, adsorption occurs in excess of that predicted by the equations. Such processes account for reported high surface areas, often not believed, in excess of 2000 $m^2\ g^{-1}$.
- *Process effects*, where, owing to heat treatment or gasification, or chemical treatments, certain pore sizes are closed off (annealed) or others are created in numbers in excess of mathematical statistical requirements in terms of size or potential energy.

Not all adsorbates respond to these structural variations equally. Some are more useful than others. This is a major source of disparity in result and conclusion. Such disparity should never be a source of argument of right versus wrong. It has to be accepted, perhaps grudgingly, that a porous carbon will never invest all of its structural secrets in a single isotherm.

This point is well illustrated by consideration of Figure 1.22, which compares

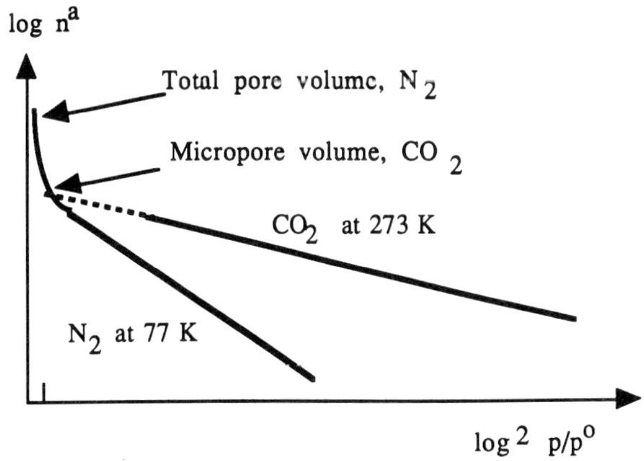

Fig. 1.22 Comparison of isotherms of nitrogen at 77 K and carbon dioxide at 273 K, on the same porous carbon, in DR coordinates to illustrate differences in adsorption behaviour

the adsorption behaviour of nitrogen at 77 K with that of carbon dioxide at 273 K, for the same microporous adsorbent. The adsorption isotherms are plotted in DR coordinates. The mechanisms of filling of porosity by the two adsorbates are different. At the lowest of relative pressures carbon dioxide is adsorbed to a larger extent than is nitrogen, within the microporosity. The larger quadrupole moment with the carbon dioxide molecule, compared with nitrogen, accounts for this behaviour. At the highest of relative pressures, nitrogen volume-fills the mesoporosity such that the extent now exceeds that of carbon dioxide. Experimental conditions obviously influence significantly how the adsorbent responds to the adsorbate. In fact, two different 'messages' are given by the carbon. The linear portions of the DR plots, at the intercept of $\log^2(p/p^0) = 0$, provide estimates of micropore volume (convertible into a surface area value) which correspond closely to each other. The difference in gradients between the two isotherms is due to differences in β, which is a shifting factor allowing superimposition of characteristic curves of different adsorbates on the same adsorbent. The curved section of the nitrogen isotherm, in the higher relative pressure region, is usually interpreted by the BET equation. The resultant, calculated surface area will be higher than that derived from the DR plot owing, of course, to the multilayers of volume filling. The nitrogen isotherm, according to BET, is thus providing a measure of the *effective surface area* (ESA). It is inappropriate to say that one value is correct and that the other is incorrect. Such an approach is not helpful. Rather, both approaches are truly valid and values are chosen according to what is required to be measured. The surface area derived from the carbon dioxide isotherm is currently a method to estimate the *total surface area* (TSA).[27–30]

1.2 Methods for measuring porosity and surface area

A range of techniques is available for estimating porosity and surface area. Comparisons are made between results obtained from isotherms and those from other methods.

Physical adsorption of gases

Adsorption isotherms

To quantify the adsorption process, extents of adsorption (mmol g^{-1}) are related to the equilibrium partial pressure p/p^0 at constant temperature to create the *isotherm*. Adsorption isotherms are studied to obtain information about the following:

- estimates of surface areas;
- estimates of pore volumes in the various porosities present, i.e. pore-size of potential energy distributions;

- assessments of the surface chemistry of the adsorbent;
- the fundamentals of adsorption processes, i.e. the nature of the adsorbed phase;
- assessments of the efficiency of industrial carbons employed in separation/purification techniques.

Interpretation of isotherms is not straightforward and this causes considerable debate among surface chemists. Adsorption processes in microporosity are the most difficult to describe accurately. Relevant theories are discussed more fully in the following chapters. The adsorption processes occurring within mesopores are more easily understood. Macroporosity behaves in the same way as an open surface to adsorption, and accounts for <1 % of the adsorption processes within microporous carbons.

Figure 1.23 sets out the six major classes of isotherm shape that are obtained from adsorption experiments. A knowledge of adsorption mechanisms in different sizes of porosity is necessary to explain the isotherm shapes.

Type I isotherms are typical of microporous solids in that micropore filling occurs significantly at relatively low partial pressures $<0.1\, p/p^0$, the adsorption process being complete at $\sim 0.5\, p/p^0$. Examples include the adsorption of nitrogen on microporous carbon at 77 K and of ammonia on charcoal at 273 K.

Type II isotherms describe physical adsorption of gases by non-porous solids. Monolayer coverage is succeeded by multilayer adsorption at higher p/p^0 values.

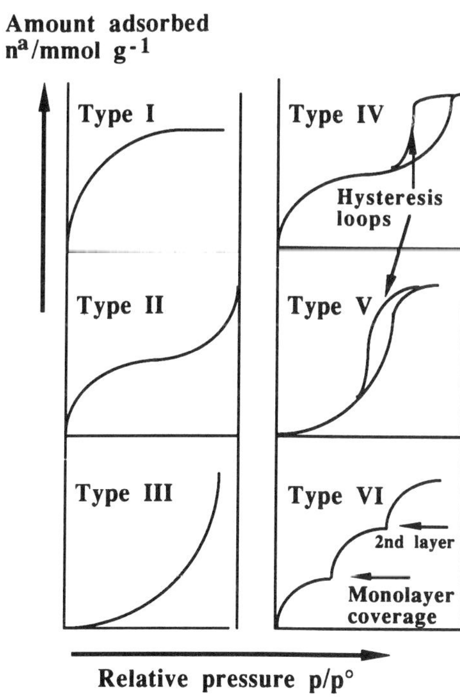

Fig. 1.23 Classification of isotherm shapes into six principal classes (Gregg and Sing[24])

Type II isotherms can also be obtained from carbons with mixed micro- and mesoporosity.

Type III and Type V isotherms are convex towards the relative pressure axis. These isotherms are characteristic of weak gas–solid interactions.[31] Type III isotherms originate from both non-porous and microporous solids and Type V isotherms from microporous or mesoporous solids. The weakness of the adsorbent–adsorbate interactions causes the uptakes at low relative pressures to be small; however, once a molecule has become adsorbed at a primary site, the adsorbate–adsorbate forces promote further adsorption in a cooperative process, which is known as the cluster theory, e.g. the adsorption of water molecules on oxides of carbon surfaces. An example is the adsorption of water vapour on graphitized carbon black.

Type IV isotherms possess a hysteresis loop, the shape of which varies from one adsorption system to another. Hysteresis loops are associated with mesoporous solids, where capillary condensation occurs.[9]

The Type VI isotherm is a stepwise isotherm and represents complete formation of successive monomolecular layers. Halsey[32] proposed that stepwise isotherms arise from extremely homogeneous, non-porous surfaces and the step height corresponds to monolayer capacity. An example is the adsorption of krypton on carbon black (graphitized at 3000 K) at 90 K.

Adsorption equations

A significant literature exists describing mathematical models to interpret shapes of isotherms obtained from porous and non-porous solids. Usually these models ignore structural features of the adsorbent, except in the most general of ways, e.g. distributions of adsorption energy (microporosity) or pore width (mesoporosity). Interpretations of isotherms from non-porous materials are relatively straightforward. Information concerning effective surface areas, pore-size distributions, micropore volumes, etc., is incorporated within the isotherm. Extracting this information is a matter for debate as direct experimental evidence is limited. Hence all equations are based on assumptions placed into models and then matched against experimental isotherms. This section is not an attempt to review exhaustively all the theories and models available for isotherm interpretation; rather, the most common are presented and commented upon.

The term 'surface area' does not describe a carbon unequivocally. Carbons with similar effective surface areas (ESA) can be very different from each other structurally. Further, values obtained using one adsorbate at a certain temperature can differ when other adsorbates at other temperatures are used. When quoting surface areas it is important to qualify them by including the technique, adsorbate and temperature used, and to state that these are effective surface areas which describe how an adsorbent is responding to a particular adsorbate.

Langmuir equation

Surface areas can be obtained from Type I (Figure 1.23) isotherms by the Langmuir equation[24,33]

$$\frac{p}{V} = \frac{p}{V_{\mathrm{m}}} + \frac{1}{bV_{\mathrm{m}}} \tag{1.4}$$

where

V = equilibrium amount (mmol g^{-1}) of gas adsorbed per unit mass of adsorbent at relative pressure p/p^0;
V_m = amount of gas required for monolayer coverage of adsorbent (mmol g^{-1});
b = constant, dependent on temperature but independent of surface coverage, and describing in some way the energetics of the surface.

Surface areas are obtained using the relationship

$$S = V_m L \sigma \quad (1.5)$$

where

S = surface area (m^2 g^{-1});
L = Avogadro's number (molecules mol^{-1});
σ = projected area of adsorbate molecule (m^2 $molecule^{-1}$), as in Figure 1.24.

The Langmuir equation is based on the following assumptions:

- only monolayer adsorption can occur;
- adsorption is localized, with no adsorbate–adsorbate interactions;
- the heat of adsorption is independent of surface coverage, i.e. the adsorbent has a homogeneous surface.

It is certain that these assumptions are invalid for any carbon. However, the equation is applicable to many isotherms and is therefore useful.

BET equation

The Brunauer, Emmett and Teller (BET) equation[34] was devised to improve the Langmuir model to account for multilayer adsorption.[24] Type II isotherms

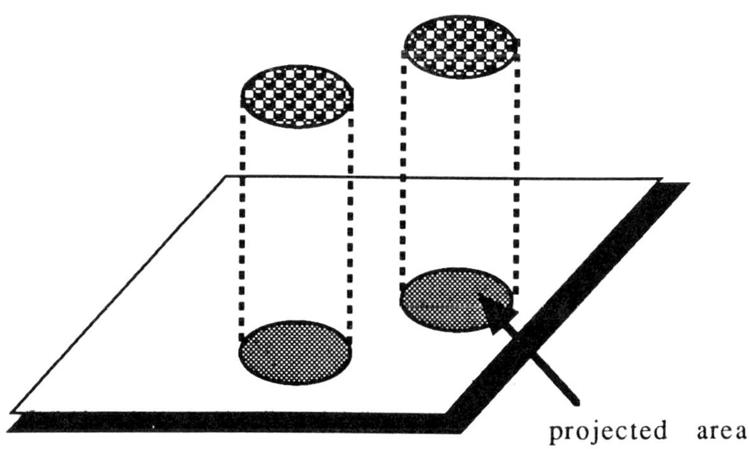

Fig. 1.24 Projection of 'area' of an admolecule to the 'surface' of adsorbent (modelled)

(Figure 1.23) describe unrestricted multilayer formation, whereas Type IV isotherms describe restricted multilayer formation.

The equation (for unrestricted multilayer formation) is as below. Equations describing restricted adsorption were discussed by Gregg and Sing.[24]

$$\frac{p}{V(p^0-p)} = \frac{1}{V_{\mathrm{m}}c} + \frac{c-1}{V_{\mathrm{m}}c} \cdot \frac{p}{p^0} \tag{1.6}$$

where

p^0 = saturation vapour pressure;
c = $\exp[(\Delta H_{\mathrm{A}} - \Delta H_{\mathrm{L}})/RT]$;
ΔH_{A} = heat of adsorption;
ΔH_{L} = heat of liquefaction.

The BET equation is widely used to interpret isotherms obtained using nitrogen at 77 K. Because the BET equation has such wide applications, researchers tend to use it indiscriminately; hence, as indicated above, carbons have been reported with surface areas as high as 5000 $m^2\,g^{-1}$. These values are misleading as the calculated surface area for an extended graphite layer, counting both sides, is ~1800 $m^2\,g^{-1}$. However, the use of the term 'effective surface area' can be applied and the values obtained indicate adsorption capacity and therefore provide useful information. The basic problem is that the modelling used for the BET equation does not describe the process of adsorption in such porous carbons.

t and α_s plots

The *t* plot was devised to detect deviations of experimental isotherms from a *standard* isotherm by testing whether the isotherm under examination was superimposable on the standard isotherm by altering the scale of ordinates[9] (Figure 1.25). The concept of a standard isotherm implies that solid surfaces,

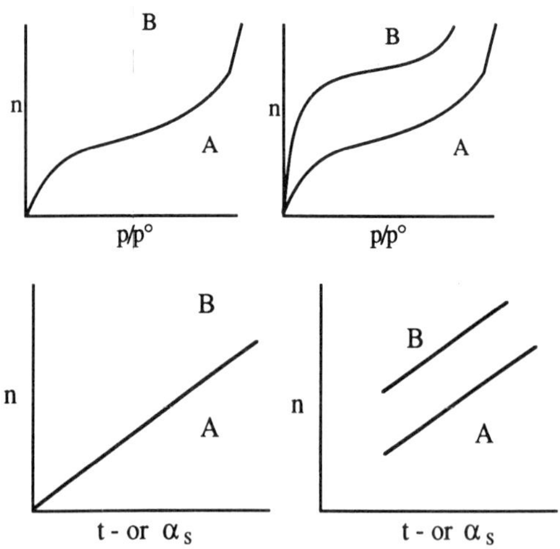

Fig. 1.25 Original isotherms and derived *t* and α_s plots

when free of the complications of adsorption in porosity, adsorb to equal extents of surface coverage at constant relative pressure. Many solids are sufficiently free from severe conditions of surface heterogeneity to make this assumption valid. The basis for the t plot is that the adsorption of nitrogen at 77 K on non-porous materials may be described by a single standard isotherm (p/p^0 plotted against number of adsorbed layers), such that a plot of amount adsorbed against the thickness (t) of these statistical layers is linear, passing through the origin. The isotherm under test is re-drawn as a curve of amount adsorbed plotted against t, rather than p/p^0, and compared with the standard t curve. If both isotherms are identical, the t plot is a straight line, as described above, with a slope equal to V_m/s, where s is the thickness of a single molecular layer. Deviations are introduced into t plots by the presence of micro- and mesoporosity within carbons.

Mesoporosity, in which capillary condensation (hysteresis) occurs, shows an upward deviation in the t plot at p/p^0 after completion of micropore filling. Micropores cause enhanced adsorption in the low p/p^0 region and this is reflected as a downward distortion. Sing *et al.*[9] pointed out that in comparing isotherm shapes, the number of molecular layers was irrelevant and that it was sufficient to normalize isotherms by taking the amount adsorbed at some fixed relative pressure ($p/p^0 = 0.4$). This value is obtained by dividing the amount obtained at a relative pressure p by the amount adsorbed at $p/p^0 = 0.4$. The advantage of the α_s method is that it can be used for adsorbates other than nitrogen and for both Type III and Type II isotherms.[35]

Dubinin–Radushkevich (DR) equation

This equation differs from the Langmuir and BET equations in that it is not based on a model process to describe the physical adsorption of gases. Rather, it is based on considerations of energies of adsorption.[11] Dubinin[36] found empirically that characteristic curves for adsorption of many microporous carbons could be linearized using the DR equation

$$W = W_0 \exp(-\varepsilon/E)^2 \quad (1.7a)$$

(terms defined after equation 1.7b). The situation is similar to that explained in Figure 1.21, that is, the adsorption isotherm is a very specifically shaped curve. Hand-drawn curves (looking superficially like isotherms) again are not linearized using DR coordinates, i.e. a plot of $\log n^a$ versus $\log^2 (p/p^0)$. The value of the DR equation is the realization that the characteristic curve of adsorption is a measure of the pore-size (or pore-energy) distribution. If these distributions follow equations for natural phenomena, Rayleigh, Gaussian, Lorentzian or log-normal, etc., then the mathematical realization of a linear plot of $\log n^a$ versus $\log^2 (p/p^0)$ is predictable.[29,37] As indicated above, the presence of about 10^{21} adsorption sites is a good statistical number to work with and hence should give smooth pore-size distribution curves. Hence the DR approach is dominantly mathematical descriptions of energies into which the adsorption process fits, in a secondary sort of way.

Dubinin[36] recognized that the adsorption energy for micropores is much higher than for supermicropores due to enhanced adsorption effects in microporosity. He proposed that micro- and supermicropores filled differently according to their respective adsorption force fields and the interaction between the adsorbed molecules. This mechanism is known as TVFM (the volume filling of micropores). Dubinin used a slit-shaped pore model as the basis for his theory. He also assumed that the distribution of pore sizes was heterogeneous and followed a Gaussian distribution.

Evidence to support Dubinin's theory was obtained from active carbons, which at low levels of burn-off have a homogeneous micropore distribution.[38] Small-angle X-ray scattering and heats of immersion[39] using adsorbates of different molecular sizes gave comparable results for micropore sizes. The derivation of the DR equation has been reported in the literature.[36,40] The equation can be written as

$$W = W_0 \exp\left[\frac{-BT}{\beta} \cdot \log^2\left(\frac{p^0}{p}\right)\right] \tag{1.7b}$$

where

W = volume of adsorbate filling micropore ($cm^3\ g^{-1}$) at relative pressure p/p^0 and temperature T (K);
W_0 = total volume of micropores ($cm^3\ g^{-1}$);
p^0 = saturation vapour pressure;
β = adsorbate affinity coefficient (C_6H_6 is employed as the standard vapour, where $\beta_{C_6H_6} = 1$);
B = constant, related to the adsorption potential of the micropores.

A plot of log W against $\log^2 (p^0/p)$ should be rectilinear. Figure 1.26 contains

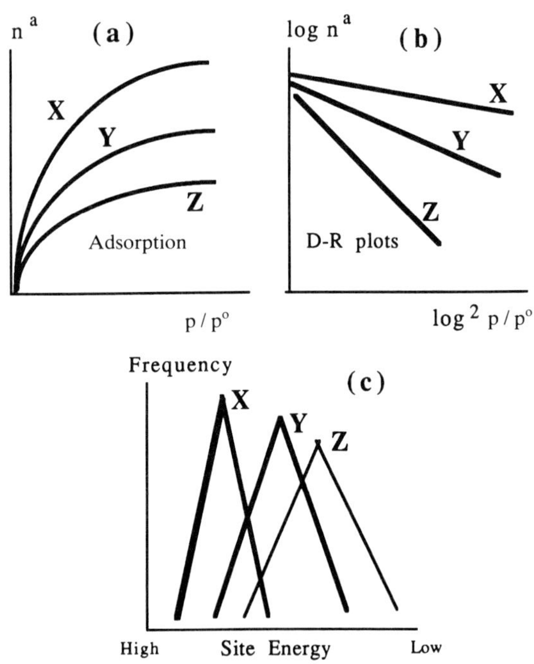

Fig. 1.26 Adsorption data as presented (a) as an isotherm, (b) in DR coordinates and (c) as a distribution of adsorption energy

plots of adsorption data, as isotherms, in DR coordinates and as distributions of adsorption energy. The micropore volume is taken from where the line intercepts the $x = 0$ axis. The parameter B can be obtained from the slope and is referred to as the structural constant of the adsorbent. The value of B is a qualitative measurement of the relative average micropore sizes of the adsorbent. The steeper the gradient, the larger the average micropore size (i.e. a wider distribution). Hence the following information can be obtained by interpreting an adsorption isotherm using the DR equation

- extents of micropore volume capacity;
- relative pore-size distributions;
- deviations of adsorption energies from a Gaussian or related type of distribution.

Langmuir and BET equations predict values of V_m, the number of mmol g^{-1} required to establish a monolayer. Conversion into a surface area value ($m^2\,g^{-1}$) requires a knowledge of the projected area of an admolecule. This is usually calculated from bulk densities of liquid adsorbates. Hence there is an estimation here. The DR equation predicts values of micropore volume (W_0). To convert this number into a surface area value requires a knowledge of the density of the adsorbed phase (presumably a liquid) in order to estimate the projected area. Equally, an estimation has to be made here. For example, adsorbed carbon dioxide, at 273 K, is assumed to have a density of 1.0 g cm^{-3}.

Stoeckli and Kraehenbuehl[37] developed further the theoretical basis for the equation, which is effective in describing microporous carbons provided that the distribution of pore sizes follows a Gaussian (or close related) distribution. The DR equation is linear over many orders of magnitude of pressure, although deviations do occur. Some of the common deviations have been explained by Marsh and Rand[40] and Marsh.[27]

Figure 1.27 indicates relationships between deviations from linearity in the DR plot and the shape of the associated pore-size distributions.

Isotherm A indicates that adsorption obeys TVFM, in that pore filling progresses from smallest to largest micropores until all porosity is filled when $\log^2 (p^0/p) = 0$.

Isotherm B occurs with microporous carbons and indicates either that the larger size of porosity is absent or the entire distribution of microporosity is filled at a relative pressure of less than unity. The position of the cut-off on the x-axis indicates the width of the pore-size distribution. The lower the relative pressure of the cut-off position, the narrower and smaller the pore-size distribution.

Isotherm C shows an upward deviation at high values of relative pressure, i.e. more adsorbate is adsorbed than expected. This 'excess' adsorption could be due to the filling of supermicro- or mesopores, or multilayer adsorption on non-porous surfaces. Extrapolation of the DR line gives the micropore volume in this case.

Isotherm D, which shows a negative deviation at low partial pressure, could be due to activated diffusion and/or molecular sieve effects. Increasing the temperature of the isotherm would distinguish between these phenomena.

Isotherm E, with total deviation from linearity, is associated with narrow micropores, isotherms not at equilibrium and mixed pore-size distributions (either mixed carbons or mixed porosities within a carbon).

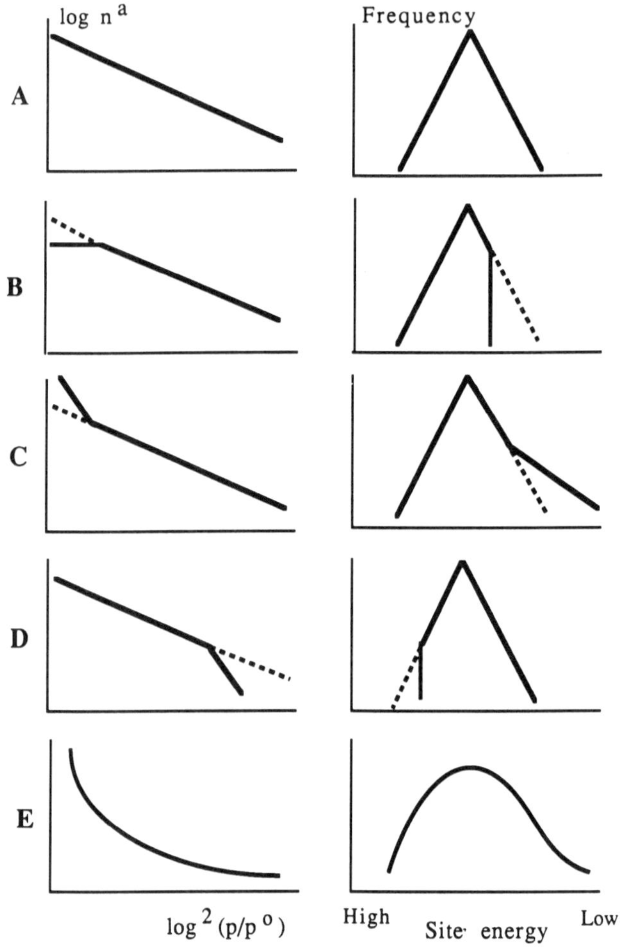

Fig. 1.27 Deviations in DR plots explained in terms of distributions of site energy

When the DR equation gives a non-linear plot owing to the distribution of site energies being very narrow, then the Dubinin–Astakhov (DA) equation:

$$W = W_0 D \log^n (p/p^0) \tag{1.8}$$

can be used.[41] The DA equation contains a variable exponent n, which replaces the square term in the DR equation. This exponent takes values between 1 and 3. Isotherms of non-porous or mesoporous materials are linearized by using $n = 1$. However, for a non-homogeneous distribution of pore sizes n values are >2; for example, molecular sieves generally have n values near to 3.

The DR equation assumes that the characteristic curve of adsorption is independent of temperature. This follows as van der Waals forces are also independent of temperature. However, for polar adsorbates this does not always hold true. For example, adsorption of water vapour does not follow the Dubinin theory of pore filling. Dubinin and Serpinski[42] proposed that water adsorption occurs at hydrophilic centres which would give the classical Type III isotherm shape. Other species, such as heavy metal ions removed from aqueous solution, also appear to be adsorbed on specific centres, representing only a fraction of the available surface.

An advantage of the DR equation is that it can be used to interpret isotherms obtained over low ranges of relative pressure (<0.03), to obtain values of micropore filling, as in the case of carbon dioxide adsorption at 273–293 K for pressures up to 0.1 MPa. The log–log^2 plot allows a short extrapolation to the value of log^2 (p/p^0) of zero. To overcome activated diffusion effects (see below), adsorptions at higher temperatures are necessary (e.g. >77 K for nitrogen). Carbon dioxide is adsorbed sufficiently at 273 K to allow the measurement of an adsorption isotherm, and this is interpreted by the DR equation.

Activated diffusion

Physical adsorption is generally a rapid process (hours); if longer equilibration times are observed, this indicates activated diffusion, to be confirmed by creating the isotherm at a higher adsorption temperature when equilibration times are reduced. Activated diffusion operates when the dimensions of the adsorbate molecules are only slightly smaller than the pore diameter. At low temperatures the gas molecules have insufficient kinetic energy to enter the pore. Figure 1.28 illustrates this effect. It describes an experiment with a carbon containing microporosity such that activated diffusion effects operate with the adsorbate. The experiment measures amounts adsorbed at constant pressure to create an isobar. The duration for each adsorption point is arbitrarily fixed as 1 hr. The experiment begins at the lowest temperatures, which are raised progressively after each hour of the experiment. Initially, little is adsorbed. With increasing temperature the amounts adsorbed increase, clearly indicating that the system after 1 hr of adsorption time is not in an equilibrium state (left curve in Figure 1.28). A stage is eventually reached where the amounts adsorbed decrease with increasing temperature, with equilibrium being established after the 1 hr period. At temperatures below the point of intersection, equilibrium would take so long to be established (days, weeks, years) that it would not be unreasonable to assume that equilibrium had been reached and, consequently, that porosity was absent from the carbon. Hence, there is a need to carry out

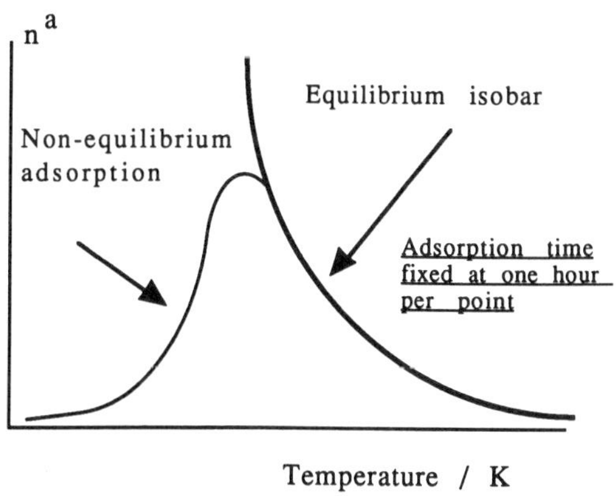

Fig. 1.28 Effect of temperature of adsorption on extents of adsorption when activated diffusion effects operate

adsorption experiments at different temperatures, e.g. N_2 at 77 K and CO_2 at 273 K, to provide information about pore-size distributions within the carbon. It is not unusual to have surface area values from these two isotherms as widely apart as 10 and 500 $m^2 g^{-1}$, respectively, indicating that the bulk of the microporosity is in the range of ultramicroporosity.

Molecular sieve effects

The models of microporosity (Figure 1.12) indicate that as an adsorbate molecule diffuses through the solid, it passes from one site to another, each of different size. Each adsorption site is unique in size, shape and potential energy. The size, configuration and frequency of these volume elements (adsorption sites or pore volumes) are a function of the development history of the carbon being studied. A critical stage exists when the diameter of one of the sites and the effective diameter of the adsorbate molecule are commensurate. When the site sizes are smaller than the adsorbate molecules they cannot occupy that particular site and adsorption is reduced. This is the basis for molecular sieving effects (Figures 1.12 and 1.13).

Molecular sieving can be used for separation processes.[43] When nitrogen and oxygen molecules are present in the gas phase, above a microporous carbon with molecular sieving properties for these gases, then oxygen molecules, with smaller kinetic diameters, diffuse initially into the micropores more rapidly (kinetic separation) than nitrogen. On the other hand, at equilibrium, at 0.1 MPa pressure, more nitrogen is adsorbed than oxygen (Figure 1.29). This selective adsorption of molecules of adsorbates by reason of faster diffusion rates and greater extents of adsorption is the basis for gas separations using the technique of pressure-swing adsorption. The critical dimension of a pore is the diameter at which the adsorbate has sufficient energy to be able to enter. In addition to the energy barrier at the entrance to the pore, diffusion rates are also dependent on the energy barriers for diffusion within the porosity.[44]

Immersion calorimetry

Immersion calorimetry is a technique complementary to adsorption methods in the characterization of surfaces. A knowledge of pore structure and adsorption

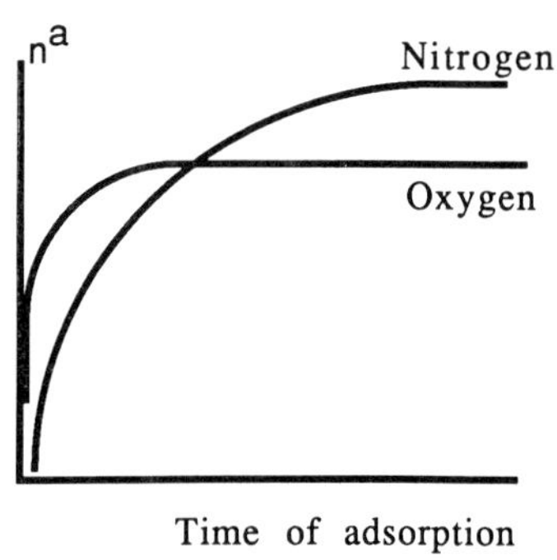

Fig. 1.29 Variation of extents of adsorption of oxygen and nitrogen with time on a microporous molecular sieve carbon

models used to interpret isotherms is necessary to assess critically immersion calorimetric results. Immersion calorimetry compares with gas adsorption in that information regarding a surface is specific to the adsorbate used. Enthalpies (ΔH, $J\,g^{-1}$ or $J\,m^{-2}$) of immersion are obtained by measuring heat release, calorimetrically, during the total immersion of an outgassed carbon sample in an appropriate medium. Immersion calorimetry describes the complete wetting/adsorption process for internal and external surfaces. When appropriate calibrations are available then $J\,g^{-1}$ values can be converted into $m^2\,g^{-1}$. However, because of differences between carbons in terms of surface chemical compositions and pore-size distributions, there is no single calibration value (conversion constant), obtained from calorimetric studies that is equally applicable to all carbons. Estimations of surface areas are therefore subjective in nature.

The procedure for the use of calorimeters in immersion experiments has been widely reported.[45,46] Values obtained for heats of immersion using organic liquids[37] can be related to the characterization energy βE_0 of the Dubinin–Astakhov equation. The structural constant B is related to E_0 (the characteristic free energy for the adsorption of a reference vapour) by the relationship

$$E_0 = 0.01915/B^{0.5} \tag{1.9}$$

The theory of volume filling of micropores (TVFM), proposed by Dubinin,[47] does not apply to water. Therefore, heats of immersion in water are not related to the DA equation. Stoeckli *et al.*[48] investigated the enthalpies of immersion of pure active carbons in water, using the Dubinin–Serpinsky relationship. Dubinin and Serpinski[42] developed an equation to account for the adsorption of water by microporous carbons (usually Type III, or Type IV isotherms)

$$\frac{p}{p^0} = \frac{a}{c(a_0 + a)(1 - ka)} \tag{1.10}$$

where

a = amount adsorbed at relative pressure p/p^0;
a_0 = number of primary sites for water adsorption;
c = ratio between the rates of adsorption and desorption (varies between 1.6 and 2.5 for activated carbons).

The term $1 - ka$ takes into account the decrease in available adsorption sites with increasing micropore filling; k is calculated from the condition that at $p/p^0 = 1$, the limiting amount adsorbed is a_s.

Differences in the adsorption behaviour of water compared with organic compounds are due to strong dispersion forces between the carbon surface and organic adsorbate and a weak interaction between the non-polar surface and the water molecules.[49] Therefore, micropore filling by water molecules occurs at a higher relative pressure than for organic compounds. At low p/p^0 values, water molecules are thought to be attracted to primary adsorption sites [(a_0), probably oxygen bonded chemically to carbon surfaces and termed 'surface oxides'], at which a form of hydrogen bonding is involved. The enthalpic contributions for the two processes are different, *primary* adsorption associated with large values ~25 $kJ\,mol^{-1}$ versus ~0.6 $kJ\,mol^{-1}$ for water adsorption in micropores.[48]

Evidence for *primary* sites has been obtained by Barton *et al.*[50] They observed a linear correlation between the amount of oxide titratable with aqueous KOH and the enthalpies of immersion in water of a mildly oxidized commercial active carbon. Stoeckli and Kraehenbuehl[37] showed that ΔH_i (H_2O: 307 K) diminishes from -140 to 40 J cm^{-3} in micropores for an active carbon reduced in hydrogen, whereas ΔH_i for benzene and other non-specific liquids remains virtually unchanged. This supports the theory that the primary site (a_0) for water adsorption is an oxide, so that under reducing conditions the number of sites (a_0) decreases. Stoeckli and Kraehenbuehl[37] also found that ΔH_i (H_2O) decreased proportionally with the amount of oxygen desorbed as CO from the carbon surface.

Dubinin[51] investigated water isotherms in detail. He proposed that the amount of water adsorbed near $p/p^0 = 0.6$ corresponded to monolayer coverage of the micropore walls by water molecules. As $a_{0.6}$ is a reasonable estimate of the surface area of the micropores [not volume (W_0)], $a_0/a_{0.6}$ represents the fraction of the surface occupied by the primary sites. This ratio is a useful quantity for the characterization and comparison of active carbons. The concept of a primary site is also applicable to the adsorption of butanol from heptane.

Immersion calorimetry is complementary to the Dubinin–Radushkevich approach (DR equation, used to estimate surface parameters of microporous carbons) and can be used to investigate micropore sizes by employing molecular probes. Kraehenbuehl and Stoeckli[45] investigated micropore size distributions in the range 0.4–0.8 nm and found close agreement between immersion calorimetry and gas adsorption results. Innes *et al.*[52] also used immersion calorimetry with liquids of various molecular dimensions and compared distributions of pore sizes with HRTEM results.

Flow microcalorimetry

The flow microcalorimeter operates on the same principle as immersion calorimetry, namely that phenomena occurring at solid/gas or solid/liquid interfaces are associated with the evolution or absorption of heat. This microcalorimeter was designed by Groszek[53] and developed by Microscal, London. It is based on a heat flow calorimeter, and uses thermistors rather than thermocouples for heat detection. This made it possible to construct a compact calorimeter with high sensitivity (4.2 mJ), able to detect changes in temperature as small as 10^{-5}°C.[54] It can be used with both gaseous and liquid adsorbates. Gaseous adsorbates pass into the flow microcalorimeter with a carrier gas, usually helium.

The following interactions can be studied using the flow microcalorimeter:

- heats of adsorption and desorption;
- rates of adsorption and desorption;
- the extent of reversibility;
- adsorption of several components, competitive adsorption and synergistic surface effects;
- preferential adsorption on specific surface sites;
- effect of temperature on heats of adsorption.

In liquid-phase operation, preferential adsorption occurs by the adsorbate displacing carrier molecules from specific sites. Preferential heats of adsorption provide information about surface functionality and specific adsorption sites. The basic principle for all adsorbate systems is to flood the sample surface with a carrier liquid; this can be organic or aqueous. A dilute solution of the adsorbate is prepared by dissolving it in the chosen carrier liquid. By using a dilute solution, heats of mixing of the adsorbate in the carrier are minimized. Flow rates of $\sim 6\ \text{ml h}^{-1}$ are used and the change-over from pure carrier to adsorbate–carrier is carried out via a Vicki valve, once thermal equilibrium in the carrier flow has been established. The resultant heats are recorded and integrated by computer. The choice of adsorbate and carrier enable specific aspects of the surface to be monitored.

Preferential heats of adsorption of *n*-butanol from *n*-heptane have been used successfully to measure polar surface areas. Because the concentration of the adsorbate in the carrier can be varied, it is possible to obtain isotherms; hence a conversion factor from J g^{-1} to $\text{m}^2\ \text{g}^{-1}$ can be calculated (the debate over the use of TSA or ESA is relevant here). Groszek[55] found for polar surfaces that the relationship between the heat of adsorption of *n*-butanol and the surface areas obtained for graphites was

$$1\ \text{J} = 6.2\ \text{m}^2 \tag{1.11}$$

Flow microcalorimetry is ideally suited to the use of molecular probes, i.e. straight- and branched-chain alcohols. Groszek[56] observed that for microporous polyvinylidene chloride (PVDC) carbons, most of the polar sites are present in pores having diameters not exceeding 0.5 nm. This was deduced from the fact that heats of adsorption of *t*-butanol were less than those for *n*-butanol.

In gas-phase studies, heats of gas adsorption provide information additional to adsorption isotherms. In comparison with other techniques (including other types of calorimetry), flow microcalorimetry provides a fast, versatile and sensitive means of characterizing surfaces. However, it must be stressed that flow microcalorimetry should always be carried out in conjunction with other techniques to confirm results. Heats of adsorption of carbon dioxide give valuable information about microporosity which is filled at very low partial pressures of <0.03 (assumed to be ultramicroporosity), but for microporous materials (active carbons) there are no correlations between Q_{CO_2} and N_2 BET surface area.[25]

Small-angle scattering of X-rays

Small-angle scattering of X-rays (SAXS) or neutrons (SANS) is used to characterize porous materials and to study their structure on a scale of 0.5–200 nm.[57,58] From the scattering data, pore sizes can be estimated, as well as surface roughness at the boundaries of pores.[59] X-rays or neutrons from a source are focused into a fine beam by slits and directed on to the sample. A small fraction of the beam is scattered and detected. The intensity of the scattered beam and the way in which the intensity varies with the scattering angle are determined by the structure of the sample. Structural information can therefore be deduced from a knowledge of scattering intensity and angle. Small-angle scattering is caused by variations of electron density within a sample, i.e. the electron density falls to zero in a pore. A 'two-phase' approximation,[58] where one phase is the solid material and the other is the pores, is used to analyse the scattering curves. Using this approximation, the intensity of SAXS is proportional to the square of the difference of the electron densities of the two phases. The boundary between the two phases is assumed to be infinitely sharp.

Renouprez and Avom[60] compared nitrogen adsorption and small-angle X-ray scattering techniques in the characterization of activated palm-tree cobs. They observed that SAXS gave larger surface areas and pore volumes in all cases. This could be because SAXS measures all interfacial situations, including ultramicroporosity and any closed porosity which is not accessible to adsorption techniques. SAXS is also able to measure pores with diameters <0.5 nm, which nitrogen is not able to penetrate. Mineral impurities may be partially responsible for the differences in results from the techniques as they can increase scattering leading to anomalously high surface areas for SAXS. Stacey[61] compared SANS with TEM and gas adsorption for measuring mesoporosity in carbons. He found that for some samples close agreement was obtained between the techniques, and that SANS gives best results when applied to systems with low levels of porosity.

High-resolution transmission electron microscopy

High-resolution transmission electron microscopy (HRTEM) has been used to study a range of carbons, including graphites, anthracites and active carbons.[13,52,62] HRTEM is a particularly useful technique in that microporosity can be observed, providing information about pore sizes, shapes and their distributions. HRTEM with the use of lattice fringe imaging and light- and dark-field imaging gives details of the local organization and orientation of the small (<1 nm) lamellar-like constituent molecules (LCM). The principle of lattice fringe-imaging consists in forming an interference image with two beams, e.g. the unscattered beam and the (002) beam; this gives (002) lattice fringes. Dark-field imaging consists of forming an image with a single (*hkl*) scattered beam.

Oberlin,[15] for a wide range of materials using the dark-field technique, produced fringe-images and measured d_{002}, i.e. the spacing between aromatic sheets of carbonaceous material. Changes are clearly seen in the structure as the parent material and heat treatment temperature (HTT) are changed. Disadvantages of HRTEM are the small size of sample observed; the shapes and size distributions of microporosity may not be statistically representative. The preparation of samples for HRTEM is also difficult in that very thin sections are required. Three-dimensional deductions of structure are not easy. However, HRTEM provides details of structure within carbons that cannot be realized by any other technique.

Mercury porosimetry

Mercury porosimetry, used to investigate meso- and macroporosity in carbons, is based on the penetration of mercury, under pressure, into porosity. Pressure is required to force the mercury into the structure because it does not wet the carbon surface owing to its high contact angle ($\theta \approx 140°$). Washburn in 1921 suggested the following equation to calculate pore radii from mercury intrusion data,[63] based on a non-intersecting cylindrical pore model.

$$r = \frac{2\gamma \cos\theta}{\Delta p} \tag{1.12}$$

where

r = pore radius (assumed to be cylindrical) (μm);
Δp = pressure required to force mercury into the pore [MPa (MN m^{-2})];
γ = surface tension (N m^{-1}).

The relationship between pore radius and pressure depends on the shape of the pore, and this necessitates the selection of a model. However, it is unlikely that meso- and macropores are all uniform in shape. Therefore, results obtained from mercury porosimetry must be viewed with caution. This does not reduce the usefulness of the technique as reasonable agreements have been obtained when comparing mercury porosimetry with other methods. Day and Fletcher[64] compared nitrogen adsorption and densitometry in helium and mercury with mercury porosimetry. There was good agreement between methods provided that the carbon was largely mesoporous. The presence of macropores led to poor correlations between techniques. Table 1.1 gives results from adsorption, mercury porosimetry and densitometry for carbons. The limits for pore radii measured by this technique are in the range ~3.5 nm–7.5 mm.[2]

Mercury porosimetry is not suitable for micropore analysis, nor can it be used to measure mesoporosity in friable carbons, as the high pressures required (up to 100 MPa) can damage pore structures. Practical difficulties also arise, such as finding contaminant-free mercury because impurities tend to lower the contact angle, producing misleading results. Surface areas and pore-size distributions can be estimated from mercury porosimetry and results are reported in good agreement with those obtained using the BET method.[65]

Image analysis

Image analysis is the quantitative assessment of surface features (generally meso- and macroporosity) created by such techniques as optical and scanning

Table 1.1 Results from adsorption, mercury porosimetry and densitometry for carbons[a]

	Mean pore size nm				Pore volume $cm^3\ g^{-1}$				
	Mesoporosity		Macroporosity		Mesoporosity		Macroporosity		Total densi-
Set	N_2	Hg	N_2	Hg	N_2	Hg	N_2	Hg	tometry
1	9.5	10.0			0.65	0.65			0.65
	16.0	17.6			0.59	0.55			0.55
2	5.5	6.5	20.0	60.0	0.12	0.10	0.05	0.15	0.22
	3.8	5.0	30.0	60.0	0.10	0.05	0.20	0.32	0.38
3	5.0	5.5		300.0	0.30	0.15		0.11	0.32
	32.0	25.0		500.0	0.10	0.06		0.04	0.10
				6000.0					

[a]The values of macropore volume were determined from nitrogen adsorption/desorption data using the procedure contained in the analytical software of the commercial instrument (Micrometrics Digisorb 2500).

electron microscopy. Image analysis provides information about pore-shape factors and spatial distributions. Originally, the cross-sectional areas and dimensions of pores were measured by hand. However, the recent development of computing facilities, enabling large quantities of data to be stored and analysed, has resulted in an upgrading in the technique of image analysis. It is now possible to grade a monochromatic or colour picture in terms of shade intensity or colour via the pixels (picture elements) generated on a television screen from the image of a microscope.

Optical and electron microscopy are used for image analysis,[66] with resolution limited by the choice of technique, i.e. the limit for optical microscopy is 1.0 μm. By appropriate manipulation of the data (number and manner of pixels selected) stored in the computer, it is possible to obtain the following information:

- total porosity as a percentage of the surface;
- pore-size distributions;
- pore-shape analysis, Feret diameters, etc.

One major advantage of image analysis is that a pore model is not necessary to analyse and interpret the data. It must be remembered that all images are two-dimensional and to obtain three-dimensional structures the sample must be sectioned. This approach allows interconnections and blind/closed pores (meso- and macro-) to be recognized, which is not the case for the majority of other techniques used to investigate porosity. Particular attention must be paid to sample preparation and sufficient sample must be analysed to give statistically representative results.

Patrick,[67] in a study of metallurgical cokes, established a good relationship between coke tensile strength and macropore structural data obtained by image analysis. More recently, a form of image analysis was used by Innes *et al.*[52] to study microporosity. They established a correlation for micropore distributions obtained from molecular probes with size analysis from fringe-imaging high-resolution electron microscopy.

Fractals

One method of characterizing surface 'roughness' is the use of surface fractal dimensions, D. For a smooth surface $D \approx 2$, and with increasing roughness D approaches 3. Values for D can be obtained by using progressively smaller and smaller probes (adsorbate molecules). The value of D is found from the increase in A_s with decrease in a_m, where A_s is the specific surface area and a_m is the effective cross-sectional area of the adsorbate molecule. For a series of adsorbate molecules of different size, but with geometrically similar cross-sections, e.g. spherical molecules, the relationship is

$$A_s \propto a_m^{-[(2-D)/2]} \tag{1.13}$$

Avnir *et al.*[68] showed that carbon blacks have fractal dimensions near to 2, whereas activated charcoals gave values closer to 3. However, on further activation, the D value for the charcoals decreased, indicating that increasing

Fig. 1.30 Model of microporosity in carbons as space between carbon atoms

activation 'smooths' carbon surfaces. The use of fractal dimensions, as applied to carbons, is still developing but does provide additional information for the characterization of surfaces.[69]

A final model

Figures 1.30 and 1.31 are models which approximate to a possible relationship (two-dimensional) between positions of carbon atoms and space between these carbon atoms to create microporosity (Figure 1.30) and occupied by nitrogen molecules (Figure 1.31). In a microporous carbon, with a surface area of $1000\ m^2\ g^{-1}$, with a monolayer coverage of the adsorbate nitrogen, the ratio of carbon atoms to nitrogen molecules is about 10:1. This is the ratio of Figure 1.31. The spaces occupied by the nitrogen molecules are interconnected in three dimensions to create the permeability and molecular sieve properties. Rudzinski and Everett[70] discussed adsorption by heterogeneous porous carbons and the application of theoretical approaches to isotherm interpretation. No serious attempt was made to discuss these approaches in terms of realistic models of

Fig. 1.31 Model of microporosity in carbons with nitrogen molecules adsorbed in spaces between carbon atoms

structure in porosity. For example, in activated carbons the activation process may not be uniform. Such carbons behave, from an adsorption point of view, like a physical mixture of several porous carbons. The difficulty of creating meaningful adsorption equations to describe these one-off systems can therefore be appreciated.

1.3 References

1. FAUST, S.D. and ALY, O.M. 1983: *Chemistry of Water Treatment.* London: Butterworths.
2. BANSAL, R.C., DONNET, J.B. and STOECKLI, H.F. 1988: *Active Carbon.* New York: Marcel Dekker.
3. MATTSON, J.S. and MARK, H.B., Jr, 1971: *Activated Carbon, Surface Chemistry and Adsorption from Solution.* New York: Marcel Dekker.
4. MARSH, H. 1986: in Figuerido, J.L. and Moulijn, J.A. (eds) *Carbon and Coal Gasification.* Martinus Nijhoff, Dordrecht, **Vol. 105**, p. 27.
5. IUPAC, International Committee for Characterization and Terminology of Carbon. *Carbon* 1982: **20**, 445; 1983: **21** 517; 1985: **23** 601.
6. DIEDERICH, F., WHETTEN, R.L., THILGEN, C., ETTL, R. CHAO, I. and ALVAREZ, M.M. 1991: *Science* **254**, 1768.
7. FRANKLIN, R.E. 1951: *Proceedings of the Royal Society London, Ser. A* **209**, 196.
8. MARSH, H. 1991: *Carbon* **29**, 703.
9. SING, K.S.W., EVERETT, D.H., HAUL, R.A.W., MOSCOU, L., PIEROTTI, R.A., ROUQUEROL, J. and SIEMIENIEWSKA, T. 1985: *Pure and Applied Chemistry* **57**, 603.
10. VOOYS, DE F. 1983: *Activated Carbon – A Fascinating Material.* Norit, Amsterdam, 2nd edn, Chapter 1.
11. McENANEY, B. and MAYS, T.J. 1989: *Introduction to Carbon Science.* London: Butterworths, Chapter 5.
12. RODRIGUEZ-REINOSO, F., LINARES-SOLANO, A. MOLINA-SABIO M. and LOPEZ-GONZALEZ, J. de D. 1984: *Adsorption Science and Technology* **1**, 211.
13. EVANS, M. and MARSH, H. 1979: in Unger, K.K., Rouquerol, J., Sing K.S.W. and Kral, H. (eds) *Characterization of Porous Solids* London: SCI, p. 53.
14. MARSH, H. and WYNNE-JONES, W.F.K. 1964: *Carbon* **1**, 269.
15. OBERLIN, A. 1984: *Carbon* **22**, 521.
16. FREEMAN, J.J., GIMBLETT, F.G.R., ROBERTS, R.A. and SING, K.S.W. 1987: *Carbon* **25**, 559.
17. WIGMANS, T. 1986: in Figuerido, J.L. and Moulijn, J.A. (eds) *Carbon and Coal Gasification* Dordrecht: Martinus Nijhoff, **Vol. 105**, p. 559.
18. WIGMANS, T. 1989: *Carbon* **27**, 13.
19. WALKER, P.L., Jr, RUSINKO, F., Jr and AUSTIN, L.G. 1959: *Advances in Catalysis* 11, 133.
20. ERGUN, S. and MENTSER, M. 1965: *Chemistry and Physics of Carbon.* **1**, Marcel Dekker, New York, 203.
21. RAND, B. and MARSH, H. 1971: *Carbon* **9**, 79.
22. RODRIGUEZ-REINOSO, F., MARTIN-MARTINEZ, J.M. and MOLINA-SABIO, M. 1985: *Carbon* **23**, 9.
23. ILEY, M., MARSH, H. and RODRIGUEZ-REINOSO, F. 1973: *Carbon* **11**, 633.
24. GREGG, S.J. and SING, K.S.W. 1982: *Adsorption, Surface Area and Porosity.* London: Academic Press, 2nd edn.
25. BYRNE, J.F. 1990: *PhD Thesis*, University of Newcastle upon Tyne.
26. FREEMAN, E.M., SIEMIENIEWSKA, T., MARSH, H. and RAND, B. 1970: *Carbon* **8**, 7.

27. MARSH, H. 1987: *Carbon* **25**, 49.
28. MARSH, H. and SIEMIENIEWSKA, T. 1965: *Fuel* **44**, 355.
29. MARSH, H. and SIEMIENIEWSKA, T. 1967: *Fuel* **46**, 441.
30. RODRIGUEZ-REINOSO, F. and LINARES-SOLANO, A. 1988: *Chemistry and Physics of Carbon* **21**, Marcel Dekker, New York, 1.
31. KISELEV, A.V. 1968: *Journal of Colloid Interface Science* **28**, 430.
32. HALSEY, G.D. 1948: *Journal of Chemical Physics* **16**, 931.
33. LANGMUIR, I. 1916: *Journal of the American Chemical Society* **38**, 2221.
34. BRUNAUER, S., EMMETT, P.H. and TELLER, E. 1938: *Journal of the American Chemical Society* **60**, 309.
35. SING, K.S.W. 1989: *Carbon* **27**, 5.
36. DUBININ, M.M. 1966: *Chemistry and Physics of Carbon* **2**, Marcel Dekker, New York, 51.
37. STOECKLI, H.F. and KRAEHENBUEHL, F. 1981: *Carbon* **19**, 353.
38. McENANEY, B. 1987: *Carbon* **25**, 69.
39. JARONIEC, M. and MADEY, R. 1989: *Carbon* **27**, 77.
40. MARSH, H. and RAND, B. 1970: *Journal of Colloid Interface Science* **33**, 101.
41. STOECKLI, H.F. 1990: *Carbon* **28**, 1.
42. DUBININ, M.M. and SERPINSKI, V.V. 1981: *Carbon* **19**, 402.
43. YANG, R.T. 1987: *Gas Separation by Adsorption Processes*. Boston: Butterworths.
44. RAO, M.B. and JENKINS, R.G. 1986: *Carbon* **24**, 236.
45. KRAEHENBUEHL, F. and STOECKLI, H.F. 1986: *Carbon* **24**, 483.
46. RADEKE, K.H. 1984: *Carbon* **22**, 473.
47. DUBININ, M.M. 1981: *Carbon* **19**, 321.
48. STOECKLI, H.F., KRAEHENBUEHL, F. and MOREL, D. 1983: *Carbon* **21**, 589.
49. BARTON, S.S., EVANS, M.J.B., HOLLAND, J. and KORESH, J.E. 1984: *Carbon* **22**, 265.
50. BARTON, S.S., DACEY, J.R. and EVANS, M.J.B. 1982: *Colloid and Polymer Science* **206**, 726.
51. DUBININ, M.M. 1980: *Carbon* **18**, 355.
52. INNES, R.W., FRYER, J.R. and STOECKLI, H.F. 1989: *Carbon* **27**, 71.
53. GROSZEK, A.J. 1970: *Proceedings of the Royal Society London, Ser. A* **314**, 476.
54. GROSZEK, A.J. 1988: *Materials Science Forum* **25/26**, 483.
55. GROSZEK, A.J. 1987: *Carbon* **25**, 717.
56. GROSZEK, A.J. 1989: *Carbon* **27**, 33.
57. JANOSI, A. and STOECKLI, H.F. 1979: *Carbon* **17**, 465.
58. GLATTER, O. and KRATKY, O. 1982: *SAXS*. London: Academic Press.
59. SCHMIDT, P.W. 1988: in Unger, K.K., Rouquerol, J., Sing, K.S.W. and Kral, H. (eds) *Characterization of Porous Solids*. Amsterdam: Elsevier, **Vol. 39**, p. 35.
60. RENOUPREZ, A. and AVOM, J. 1988: in Unger, K.K., Rouquerol, J., Sing, K.S.W. and Kral, H. (eds) *Characterization of Porous Solids* Amsterdam: Elsevier, Vol. **39**, p. 49.
61. STACEY, M.H. 1988: in *Characterization of Porous Solids*. Amsterdam: Elsevier, Vol. 39, p. 55.
62. MARSH, H., CRAWFORD, D., O'GRADY, T.M. and WENNERBERG, A. 1982: *Carbon* **20**, 419.
63. GREGG, S.J. and SING, K.S.W. 1967: *Adsorption, Surface Area and Porosity*. London: Academic Press.
64. DAY, M. and FLETCHER, R. 1988: in Unger, K.K., Rouquerol, J., Sing, K.S.W. and Kral, H. (eds) *Characterization of Porous Solids*, Amsterdam: Elsevier, **Vol. 39**, p. 491.
65. ROOTARE, H.M. and PRESZLOW, C.F. 1967: *Journal of Physical Chemistry* **71**, 2733.
66. PICKUP, I.M., MAYS, T.J. and McENANEY, B. 1986: in *Proceedings of 4th International Carbon Conference*. Baden-Baden, Germany, June. Arbeitreiss Kohlenstoff der Deutschen Keramischen Gesselschaft E.B., p. 240.

67. PATRICK, J. 1983: *Journal of Microscopy* **132**, 333.
68. AVNIR, D., FARIN, D. and PFEIFER, P. 1983: *Journal of Chemical Physics* **79**, 3566.
69. FARIN, D. and AVNIR, D. 1988: in Unger, K.K., Rouquerol, J., Sing, K.S.W. and Kral, H. (eds) *Characterization of Porous Solids* Amsterdam: Elsevier, Vol. 39, p. 421.
70. RUDZINSKI, W. and EVERETT, D.H. 1992: *Adsorption of Gases on Heterogeneous Surfaces.* London: Academic Press.

2 Physisorption of Gases by Porous Carbons

Kenneth S.W. Sing
Department of Chemistry
Exeter University
Exeter, Devon
United Kingdom

2.1 Introduction

Gas adsorption is one of the most widely used techniques for the characterization of porous carbons. The aim of most routine work is to determine the surface area and pore size distribution, but this is not always easy to achieve. The complexity of both the pore structure and the mechanisms of adsorption makes it necessary to adopt a cautious approach in the interpretation of experimental data. On the other hand, adsorption measurements provide a useful 'fingerprint' of the microstructure and, of course, are essential if the carbon is to be used as an adsorbent or catalyst support.

A characteristic feature of activated carbons is the large internal surface located within a network of narrow pores. In the context of adsorption, it is expedient to classify pores according to their size[1]: *micropores* have widths up to about 2 nm; *mesopores* are in the range 2–50 nm; *macropores* have widths greater than about 50 nm (0.05 μm). However, these limits are necessarily somewhat arbitrary because, as will be explained later, the mechanisms of adsorption are affected by pore shape, the relative size of the gas molecules and the nature of the gas–solid interactions.

In this chapter, we are concerned with physical adsorption (physisorption), which, in contrast to chemisorption, occurs to some extent whenever a clean solid surface is exposed to a gas. The amount adsorbed, n, by unit mass of solid (the adsorbent) is dependent on the gas (adsorptive) pressure, p, the temperature, T, the properties of the adsorbent and the nature of the gas–solid interactions. Thus, for the adsorption of a given adsorptive by a particular adsorbent, we may write

$$n = f(p, T)_{\text{system}} \tag{2.1}$$

and at constant temperature,

$$n = f(p)_{T,\text{system}} \tag{2.2}$$

This equation represents the adsorption isotherm, which is generally presented in the graphical form of n plotted against p or p/p^0, where p^0 is the saturation pressure of the adsorptive at the temperature of measurement.

Physisorption is always accompanied by the evolution of heat which, if measured under properly controlled conditions,[2] may be equated to a change in the thermodynamic state of the system. Thus, if we specify constant temperature and volume, we may write

$$Q_{T,V} = -\Delta_a u \tag{2.3}$$

where $Q_{T,V}$ may be regarded as an integral heat of adsorption and $\Delta_a u$ is the molar integral energy of adsorption. If we make the usual assumption that the adsorbent remains unchanged, we may define $\Delta_a u$ as the difference between the molar internal energies of the adsorbed state (i.e. the adsorbate) and the gaseous adsorptive (i.e. $u_a - u_g$).

If the mechanism of physisorption is to be investigated, it is useful to consider the differential molar energy of adsorption

$$\Delta_a \bar{u} = \bar{u}_a - u_g \tag{2.4}$$

and the differential molar enthalpy of adsorption

$$\Delta_a \bar{h} = \bar{h}_a - h_g \tag{2.5}$$

Thermodynamic consistency is maintained if we define

$$\bar{u}_a = \left(\frac{\partial u_a}{\partial n}\right)_{T,A} \tag{2.6}$$

and

$$\bar{h}_a = \left(\frac{\partial h_a}{\partial n}\right)_{T,p} \tag{2.7}$$

where A is the surface area of the adsorbent. Further, by extending the principles of thermodynamic equilibrium,[3] we may derive the equation

$$\left(\frac{\partial \ln p}{\partial T}\right)_n = -\frac{\Delta_a \bar{h}}{RT^2} \tag{2.8}$$

or

$$\left(\frac{\partial \ln p}{\partial T}\right)_n = \frac{q^{st}}{RT^2} \tag{2.9}$$

where q^{st} is a form of differential heat of adsorption, which is generally known as the isosteric heat. Integration of equation 2.9 over a limited range of temperature (assuming q^{st} is to remain constant) gives

$$\ln (p)_n = -q^{st}/R + \text{constant} \tag{2.10}$$

which allows us to calculate q^{st} from a series of isotherms determined at neighbouring temperatures.

It is of particular interest to determine q^{st}, the isosteric heat of adsorption at zero surface coverage (or zero fractional pore filling), since it provides an indication of the magnitude of the adsorbent–adsorbate interaction energy. One approach is to apply a virial-type expansion to analyse the isotherm data obtained at very low surface coverage.[4] The derived values of the Henry's law constant, k_H, are then used in place of $(p)_n$ in equation 2.10; thus, log k_H is plotted against $1/T$ and q^{st} is obtained from the slope.[5] An alternative procedure is to use a chromatographic technique to determine the specific retention volume of the adsorptive as a function of the adsorbent column temperature.[6] These two procedures are capable of giving q^{st} values in good agreement,[5] provided that certain conditions are fulfilled (e.g. that the curvature of the isotherm is not too great and that thermodynamic equilibrium is established).

Many different procedures have been developed for determining the amount of gas adsorbed.[1] Gas volumetric methods, which are generally preferred for low-temperature measurements with gases such as argon and nitrogen, depend on the measurement of the quantity of gas removed from the gas phase. The amount adsorbed at equilibrium pressure is given as the difference between the amount of gas admitted and the amount remaining in the space around the adsorbent. The adsorption isotherm is usually constructed point-by-point, but a quasi-equilibrium procedure[7] can be used provided that the rate of adsorption is not too slow. Chromatographic techniques may be employed,[8,9] but it is essential to ensure that the results are not affected by changes in flow rate.

A gravimetric technique is generally a more convenient way of determining the amount adsorbed if the adsorptive is in the form of a condensable vapour such as water[10] or benzene.[11] In this case the change in weight of the adsorbent can be followed directly during the outgassing and adsorption stages.

All physisorbed material must be removed from the adsorbent surface before the isotherm is determined. This is usually achieved by outgassing the adsorbent at elevated temperature. To obtain reproducible isotherms, microporous carbons should be outgassed at ~300°C. In some cases, to obtain reproducible isotherms it is also necessary to flush several times with the adsorptive.

2.2 Energetics of physisorption

The potential energy of an isolated gas molecule in the vicinity of a solid surface is dependent on both its distance, z, from the surface and its location (x,y) relative to the surface structure. Unless some form of chemical bonding is involved, it is usually assumed that the surface is inert and therefore that the gas–solid interaction energy, ϕ, is controlled by the change in potential energy of the gas molecule, i.e. by the depth of the potential energy well corresponding to the most favourably physisorbed state. The further assumption is generally made that the interaction energy may be computed by summation of the pairwise attractive and repulsive interaction.[12]

For the simplest type of adsorption system, i.e. a small non-polar molecule on a uniform non-polar surface, the interaction energy may be expressed in the approximate Lennard–Jones (12:6) form[13] as

$$\phi(z) = B_{ij} \sum r^{-12} - c_{ij} \sum r^{-6} \tag{2.11}$$

where z is the distance of the adsorbed molecule, i, from the surface plane, r_{ij} is the distance between i and an atom, j, in the solid and B_{ij} and C_{ij} are characteristic repulsion and attraction constants for the gas–solid system.

The r^{-12} and r^{-6} terms in equation 2.11 are associated with the short-range repulsion and the dispersion attractive forces respectively, which always contribute to physisorption and are therefore known as the non-specific interactions.[14] In addition, specific interactions come into play when polar molecules are adsorbed on certain surfaces. Barrer[15,16] has shown that the overall interaction energy may be represented by a general expression of the form

$$\phi = \phi_D + \phi_R + \phi_p + \phi_{F\mu} + \phi_{\dot{F}Q} + \phi_{sp} \tag{2.12}$$

where $(\phi_D + \phi_R)$ is the sum of the non-specific attraction and repulsion terms and $(\phi_p + \phi_{F\mu} + \phi\phi_{\dot{F}Q})$ is the sum of the specific adsorbent–adsorbate interactions (ϕ_p, the polarization energy; $\phi_{F\mu}$, the field–dipole energy; $\phi_{\dot{F}Q}$, the field gradient–quadrupole energy). The term ϕ_{sp}, the adsorbate–adsorbate energy (or self-potential term[16]), becomes increasingly important as the surface coverage is increased. Generally, ϕ_p is relatively small, but the terms $\phi_{F\mu}$ and $\phi_{\dot{F}Q}$ become important when polar molecules interact with ionic or polar sites.

Extensive calculations of the non-specific interaction energies have been reported.[4,12,13,17,18] By replacing the (12:6) potential in equation 2.11 by a (12:6:8:10) form of equation, Avgul and Kiselev[17] were able to obtain good agreement with experimental heat data for a wide range of molecules adsorbed on graphitized carbon blacks. Ungraphitized carbons exhibit a high degree of energetic heterogeneity,[12,13,17] as indicated by the steady decrease in the differential heat of adsorption of argon, nitrogen and other gases with surface coverage, but heat treatment at temperatures >2000°C brings about a change in shape of the particles with the development of a remarkably uniform surface

Table 2.1 Low-coverage isosteric heats of adsorption

Adsorptive	q^{st}/kJ mol^{-1} [a] Non-porous carbons	Molecular sieve carbons	Ref.
Ar	10	20	22
N_2	10	21	22
CH_4	13	23	23, 24
C_2H_6	20	37	23
C_3H_8	26	52	23
n-C_4H_{10}	32	66	23
n-C_5H_{12}	38	80	23
n-C_6H_{14}	43	94	23

[a] Rounded to the nearest unit.

associated with the exposure of the graphitic basal plane and the removal of polar groups.[3,17,19] After such treatment, the differential heat of adsorption (e.g. for N_2 or Ar) undergoes little change at low surface coverage, but exhibits a sharp maximum as the monolayer capacity is approached.[19] This increase in q^{st} is due to the increasingly important contribution of ϕ_{sp}.

It is not surprising to find that the adsorbent–adsorbate interaction energy is considerably enhanced when physisorption occurs in very narrow pores. Calculations based on equation 2.11, or more refined expressions, for the non-specific interactions have indicated[20,21] that the overlap of the adsorption forces in slit-shaped pores results in a maximum enhancement of adsorption energy of around twofold when the slit width, d, is slightly larger than the molecular diameter, σ. Further, it appears that the enhancement is insignificant when d is only slightly larger than 2σ. The experimental values of q^{st} in Table 2.1 were obtained by the application of various techniques (calorimetry,[22] gas chromatography[23] and virial analysis[24]). It will be noted that the ratio of the corresponding values for each adsorptive is remarkably close to 2, thus confirming the theoretically predicted enhancement of the adsorbent–adsorbate interaction energy. Further, for each type of carbon the n-alkanes exhibit a regular increase of q^{st} as the hydrocarbon chain is lengthened. The significance of these results is discussed later in the context of the assessment of microporosity.

2.3 Types of physisorption isotherms

The first step in the interpretation of a physisorption isotherm is the visual inspection of its shape and hence identification of the principal mechanism of adsorption.[3] Most physisorption isotherms may be grouped into the six types[1] shown in Figure 2.1.

The reversible Type II isotherm is the normal form of isotherm given by a non-porous or macroporous adsorbent on which unrestricted monolayer–multilayer adsorption can occur. The appearance of well-defined point B indicates completion of monolayer coverage and the onset of multilayer development.[3]

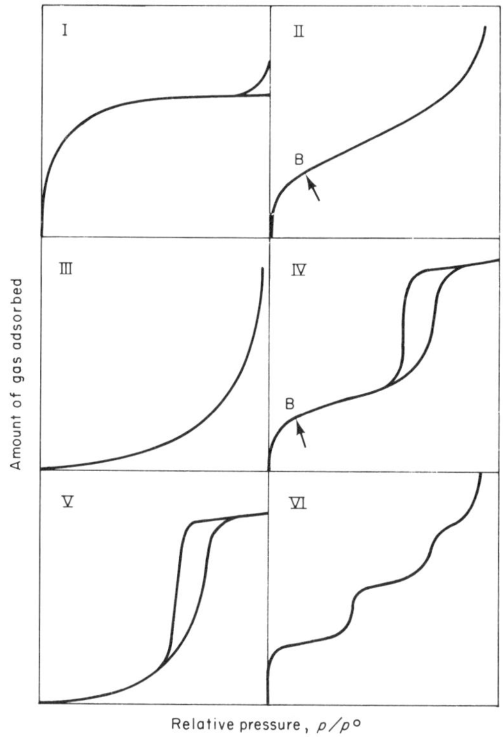

Fig. 2.1 Types of physisorption isotherms

The Type III isotherm is generally associated with weak adsorbent–adsorbate and relatively strong adsorbate–adsorbate interactions. In this case cooperative effects lead to the development of patches of multilayer before a uniform monolayer has been formed. The adsorption of water vapour on a non-porous carbon surface is probably the best example of this type of system.[25]

The most striking features of the Type IV isotherm are its hysteresis loop, which is associated with capillary condensation in mesopores,[26] and the plateau at high p/p^0. The initial part of the isotherm follows the same path as the corresponding Type II isotherm and is therefore the result of monolayer–multilayer adsorption on the mesopore walls.

The Type I isotherm is given by a microporous solid having a small external area. The very steep region at low p/p^0 is due to the filling of very narrow pores and the limiting uptake is dependent on the accessible micropore volume.[27]

Some microporous carbons are amongst the few adsorbents to give Type V water isotherms.[25] As with Type III isotherms, the adsorbent–adsorbate interactions are exceptionally weak in comparison with the strength of the adsorbent–adsorbate interactions, which of course involve hydrogen bonding in the case of water.

The Type VI isotherm is also uncommon: it represents stepwise multilayer adsorption on a uniform non-porous surface such as graphitized carbon.[28] The step height provides a measure of the capacity for each adsorbed layer and in the simplest case it remains nearly constant over two or three adsorbed layers.

2.4 Determination of surface area

Over the past 50 years, the Brunauer–Emmett–Teller (BET) method[29] has remained the most widely used procedure for the determination of the surface area of highly porous materials. It has become increasingly evident,[30] however, that the method has serious limitations, especially when applied to materials such as activated carbons.[31]

The main assumptions made by Brunauer, Emmett and Teller[29] in their extension of the Langmuir mechanism to multilayer adsorption were that localized adsorption occurs on an array of surface sites of uniform energy, molecules in the first layer act as sites for molecules in the second and higher layers and the evaporation and condensation properties of all layers above the first are very similar to those of the liquid adsorptive.

On the basis of these assumptions and under steady-state conditions, the rates of condensation and evaporation were equated for each adsorbed layer. Summation of the amounts adsorbed gave the well-known BET equation

$$\frac{p}{n(p^0-p)} = \frac{1}{n_m C} + \frac{C-1}{n_m C} \cdot \frac{p}{p^0} \tag{2.13}$$

where n_m is the monolayer capacity and C is a constant which, according to the theory, is related exponentially to the heat of first-layer adsorption.

According to equation 2.13, the plot of $p/n(p^0-p)$ against p/p^0 should be linear, and from the slope and intercept it should be easy to calculate both n_m and C. In practice, the linearity of this 'BET plot' is always restricted to a part of the isotherm.[30] In many cases the range of linearity extends over $p/p^0 \approx 0.05$–0.3, but with graphitized and microporous carbons linear BET plots are given only at $p/p^0 < 0.1$.

In support of their theory, Brunauer, Emmett and Teller claimed[29] that close agreement is generally obtained between n_m and the amount adsorbed at point B. This is indeed true for isotherms of Types II or IV, provided that the value of C is relatively high (i.e. $C \gtrsim 100$). If the C value is low ($C \lesssim 20$) the isotherm shape is such that point B cannot be located[30] and then the validity of n_m is questionable.[1] The significance of a high value of C associated with a Type I isotherm is discussed later.

The second stage in the application of the BET method is the calculation of the surface area, A_{BET}, from n_m. This step obviously requires prior knowledge of the average area, a_m, occupied by the adsorbate molecule in the completed monolayer. Thus

$$A_{BET} = n_m L a_m \tag{2.14}$$

where L is the Avogadro constant.

For several reasons, nitrogen (at 77 K) is generally considered to be the most suitable adsorptive for surface area determination[1,13,30] and it is usually assumed that the nitrogen monolayer is close-packed[29] so that $a_m(N_2) = 0.162\ \mathrm{nm}^2$. Comparisons made[32,33] with the geometric areas of some finely divided solids appear to confirm the correctness of A_{BET} and $a_m(N_2)$ to within about 10%. With other apparently non-porous materials much larger differences have been reported between BET areas and surface areas determined by electron microscopy or other techniques. Many attempts have been made to improve the accuracy of the BET method by arbitrary adjustment of the adsorbate cross-section area.[33] Unless it is based on other evidence, this approach cannot be regarded as satisfactory since it necessarily assumes the validity of n_m and the constancy of a_m.

An alternative approach is more empirical and is based on the concept of the standard isotherm.[30] Thus, by plotting isotherms determined on well-defined non-porous solids in a dimensionless form one should be able to establish a 'standard' monolayer–multilayer curve at a certain temperature for a given gas–solid system. It has been found[34] expedient to express the 'reduced' adsorption as n/n_s ($=\alpha_s$), where n_s is the amount adsorbed at a pre-selected fixed relative pressure $(p/p^0)_s$. Such normalized isotherms have been constructed for a number of different systems including adsorption of nitrogen and hydrocarbons by non-porous carbons.[35–38] In principle, α_s could be set equal to unity at any chosen point on the standard isotherm, but in practice it is usually convenient[34] to take $\alpha_s = 1$ at $(p/p^0)_s = 0.4$.

Since the reduced form of the standard isotherm for a given gas–solid system has been arrived at empirically, it can be adopted as a basis for the analysis of other isotherms and as a means of checking the consistency of BET areas. The α_s method[34] provides a simple and effective way of making this analysis. It consists of plotting the amount adsorbed by the adsorbent under investigation against α_s (i.e. interpolated from the appropriate standard data).

Three hypothetical α_s plots are shown in (d), (e) and (f) in Figure 2.2 along with the corresponding isotherms in (a), (b) and (c). The linear plot in (d) is the result of unrestricted monolayer–multilayer adsorption on a non-porous or macroporous solid giving similar adsorbent–adsorbate interactions to those of the reference material.[35] The original isotherm may be of Type II, III or even VI, provided that the appropriate standard is adopted; in the case of carbon this means that the degree of graphitization should be taken into account.[35] The α_s plot in (e) in Figure 2.2, which provides clear evidence of capillary condensation

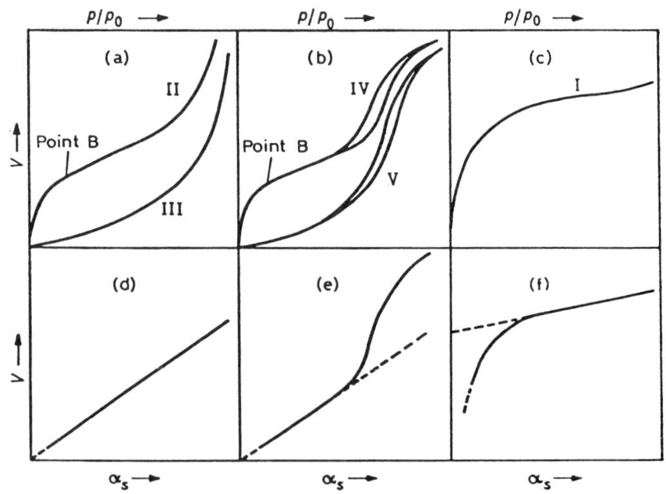

Fig. 2.2 Types of α_s plots

mesopores, is usually derived from a Type IV isotherm, but if the adsorbent–adsorbate interaction is very weak the isotherm may be close to Type V. In the simplest form of (e) the total surface area (i.e. of the mesopore walls) may be calculated from the slope of the initial linear section.[34] The α_s plot in (f) is derived from a Type I isotherm or composite of Types I and II isotherms; here micropore filling is followed by multilayer adsorption on a small external surface.[35]

It is evident that the values of surface area are calculated from the slopes of the linear regions of the α_s plots by using a proportionality (or normalizing) factor, which has been obtained from a standard isotherm determined on a non-porous reference solid of known surface area. In principle, therefore, the α_s method can be made completely independent of the BET method, provided, of course, that the reference area has been determined by electron microscopy or by some other independent technique. For nitrogen adsorption at 77 K on non-porous carbons the equation

$$A_s = 2.86\, V/\alpha_s \qquad (2.15)$$

can be applied[35] when the adsorption, V, is given in cm^3 (stp) g^{-1}. With ungraphitized carbons the values of area, A_s, are in excellent agreement with the corresponding BET areas, but this is not the case with graphitized carbons. The discrepancy has been attributed to the localized nature of the monolayer and hence $a_m(N_2) = 0.195\ nm^2$. When equation 2.15 is applied to the linear multilayer section of an α_s plot of the type shown in (f) in Figure 2.2, it gives the external surface area, which is much lower than the corresponding BET value.

As pointed out earlier, physisorption in narrow micropores of molecular dimensions is associated with enhanced adsorbent–adsorbate interactions.[20,21,39] This, in turn, leads to a distortion in the shape of the isotherm at low p/p^0, which can be detected as an increase in the BET C value[39] and also as a marked deviation from linearity of the α_s plot. It is evident that this form of physisorption involves pore filling rather than surface coverage and it follows that application of the BET method cannot yield a reliable value of the internal surface area. If the BET method is employed to characterize a molecular sieve carbon, it is recommended that the derived value of area should be designated as the *BET area* and that the experimental conditions and range of linearity of the BET plot should be indicated.[1]

There remains the question of the validity of the BET area of a microporous solid which does not possess any narrow micropores which would give rise to enhanced adsorbent–adsorbate interactions. The results of recent nonane preadsorption work[40] support the view that under certain circumstances the BET method may be used to assess the internal area of the micropores in the range ~1–2 nm. We shall return to this aspect in the context of the assessment of microporosity.

2.5 Assessment of mesoporosity

It is evident from the above discussion that the monolayer section of a simple Type IV isotherm follows the same path as the corresponding Type II isotherm and that the onset of capillary condensation is shown by an upward deviation. However, there is always the possibility that the pore size distribution is not restricted to the mesopore range but that it extends down into the micropore region. A relatively small micropore-filling contribution cannot be easily identified by visual inspection of the isotherm shape and therefore the α_s method has been found to be a useful procedure for checking the validity of the BET area and detecting any significant micropore-filling contribution.

The evaluation of mesopore size is usually based on the application of the Kelvin equation in the form[1]

$$\frac{1}{r_1} + \frac{1}{r_2} = -\frac{RT}{\gamma^l v^l} \ln \frac{p}{p^0} \tag{2.16}$$

which relates the principal radii, r_1 and r_2, of curvature of the liquid meniscus in the pore to the p/p^0 at which equilibrium is established with the vapour phase; γ^l is the surface tension of the liquid condensate and v^l is its molar volume. To apply equation 2.16) it is necessary to assume[41] that the curvature and shape of the meniscus are controlled by the pore size and shape, the pores are of well-defined shape and the initial monolayer–multilayer development on the pore walls follows the same isotherm path as on the corresponding non-porous surface.

If the width of a parallel-sided slit in an activated carbon is d_p and a correction made for the multilayer thickness, t, equation 2.16 becomes[1]

$$d_p = \frac{2\gamma^l v^l}{RT \ln(p^0/p)} + 2t. \tag{2.17}$$

Various mathematical procedures have been developed[42] for the computation of the pore-size distribution, which is usually expressed in the graphical form $\Delta V_p/\Delta d_p$ vs d_p, allowance being made for the effect of multilayer adsorption in progressively reducing the dimensions of the free pore space available for capillary condensation. In addition, it is necessary to make the further assumptions that the pores are rigid and of uniform shape, the pore size is confined to the mesopore range, there are no pore blocking effects and *either* the adsorption *or* the desorption branch of the hysteresis loop represents true equilibrium with respect to capillary condensation.[43]

In practice, none of these assumptions is entirely safe.[41–43] Special problems arise with many activated carbons because of the wide distribution of pore size, which often extends from the molecular scale into the macropore range. Further, a growing amount of evidence suggests that the slits in porous carbons are not entirely rigid.[44]

The total mesopore volume, V_p, is often derived from the amount of vapour adsorbed at a relative pressure close to unity (say, at $p/p^0 = 0.95$) by assuming that the pores have been filled by the condensed adsorptive in its normal liquid state. This procedure is justifiable[42] only if the Type IV isotherm has a distinct plateau at high p/p^0. With many activated carbons the plateau is either non-existent or very indistinct so that the *adsorption* branch of the isotherm is virtually Type II. In this case, meniscus formation is likely to be delayed beyond the p/p^0 at which capillary condensation would be expected to occur and equation 2.17 is then no longer applicable.

It is evident that the computation of mesopore-size distribution is subject to a number of uncertainties. With some activated carbons the derived pore-size distribution may give a completely misleading picture of the pore structure and it is recommended, therefore, that the computation should be restricted to Type IV isotherms having narrow and nearly vertical hysteresis loops.[1,42] The appearance of this type of isotherm is generally indicative of a relatively narrow distribution of fairly uniform mesopores. On the other hand, wide hysteresis loops are usually associated with more complex and non-uniform pore structures. With such adsorbents the pore filling–emptying mechanisms are likely to be complicated by network effects and by pore blocking and the trapping of adsorbed molecules.

Many activated carbons, particularly those prepared at high burn-off, exhibit some mesopore character, but very few give well-defined Type IV isotherms. These findings are consistent with the ill-defined nature of their mesopore structure as revealed by electron microscopy.

2.6 Assessment of microporosity

It is now generally recognized that most activated carbons are highly microporous and that the distribution of pore size usually extends down to molecular dimensions.[45] It has been pointed out above that micropore-filling occurs as a distinctive mechanism over a lower range of p/p^0 than capillary condensation. It has been noted also that the enhancement of the gas–solid interaction energy probably becomes insignificant[21] when the slit width is larger than $\sim 2\sigma$. This limit would appear to restrict the micropore-filling of small molecules such as Ar, N_2, CO_2 and the lower hydrocarbons to pores of maximum width 0.8 nm. In consequence, it might seem to be necessary to modify the IUPAC classification of pore size.[1]

To overcome this apparent discrepancy, Dubinin[46] introduced the term 'supermicropores', which he applied to the intermediate group of pores between micropores and mesopores. It was suggested by Dubinin[47] that in this intermediate range the micropore properties gradually decline whilst the mesopore character is developed.

An alternative approach[48] is to place more emphasis on the pore-filling mechanisms than on the absolute pore width. Recent work[27] has indicated that there are indeed two distinctive micropore-filling mechanisms. The first, termed 'primary micropore filling', takes place at very low p/p^0 (e.g. 10^{-5}–10^{-2}) and involves the entry of individual molecules into pores of molecular dimensions (in the range $d_p = \sigma$–2σ). Secondary micropore filling, on the other hand, is believed to be a cooperative process, occurring in the wider micropores (e.g. in slits of $d_p \approx 2\sigma$–5σ) and involving adsorbate–adsorbate interaction or quasi-multilayer adsorption. This process takes place over a much wider range of the isotherm (e.g. $p/p^0 \approx 0.01$–0.3), but still before the onset of capillary condensation.

If the plateau of a reversible Type I isotherm is virtually horizontal up to high p/p^0, the limiting uptake can be taken as a direct measure of the micropore capacity, n_p, since the external area must be very small and the multilayer contribution negligible. In the past it has been customary to convert n_p, into the micropore volume, V_p, the assumption being made that the adsorbate is in the liquid state, as in the case of mesopore filling. The soundness of this assumption becomes more and more doubtful, however, as the pore width is decreased to molecular dimensions.[49]

It can readily be shown[49] that the degree of molecular packing is dependent on both the ratio d_p/σ and the pore shape. The effect is very pronounced in the case of cylindrical pores and, although a higher degree of packing is possible in narrow slits, the overall packing density is still significantly lower than that in the bulk liquid (by up to ~30%). To allow for these differences, it is suggested that the term *effective micropore volume* should be adopted and an indication given of the particular adsorptive and conditions employed.

Much of the early work on micropore filling was carried out by Dubinin and co-workers.[50,51] indeed, their pioneering studies were largely responsible for the formulation of the concept of volume-filling as a distinctive mechanism of physisorption.[52] Dubinin[53] adapted the characteristic curve principle of Polanyi and in the first place assumed that the fractional degree of micropore-filling may be expressed in the form

$$n/n_p = \exp[-k(A/\beta)^2] \tag{2.18}$$

where A is the affinity or differential free energy of adsorption, i.e.

$$A = RT\ln(p/p^0) \tag{2.19}$$

(originally termed the adsorption potential), k is a characteristic constant and β is a scaling factor. By combining equations 2.18 and 2.19, Dubinin and Radushkevich[54] obtained an equation for the micropore-filling isotherm, which is usually given in the linear form [i.e. the Dubinin–Radushkevich (DR) equation]

$$\log n = \log n_p - B(T/\beta)^2 \log^2(p^0/p) \tag{2.20}$$

where

$$B = 2.303\,R^2/k$$

and is considered to be a structural constant, which characterizes the pore structure of the adsorbent.

According to equation 2.20, the DR plot of $\log n$ against $\log^2(p^0/p)$ should be linear with intercept $\log n_p$ and slope $-B(T/\beta)$.[2] Dubinin selected benzene as the standard vapour so that in this case $\beta = 1$. In this manner it was possible to calculate the value of B for a given activated carbon.

Some microporous carbons are found[35,39,55] to give linear DR plots over an appreciable range of p/p^0, whereas in other cases the initial linear region is restricted to $p/p^0 < 0.01$. Generally, the overall range of linearity of the DR plot is reduced as the micropore-size distribution is widened.[35,45] To overcome this problem, Dubinin first modified equation 2.20 to include an additional term to take account of adsorption in both narrow and wide micropores. At a later stage, Dubinin and Astakhov[56] proposed a more general equation (DA equation) for the characteristic curve

$$n/n_p = \exp[-(A/\beta E_0)]^m \tag{2.21}$$

where m is a small integer and the parameter E_0 is termed the characteristic energy of adsorption. Obviously, equations 2.18 and 2.21 are essentially identical if $m = 2$.

A different approach was adopted by Stoeckli,[57] who suggested that for a carbon with a heterogeneous collection of micropores the overall isotherm is best expressed as the sum of the DR contributions from the individual pore groups so that

$$n = \sum n_{p,j} \exp[-B_j(T/\beta)^2 \log^2(p^0/p)] \tag{2.22}$$

each pore group being characterized by its own $n_{p,j}$ and B_j. For a continuous Gaussian distribution, Stoeckli replaced the summation by integration and obtained

$$n = \int_0^\infty f(B)\exp[-B(T/\beta)^2 \log^2(p^0/p)]\,\mathrm{d}B \tag{2.23}$$

where $f(B)$ is dependent on the half-width of a normalized Gaussian distribution of pore size. Stoeckli and co-workers[57,58] have shown that this extended DRS equation can be applied to the isotherms of a variety of simple gases and vapours determined over a wide range of T and p/p^0 on a number of typical activated carbons.[45]

It must be kept in mind that the range over which a particular mathematical expression can be fitted to experimental isotherm data does not by itself provide sufficient evidence for a particular mechanism of adsorption.[3,13] In the present context we must endeavour to obtain an independent assessment of the micropore capacity. For this purpose the α_s method is helpful because it does not rely on any assumptions concerning the mechanisms of pore filling or surface coverage.[34]

As discussed earlier, to apply the α_s method it is first necessary to determine a standard isotherm for the particular adsorptive on a suitable non-porous reference solid. A particular difficulty arises with porous carbons because the shape of the initial region of the isotherm is dependent on both microporosity and surface structure[37] (e.g. degree of graphitization). It follows that the latter must be taken into account if the micropore filling effects are to be properly assessed. When micropore filling is succeeded by a range of multilayer adsorption before the onset of any capillary condensation, the α_s plot has the features of (f) in Figure 2.2 and then the intercept obtained by backward extrapolation of the linear multilayer section[35] (i.e. to $\alpha_s = 0$) may be taken as a direct measure of the micropore capacity, n_p. By making the usual assumption of liquid density, we may thus obtain the effective micropore volume, V_p.

Table 2.2 gives a comparison of the corresponding values of pore volume obtained by application of the DR (i.e. V_p^D) and α_s (i.e. V_p^α) methods for the analysis of nitrogen isotherms on a set of typical microporous carbons.[48,59] It is evident that the corresponding values of V_p^D and V_p^α are in good agreement provided that micropore filling is restricted to the primary range. The lack of agreement in the case of secondary micropore filling is not surprising in view of the restricted range of linearity of the DR plot. Similar results have been

Table 2.2 Effective pore volumes of microporous carbons[48] as determined by DR and α_s methods

Carbon	Micropore-filling mechanism[a]	Pore volume $cm^3\ g^{-1}$	
		V_p^D	V_p^α
XR2	P	0.27	0.27
JF002	P	0.35	0.33
Carbosieve	P	0.45	0.43
AM4	P+S	0.43	0.49
JF517	P+S	0.47	0.76
PX21	P+S	0.99	1.57
AX21	P+S	1.00	1.52

[a] P = primary micropore filling; S = secondary micropore filling.

reported by Rodríguez-Reinoso,[60] who also concludes that the α_s method provides the more reliable assessment of the *total* micropore volume. The application of either the DA equation or the DRS equation is likely to yield closer agreement between V_p^D and V_p^α, but this aspect does not appear to have been studied.

Dubinin[61] and Dubinin and Stoeckli[62] have discussed the significance of the structural constant, B, and the characteristic energy, E_0. Small-angle scattering of X-rays has been used as an independent technique to assess the micropore width, d_p, and a general relationship of the form

$$d_p E_0 = \text{constant} \tag{2.24}$$

has been found to hold for a series of activated carbons. Thus, as the level of burn-off is increased, the pores are widened and the value of E_0 decreases; the pore system then becomes more accessible to larger molecules.[45]

The status and validity of the DR equation have been discussed by a number of workers.[35,55,62–64] There appears to be general agreement that Dubinin's original approach was essentially empirical, but recently McEnaney[63] and Kadlec[64] have argued that the mathematical form of the DA equation can be justified on theoretical grounds, the parameters m and E_0 being dependent on adsorbent heterogeneity. It remains to be seen whether it will ever be possible to verify the theoretical significance of these terms, but for the time being it is likely that the DR, DA and DRS equations will continue to be applied in a useful and empirical manner.

With the aid of adsorption microcalorimetry, Rouquerol and co-workers[65] carried out a systematic investigation of the adsorption of nitrogen and argon by a series of activated charcoals. The results are consistent with the hypothesis of primary and secondary micropore filling, and also provide additional evidence to support the choice of untreated Vulcan carbon black as the most appropriate non-porous reference adsorbent.

Some microporous carbons have been found to adsorb small amounts of nitrogen at 77 K whilst exhibiting much larger adsorption capacities at higher temperatures. These effects are associated with the 'activated entry' of adsorbate molecules through very narrow pore entrances or constrictions.[66] Thus, with some activated carbons it has been found that the uptake of carbon dioxide (e.g. at 273 K) is much larger than that of nitrogen at 77 K, although the minimum dimensions of the two molecules are not very different (CO_2, 0.28 nm;

N_2, 0.30 nm). The most important factor in allowing the larger uptake of carbon dioxide is the higher temperature of measurement, which provides the adsorbate molecules with sufficient kinetic energy to overcome the energy barrier at the pore entrance.[66] It has been pointed out by Rodríguez-Reinoso and co-workers[60,67] that it is possible to obtain a more complete picture of the micropore structure by the use of both nitrogen and carbon dioxide. It must be kept in mind, however, that the CO_2 molecule has a high quadrupole moment and that the contribution of the $\phi_{\dot{F}Q}$ term in equation 2.12 may lead to an unwelcome sensitivity of the isotherm shape to the presence of exposed ions or polar groups.[68]

Various attempts have been made to employ physisorption measurements for the evaluation of the micropore-size distribution. The most direct approach is to use a selection of molecular probes of different size and shape. This procedure, as illustrated by the work of Dacey and Thomas[69] and Barton *et al.*,[70] is eminently suitable for the characterization of molecular sieve carbons, but obviously cannot be used as the micropore width becomes larger than the dimensions of the largest available adsorptive molecules (say, 0.8 nm). Of course, this range can be extended by the use of heat of immersion or adsorption from solution measurements, but the interpretation of the experimental data is then less straightforward.

Stoeckli's generalized treatment[58] of the DR equation provides an elegant method for the determination of micropore-size distribution. As noted earlier, Stoeckli proposed that $f(B)$ in equation 2.23 could be expressed in terms of the half-width, Δ, of a normalized Gaussian distribution of pore size. According to this view, Δ gives a measure of the spread of B and hence of the micropore width. However, McEnaney[63] and others[39,64,71] have drawn attention to the difficulty in arriving at an unambiguous interpretation of a distribution function such as $f(B)$ since the observed heterogeneity may be the result of a distribution of micropore potentials and/or depend on the presence of defects or various functional groups on the carbon surface.

A similar problem is encountered in the application of an alternative procedure proposed by Horvath and Kawazoe[72] in which use is made of the calculated enhancement of adsorption energy to provide an approximate indication of the relative pressures at which the adsorption of N_2 is expected to occur in molecular-sized pores of different widths. This method, which entails high-resolution (HRADS) measurements at very low relative pressures, seems promising provided that it can be calibrated with the aid of well-defined micropore structures. Recently, Kaneko and co-workers[73] have also attempted to characterize the p/p^0 range of filling of micropores of different dimensions. They noted that some DR plots appear to exhibit a number of linear regions and interpreted these features in terms of a multistage mechanism of micropore filling. Again, it is too early to judge the success of this interpretation, but for the reasons given above it is unlikely to be of general validity.

In view of the uncertainties involved in the interpretation of isotherms given by a particular adsorptive, we are led to the conclusion that if is preferable to use a number of adsorptives for the determination of the effective micropore-size distribution.[74] Preliminary work[59,68,75] has indicated that if account is taken of the two stages of micropore filling in relation to the pore width/molecular diameter ratio, it is possible to study the complete range of micropore size with the aid of non-polar molecules such as propane, isobutane and neopentane.

Finally, we must refer briefly to the special case of the adsorption of water vapour by microporous carbons. It is well known that activated carbons generally have a very low affinity for water, i.e. the overall adsorbent–adsorbate interaction is relatively weak.[15] However, there are some polar groups on the internal carbon surface which intereact more strongly (i.e. specifically) with water molecules and furthermore the adsorbate–adsorbate interactions (i.e. hydrogen bonding) come into play as p/p^0 is increased. Clustering of the water molecules around particular active sites thus leads to micropore filling. The most striking features of the associated Type V isotherm are the pronounced upward swing (usually at $p/p^0 \approx 0.4–0.6$) and the well-defined plateau which corresponds to quantitative pore filling.[76] The latter is in contrast to the behaviour of the hydrophobic zeolite Silicalite®,[77] which adsorbs very little water at high p/p^0. It is believed [78] that this difference is mainly due to the effect of pore shape. Thus, the hydrogen-bonded water structure can be accommodated within the slit-shaped pores of molecular sieve carbons (of $d \approx 0.5$ nm) in a manner which is not possible in the tubular pore structure of Silicalite.

2.7 References

1. SING, K.S.W., EVERETT, D.H., HAUL, R.A.W., MOSCOU, L. PIEROTTI, R.A., ROUQUEROL, J. and SIEMIENIEWSKA, T. 1985: *Pure and Applied Chemistry* **57**, 603.
2. LÉTOQUART, C., ROUQUEROL, F. and ROUQUEROL, J. 1983: *Journal de Chimie Physique* **70**, 559.
3. GREGG, S.J. and SING, K.S.W. 1982: *Adsorption, Surface Area and Porosity.* London: Academic Press, Chapter 1.
4. AVGUL, N.N., BEZUS, A.G., DOBROVA, E.S. and KISELEV, A.V. 1973: *Journal of Colloid Interface Science* **42**, 486.
5. BARRER, R.M. 1978: *Zeolites and Clay Minerals as Sorbents and Molecular Sieves.* London: Academic Press, p. 131.
6. KISELEV, A.V. 1967: *Advances in Chromatography* **4**, 113.
7. ROUQUEROL, J., ROUQUEROL, F., GRILLET, Y. and WARD, R.J. 1988: In Unger, K.K., Rouquerol, J., Sing, K.S.W. and Kral, H. (eds) *Characterization of Porous Solids.* Amsterdam: Elsevier, p. 67.
8. CONDER, J.R. and YOUNG, C.L. 1979: *Physicochemical measurement by Gas Chromatography.* Chichester, Wiley: p. 437.
9. VIDAL-MADJAR, C. and JAULMES, A. 1989: *Pure and Applied Chemistry* **61**, 2005.
10. POIRIER, J.E., FRANCOIS, M., CASES, J.M. and ROUQUEROL, J. 1987: in Liapis, A.I (ed.) *Fundamentals of Adsorption.* New York: Engineering Foundation, p. 473.
11. MIKHAIL, R. Sh. and ROBENS, E. 1983: *Microstructure and Thermal Analysis of Solid Surfaces.* Chichester: Wiley–Heyden, p. 41.
12. STEELE, W.A. 1974: *The Interaction of Gases with Solid Surfaces.* Oxford: Pergamon Press, p. 13.
13. YOUNG, D.M. and CROWELL, A.D. 1962: *Physical Adsorption of Gases.* London: Butterworth, p. 25.
14. KISELEV, A.V. 1965: *Discuss. Faraday Society* **40**, 205.
15. BARRER, R.M. 1966: *Journal of Colloid Interface Science* **21**, 415.
16. BARRER, R.M. 1978: *Zeolites and Clay Minerals as Sorbents and Molecular Sieves.* London: Academic Press, p. 174.
17. AVGUL, N.N. and KISELEV, A.V. 1970: *Chemistry and Physics of Carbon* 1.

18. PARKYNS, N.D. and SING, K.S.W. 1975: in *Colloid Science, Specialist Periodical Report*. London: Chemical Society, Vol. 2, p. 18.
19. GRILLET, Y., ROUQUEROL, F. and ROUQUEROL, J. 1979: *Journal of Colloid Interface Science* **70**, 239.
20. STOECKLI, H.F. 1974: *Helvetica Chimica Acta* **57**, 2195.
21. EVERETT, D.H. and POWL, J.C. 1976: *Journal of the Chemical Society, Faraday Transaction 1* **72**, 619.
22. ATKINSON, D., CARROTT, P.J.M., GRILLET, Y., ROUQUEROL, J. and SING, K.S.W. 1987: in Liapis, A.I. (ed.) *Fundamentals of Adsorption*. New York: Engineering Foundation, p. 89.
23. CARROTT, P.J.M. and SING, K.S.W. 1987: *Journal of Chromatography* **406**, 139.
24. CARROTT, P.J.M. and SING, K.S.W. 1989: *Adsorption Science Technology* **6**, 136.
25. GREGG, S.J. and SING, K.S.W. 1982: *Adsorption, Surface Area and Porosity*. London: Academic Press, p. 262.
26. GREGG, S.J. and SING, K.S.W. 1982: *Adsorption, Surface Area and Porosity*. London: Academic Press, Chapter 3.
27. SING, K.S.W. 1989: *Colloids and Surfaces* **38**, 113.
28. DASH, J.G. 1975: *Films on Solid Surfaces*. New York: Academic Press, p. 216.
29. BRUNAUER, S., EMMETT, P.H. and TELLER, E. 1938: *Journal of the American Chemical Society* **60**, 309.
30. GREGG, S.J. and SING, K.S.W. 1982: *Adsorption, Surface Area and Porosity*. London: Academic Press, p. 49.
31. DUBININ, M.M. 1970: in Everett, D.H. and Ottewill, R.H. (eds) *Surface Area Determination*. London: Butterworth, p. 123.
32. YOUNG, D.M. and CROWELL, A.D. 1962: *Physical Adsorption of Gases*. London: Butterworth, p. 234.
33. MCCLELLAN, A.L. and HARNSBERGER, H.F. 1967: *Journal of Colloid Interface Science* **23**, 577.
34. SING, K.S.W. 1970: in Everett, D.H. and Ottewill, R.H. (eds) *Surface Area Determination*. London: Butterworths, p. 25.
35. CARROTT, P.J.M., ROBERTS, R.A. and SING, K.S.W. 1987: *Carbon* **25**, 59.
36. CARROTT, P.J.M., ROBERTS, R.A. and SING, K.S.W. 1988: *Langmuir* **4**, 740.
37. CARROTT, P.J.M. and SING, K.S.W. 1989: *Pure and Applied Chemistry* **61**, 1835.
38. CARROTT, P.J.M., ROBERTS, R.A. and SING, K.S.W. 1988: in Lloyd, P.J. (ed.) *Particle Size Analysis*. Chichester: John Wiley, p. 217.
39. GREGG, S.J. and SING, K.S.W. 1982: *Adsorption, Surface Area and Porosity*. London: Academic Press, p. 207.
40. CARROTT, P.J.M., DRUMMOND, F.C., KENNY, M.B., ROBERTS, R.A. and SING, K.S.W. 1989: *Colloids and Surfaces* **37**, 1.
41. EVERETT, D.H. 1988: in Unger, K.K., Rouquerol, J., Sing, K.S.W. and Kral, H. (eds) *Characterization of Porous Solids*. Amsterdam: Elsevier, p. 13.
42. GREGG, S.J. and SING, K.S.W. 1982: *Adsorption, Surface Area and Porosity*. London: Academic Press, p. 132.
43. EVERETT, D.H. 1979: in Gregg, S.J., Sing, K.S.W. and Stoeckli, H.F. (eds) *Characterization of Porous Solids*. London: Society of Chemical Industry, p. 249.
44. KORESH, J.E., KIM, T.H., WALKER, D.R.B. and KOROS, W.J. 1989: *Journal of the Chemical Society, Faraday Transactions 1* **85**, 4311.
45. BANSAL, R.C., DONNET, J.B. and STOECKLI, F. 1988: *Active Carbon*. New York: Marcel Dekker, p. 119.
46. DUBININ, M.M. 1974: *Journal of Colloid Interface Science* **46**, 351.
47. DUBININ, M.M. 1983: *Carbon* **21**, 359.
48. CARROTT, P.J.M. and SING, K.S.W. 1988: in Unger, K.K., Rouquerol, J., Sing, K.S.W. and Kral, H. (eds) *Characterization of Porous Solids*. Amsterdam: Elsevier, p. 77.
49. CARROTT, P.J.M., ROBERTS, R.A. and SING, K.S.W. 1987: *Chemical Industries Newsletter (London)* 855.

50. DUBININ, M.M. 1960: *Chem. Rev.* **60**, 235.
51. BERING, B.P., DUBININ, M.M. and SERPINSKY, V.V. 1966: *Journal of Colloid Interface Science* **21**, 378.
52. DUBININ, M.M. 1967: *Journal of Colloid Interface Science* **23** 487.
53. DUBININ, M.M. 1958: in *1st International Conference on Industrial Carbon and Graphite.* London: Society of Chemical Industry, p. 219.
54. DUBININ, M.M. and RADUSHKEVICH, L.V. 1947: *Proceedings. Academy of Sciences USSR* **55**, 331.
55. RODRIGUEZ-REINOSO, F., GARRIDO, J., MARTIN-MARTINEZ, J.M., MOLINA-SABIO, M. and TORREGROSA, R. 1989: *Carbon* **27**, 23.
56. DUBININ, M.M. and ASTAKHOV, V.A. 1971: *Advances in Chemistry Series*, No. **102**, 69.
57. STOECKLI, H.F. 1977: *Journal of Colloid Interface Science* **59**, 184.
58. HUBER, U., STOECKLI, H.F. and HOURIET, J.-P. 1978: *Journal of Colloid Interface Science* **67**, 195.
59. CARROTT, P.J.M., ROBERTS, R.A. and SING, K.S.W. 1988: in Unger, K.K., Rouquerol, J., Sing, K.S.W. and Kral, H. (eds) *Characterization of Porous Solids.* Amsterdam: Elsevier, p. 89.
60. RODRIGUEZ-REINOSO, F. 1989: *Pure Applied Chemistry* **61**, 1859.
61. DUBININ, M.M. 1989: *Carbon* **27**, 457.
62. DUBININ, M.M. and STOECKLI, H.F. 1980: *Journal of Colloid Interface Science* **75**, 34.
63. MCENANEY, B. 1988: *Carbon* **26**, 267.
64. KADLEC, O. 1991: in Rodriguez-Reinoso, F., Rouquerol, J., Sing, K.S.W. and Unger, K.K. (eds) *Characterization of Porous Solids—II.* Amsterdam: Elsevier, p. 759.
65. FERNANDEZ-COLINAS, J., DENOYEL, R., GRILLET, Y., ROUQUEROL, F. and ROUQUEROL, J. 1989: *Langmuir* **5**, 1205.
66. GREGG, S.J. and SING, K.S.W. 1982: *Adsorption, Surface Area and Porosity.* London: Academic Press, p. 228.
67. GARRIDO, J., LINARES-SOLANO, A., MARTIN-MARTINEZ, J.M., MOLINA-SABIO, M., RODRIGUEZ-REINOSO, F. and TORREGROSA, R. 1987: *Langmuir* **3**, 76.
68. SING, K.S.W. 1989: *Carbon* **27**, 5.
69. DACEY, J.R. and THOMAS, D.G. 1954: *Transactions of the Faraday Society* **50**, 740.
70. BARTON, S.S., EVANS, M.J.B. and HARRISON, B.H. 1974: *Journal of Colloid Interface Science* **40**, 462.
71. JARONIEC, M. and CHOMA, J. 1989: *Colloids and Surfaces* **37**, 183.
72. HORVATH, G. and KAWAZOE, K. 1983: *Journal of Chemical Engineering of Japan* **16**, 470.
73. KAKEI, K., OZEKI, S., SUZUKI, T. and KANEKO, K. 1990: *Journal of the Chemical Society, Faraday Transactions* **86**, 371.
74. LU, X., JARONIEC, M. and MADEY, R. 1991: *Langmuir* **7**, 173.
75. ROBERTS, R.A., SING, K.S.W. and TRIPATHI, V. 1987: *Langmuir* **3**, 331.
76. GREGG, S.J. and SING, K.S.W. 1982: *Adsorption, Surface Area and Porosity.* London: Academic Press, p. 262.
77. KENNY, M.B. and SING, K.S.W. 1990: *Chemistry and Industry (London)*, 39
78. CARROTT, P.J.M., KENNY, M.B., ROBERTS, R.A., SING, K.S.W. and THEOCHASIS, C.R. 1991: in Rodriguez-Reinoso, F., Rouquerol, J., Sing, K.S.W. and Unger, K.K. (eds) *Characterization of Porous Solids—II.* Amsterdam: Elsevier, p. 685.

3 Characterization of Microporous Carbons by Adsorption and Immersion Techniques

H. Fritz Stoeckli
Institut de Chimie
University of Neuchâtel
Switzerland

3.1 Introduction

As discussed in the previous chapter, adsorption of vapours by microporous carbons may be divided into a process of micropore filling, well described by Dubinin's theory, and the physical adsorption on the surface of the meso- and macropores.

In spite of the fact that relatively large external or non-microporous areas can be found in some materials, the bulk of the adsorption takes place in the micropores whose volume can be as high as 0.6–0.8 $cm^3 g^{-1}$ for certain active carbons (active carbons are industrial carbons with a well-developed micropore system). By definition,[1] microporosity covers a relatively narrow range of dimensions (0.35–2 nm), where adsorption is reversible, in principle at least. However, in view of the dimensions of the adsorbates, energy and geometrical factors play an important role, in particular in micropores of less than two molecular diameters in width.

As shown by experimental and theoretical studies, the adsorption energy is greatly enhanced in narrow pores and entry may even require an activation energy. For ideally slit-shaped pores, having a width of two molecular diameters, the potential energy of adsorption is approximately twice as large as on a flat reference surface (graphite or graphitized carbon blacks).[2–4] The size of the micropores, in particular below 0.6–0.8 nm, also plays a role in the selectivity of adsorption, through molecular sieve and gate effects. Like other carbons, microporous carbons also contain chemically active centres, responsible for specific interactions, and whose distribution within the structure is of great importance. It will be shown that their accessibility to a first or a second adsorbate depends on their location. It is therefore important to characterize microporous carbons and in particular active carbons, from a structural and a chemical point of view. In this chapter we show that this can be achieved by a combination of techniques based on the adsorption of vapours and gases, on immersion calorimetry and on high-resolution transmission electron microscopy (HRTEM). This combined approach leads to useful correlations between parameters of the adsorption isotherms and structural properties of typical active carbons. As far as the chemistry of these solids is concerned, we shall deal essentially with the problem of the hydrophilic centres and their distribution within the structure, since water is often present in real systems.

Keeping in mind the practical requirements of the potential users of active carbons, we shall examine experimental techniques and correlations of variable degrees of sophistication, starting with the information derived from a single adsorption isotherm, provided that it covers a sufficiently wide range of relative pressures.

3.2 Experimental techniques

In the present context, the main techniques used for the characterization of microporous carbons are based on the adsorption of gases and vapours and on immersion calorimetry.[5–8]

The adsorption isotherms of gases and vapours can be determined volumetrically or gravimetrically, but in view of the high sorptive capacity of active carbons, the latter technique is preferred for these carbons. Adsorption is measured directly by monitoring the weight increase of the sample (usually 0.1–0.5 g) contained in a bucket at the end of a quartz spring. Weight changes can be measured within ±0.1 mg, which corresponds to ±0.001 cm^3 or ±0.3 m^2 for nitrogen adsorbed at 78 K. Micropore volumes of typical active carbons are therefore known to within ±0.01 $cm^3\,g^{-1}$.

The carbon sample and the spring are contained in a vacuum system, allowing it to be treated *in situ* and to be outgassed prior to adsorption experiments. Normally, outgassing under 133×10^{-5} Pa for 15–20 hrs at 300°C and for 1–2 hrs at 400–500°C is sufficient for most purposes. However, in experiments with water, a standard procedure must be adopted, since the various chemical groups which can be present at the surface are influenced by the heat treatment. Between adsorption runs on a given sample, the time of outgassing depends on the nature of the adsorbate, but in any case the temperature should not be raised too quickly, in order to avoid the decomposition of organic molecules in the micropores.

Unless specific systems are investigated, the characterization of microporous carbons can be carried out with simple vapours. Although nitrogen at 78 K and CO_2 at 273–298 K are popular adsorbates, vapours such as methylene chloride and benzene, or other simple organic compounds used near room temperature, present a number of advantages. First, the corresponding liquids can be used at the same temperature for experiments in immersion calorimetry, which provides complementary information on the carbon. Second, a simple molecule such as methylene chloride is of approximately the same size as nitrogen, but it is not affected as strongly as nitrogen by activated diffusion into narrow pores.

Since a number of simple organic compounds have vapour pressures in the range of 1.3–13.3 kPa at room temperature, it is possible to cover a relatively large range of relative pressures by using standard pressure gauges. In the course of a single experiment it is possible to follow most of the micropore filling ($\theta > 0.1$), in addition to adsorption on the external surface area, which begins in the region of $p/p^0 \approx 0.05$. This means that with simple organic vapours a great deal of information can already be obtained from a single isotherm. Moreover, the adsorption isotherm can also be compared with the enthalpy of immersion into the corresponding liquid. In the case of CO_2 at 273–293 K, on the other hand, information is limited to low relative pressures, or high values of $\log^2(p^0/p)$, and their interpretation may lead to inconsistent results.

Immersion calorimetry can be carried out at room temperature, with a simple calorimeter of the Tian–Calvet type, described in detail elsewhere,[7] and designed specifically for the energies evolved in the case of active carbons (typically 10–150 J g^{-1}, corresponding to 2–50 J released in the calorimeter). The reproducibility of the calorimeter used here is ±1% for typical experiments. The solid (usually 0.3 g) is outgassed in a glass bulb attached to a vacuum line and sealed off before being transferred to the calorimeter. In the case of preadsorption with variable amounts of *n*-nonane or *n*-undecane, the bulb is surrounded by liquid nitrogen and the degree of micropore filling is determined by weighing (this requires some practice, but excellent results can be obtained, the filling being a function of time).

Prior to immersion at 293 K, the sealed bulbs containing the prefilled carbon are left at room temperature for approximately 24 hrs. Tests carried out gravimetrically, by adsorbing N_2 at 78 K on samples prefilled with *n*-nonane, showed that in the case of typical industrial carbons no major parts of the micropore system are blocked off, provided that prefilling is not carried out too rapidly.

Table 3.1 lists the main physical properties of typical adsorbates used in adsorption and immersion experiments at 293 K. For nitrogen at 78 K, V_m = 34.6 cm^3 mol^{-1} and $\beta = 0.34$. For *n*-nonane, the extrapolated density at 78 K is 0.89 g cm^{-3} and V_m = 144 cm^3 mol^{-1}. The critical dimensions L of the adsorbates are based on their van der Waals diameters and on the liquid densities.

Table 3.1 List of organic liquids recommended for immersion and pre-adsorption experiments at 293 K (the first adsorptives, up to CCl_4, can also be used easily for vapour adsorption)

Liquid	L nm	α 10^{-3} K^{-1}	β	$-h_i$ J m^{-2}	V_m cm^3 mol^{-1}
Methylene chloride	0.33	1.34	0.66	0.152	64.02
Benzene	0.41	1.24	1.00	0.114	88.91
Cyclohexane	0.54	0.96	1.04	0.101	108.10
Carbon tetrachloride	0.63	1.22	1.05	0.115	96.50
Norborna-2,5-diene	0.66	1.20	1.62	0.110	101.63
Cyclododeca-1,5,9-triene	0.76	0.76	1.90	0.103	182.02
α-Pinene	0.75	1.02	1.70	0.110	158.75
Perchlorocyclopentadiene	0.80	1.17	1.91	0.110	159.30
Tetrabutylurea	0.93	0.85	3.50	0.118	282.34
Tetraisopropyl orthotitanate	1.05	2.23	3.31	0.205	294.57
Tetrabutyl orthotitanate monomer	1.30	0.52	3.83	0.167	340.36
Tri-2,4-xylyl phosphate	1.50	0.66	4.05	0.160	360.04
n-Nonane		1.55	1.90	0.136	178.73
n-Undecane		1.43	2.53	0.137	213.86

3.3 Dubinin's theory for the volume filling of micropores

A convenient background for the description of adsorption by microporous solids in general is provided by Dubinin's theory,[7,9,10] developed in successive stages since the late 1940s and applied mainly to active carbons. This theory is certainly not exclusive, but it provides a good description of the adsorption of a variety of vapours by microporous carbons. On the other hand, it does not describe adsorption in the region of Henry's law and a different theoretical framework has to be used for this particular domain.

In its present formulation, Dubinin's theory is expressed by the Dubinin–Astakhov (DA) equation

$$W = W_0 \exp[-(A/\beta E_0)^n]. \tag{3.1}$$

where W represents the volume filled at temperature T and relative pressure p/p^0, W_0 is the total volume of the micropores, $A = RT \ln(p^0/p)$ and n, E_0 and β are parameters of the system under investigation.

For a variety of microporous carbons (in particular active carbons of industrial origin), the adsorption data can be fitted over a variable range of relative pressures to equation 3.1 with exponent $n = 2$. This corresponds to the original equation of Dubinin and Radushkevich (DR) postulated in 1947[11]

$$W = W_0 \exp[-B(T/\beta)^2 \log^2(p^0/p)]. \tag{3.2}$$

In the modern formulation, the DR equation becomes

$$W = W_0 \exp[-(A/\beta E_0)^2]. \tag{3.3}$$

Parameter B, known as the structural constant, is related to the so-called characteristic energy E_0 in equation 3.3 by[9]

$$E_0\ (\text{kJ mol}^{-1}) = 0.01914/B^{0.5} \tag{3.4}$$

It was noted at an early stage that B increases with increasing degree of activation (burn-off) and with the accessibility of the micropore system to larger molecules. This means that decreasing values of the structural constant E_0 correspond to increasing micropore sizes. The practical range of 30–35 to 17–18 kJ mol^{-1} for E_0 corresponds to average pore widths of approximately 0.4–2 nm.

The usefulness of the DR equation is illustrated by the fact that it applies to a variety of vapours, each being characterized by a specific value of β, the so-called affinity coefficient. By convention, benzene is the reference vapour and $\beta(C_6H_6) = 1$. Experimental values of β are given in Table 3.1. This interesting property holds over wide pressure and temperature ranges, as long as molecular sieve effects are absent. Consequently, the data for different vapours can be fitted to a single curve.

The DR equation corresponds to a Type I isotherm[5] and therefore Type II isotherms, often observed with active carbons, indicate the presence of a non-negligible external surface area, S_e. On this surface, multilayer adsorption becomes significant at relative pressures $p/p^0 \geqslant 0.05$, where the filling of the micropore system is nearly completed. With the exception of a transition region, corresponding to so-called supermicropores, the two domains are distinct. The experimental adsorption isotherm therefore consists of two contributions.

$$V_t(T;p/p^0) = W(T;p/p^0) + S_e t(p/p^0)$$

where V_t is the total volume adsorbed by the solid, $W(T;p)$ is the DR equation 3.3 and $t(p/p^0)$ represents the thickness of the layer adsorbed on S_e. Various techniques are available to determine the micropore volume and the external surface area.[5,6,12,13] Basically, one may either use a standard reference isotherm $t(p/p^0)$ for N_2 (77 K) or C_6H_6 (293 K), or subject the residual isotherm (overall isotherm minus the DR equation) to a BET analysis.[14–16] Table 3.2 shows the values of S_e obtained by different techniques for carbons N–125 and U–02.

At sufficiently low relative pressures, adsorption on the external surface is negligible and the DR equation can be represented in its linearized form, by plotting $\ln W$ against A^2, $\log^2(p^0/p)$ or the quantity $y = (T/\beta)^2 \log^2(p^0/p)$. The last approach is very convenient if several adsorptives and different working temperatures are considered, and ideally a unique straight line should be obtained. The slope and the intercept lead to constant B or E_0 and to the limiting micropore volume W_0.

Figure 3.1 illustrates the case of adsorption by carbon CEP-35, a barbecue charcoal activated in CO_2 at 840°C, to a burn-off (weight loss) of 35%.[17,18] The adsorption of small molecules such as N_2 (78 K), N_2O (293 K) and C_3H_8 (293 K) shows a common linear section leading to $W_0 = 0.34$ cm^3 g^{-1} and $B = 0.41 \times 10^{-6}$ K^{-2}, which corresponds to $E_0 = 30$ kJ mol^{-1}. For CCl_4, on the other hand, a larger molecule (0.63 nm), the smaller value of $W_0 = 0.16$ cm^3 g^{-1} clearly reflects molecular sieve effects. For this adsorbate, the linear section of the DR plot leads to $E_0 = 27.5$ kJ mol^{-1} and since CCl_4 fills only the upper part of the micropore system, it appears that E_0 (or parameter B) is related to the

Table 3.2 External surface area, S_e ($m^2\ g^{-1}$), of active carbons N-125 and U-02 obtained by different techniques[a]

Carbon	Immersion		Isotherm analysis	Saturation $+N_2$ (78 K)		t Plot (N_2)	t/F(N_2)	Kelvin eqn (mesopores)
	C_6H_6	CCl_4		C_6H_{12}	C_9H_{20}			
N-125	169	153	155	155	141	128	175	145
U-02	106	116	131	93	115	110	108	96

[a] The average values are 157 ± 20 and $105 \pm 16\ m^2\ g^{-1}$. The t/F technique of Dubinin and Kadlec is described in ref. 13.

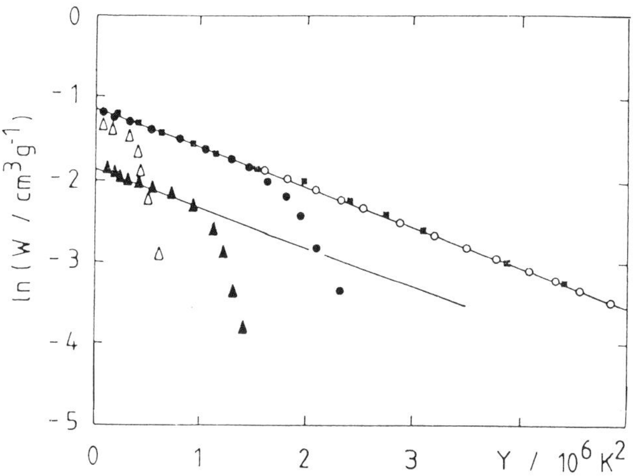

Fig. 3.1 DR plot for the adsorption of different vapours by active carbon CEP-35.[17] The adsorptions are N_2 at 78 K (●) and N_2O (○), C_3H_8 (■), C_6H_6 (△) and CCl_4 (▲) at 293 K. Variable $y = (T/\beta)^2 \log^2(p^0/p)$

average micropore width. The difference in adsorption between N_2O (293 K) and N_2 (78 K) also suggests activated entry into the smaller pores. This is further illustrated by the behaviour of benzene and carbon tetrachloride at low pressures.

Dubinin's theory postulates the filling of the micropore volume by the condensed adsorbate, but its true state is still open to discussion. To a first approximation, one may postulate that the adsorbed state is not too different from the liquid state at the same pressure and temperature. This is suggested by the study of relatively large micropore systems, without molecular sieve effects, and where similar values of W_0 are obtained by using the limiting amounts absorbed, N_{a0}, and the molar volume of the liquid adsorbates. On the other hand, in the case of narrow micropores of 1–3 molecular diameters in width, it is possible that the density and the packing of the adsorbate may be different, owing to geometrical constraints. As a working hypothesis, we suggest systematically using the liquid state as a reference (Table 3.1), and

$$W_0(\text{cm}^3\,\text{g}^{-1}) = N_{a0}(\text{mol/g}) \times V_m(\text{cm}^3\,\text{mol}^{-1}).$$

3.4 Determination of micropore distribution

The DR equation is valid for typical microporous carbons and it has provided a basis for more refined descriptions. From a historical point of view, it appeared that for strongly activated and therefore very heterogeneous carbons, the traditional DR plot was no longer linear, but showed an upward curvature (see Figure 3.2). This led Izotova and Dubinin[9,19] to postulate the existence of a bimodal system of micropores, corresponding to a sum of two contributions. In the modern formulation, the overall isotherm is given by

$$W_t = W_{01} \exp[-(A/\beta E_{01})^2] + W_{02} \exp[-(A/\beta E_{02})^2] \tag{3.5}$$

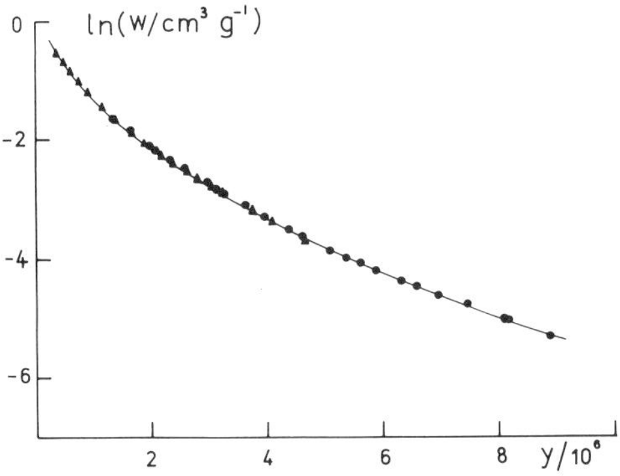

Fig. 3.2 DR plot for carbon F-85, activated to 85 wt% burn-off. The adsorbate is N_2O at 243 K (▲) and 293 K (●). Variable $y = (T/\beta)^2 \log^2(p^0/p)$

E_{01} and E_{02} are usually found in the range 25–35 and 15–20 kJ mol^{-1} respectively, corresponding formally to sets of narrow and wide micropores (0.4–0.7 and 1.4–2 nm, approximately). This approach was also based on simultaneous data from the small-angle scattering of X-rays (SAXS),[20] suggesting the presence of heterogeneous (but not necessarily bimodal) micropore systems. This leads to a first quantitative correlation between parameter B (now E_0) of the adsorption isotherm and an average micropore dimension, the so-called gyration radius R_g, used in X-ray theory.[21] Its value is close to the half-width of disc-shaped micropores.

Later, two generalizations of Dubinin and Izotova's approach were proposed successively by Stoeckli and co-workers[22] and by Dubinin.[10] The former postulated that a Gaussian distribution of parameter B, and a standard integral transform based on the DR equation 3.2 leads to the following generalized equation.

$$W_t = W_0 \exp[-B_0 y] \exp[y^2\Delta^2/2]\{0.5[1-\text{erf}(z)]\} \tag{3.6}$$

where $y = (T/\beta)^2 \log(p^0/p)$, $z = (y - B_0/\Delta^2)\Delta/2^{0.5}$ and Δ is a measure of the dispersion of B around B_0. B and B_0 being related to pore dimensions, like E_0, it is possible to obtain a micropore distribution.[7] Table 3.3 gives typical values obtained for strongly activated carbons. For these carbons, one also finds values of exponent $n<2$ in the DA equation 3.1. Dubinin's generalization, on the other hand, is based on a Gaussian distribution $f(x) = \mathrm{d}W/\mathrm{d}x$ of the half-width x of the micropores, a sequel of the earlier work based on the gyration radius R_g,

$$\frac{\mathrm{d}W}{\mathrm{d}x} = \frac{W_0}{\delta(2\pi)^{0.5}} \cdot \exp\left[\frac{-(x_0-x)^2}{2\delta^2}\right] \tag{3.7}$$

It was known that E_0 is an inverse function of x and to a first approximation[23] $E_0 = k_0/x$ (the choice of $k_0 = 12$ nm kJ mol^{-1} postulated by Dubinin is valid only for carbons with E_0 near 20–22 kJ mol^{-1}).[7] Provided that k_0 is a constant, the integral transform

$$W_t(A) = \int_0^\infty f(x) \exp[-(Ax/\beta k_o)^2]\,\mathrm{d}x \tag{3.8}$$

Table 3.3 Structural parameters of active carbons CEP-59, U-02 and F-02, derived from different models[a]

Parameter	Carbon		
	CEP-59	U-02	F-02
W_0 cm^3 g^{-1}	0.48	0.43	0.64
S_e m^2 g^{-1}	31.00	105.00	90.00
E_0 kJ mol^{-1}	25.00	20.00	18.70
$\bar{L}$ nm	0.79	1.30	1.90
B_0 10^{-6} K^{-2}	0.66	0.92	1.03
Δ 10^{-6} K^{-2}	0.18	0.21	0.29
$2x_0$ nm	0.80	1.50	1.90
δ nm	0.10	0.15	0.20
n	1.67	1.65	1.40
a nm^{-3}	2.99	0.22	0.087
$a^{-1/3}$ nm	0.69	1.65	2.20
ν	1.85	1.32	1.21

[a] In equation 3.9, $k_0 = K_0/2$.

leads to the overall adsorption isotherm

$$W_t(A) = \frac{W_0}{2(1+m\delta^2A^2)^{0.5}} \exp\left(\frac{-mx_0{}^2A^2}{1+2m\delta^2A^2}\right)\left\{1+\mathrm{erf}\left[\frac{x_0}{\delta[2(1+2m\delta^2A^2)]^{0.5}}\right]\right\} \quad (3.9)$$

where $m = (1/\beta k_0)^2$. Typical values of x_0 and δ, obtained from the best fit of the adsorption data to equation 3.9, are given in Table 3.3. It is obvious that equations 3.8 and 3.9 can also be expressed in terms of $L = 2x$, the actual width of the ideally slit-shaped micropores. This quantity is more convenient than x.

In the 1980s, work based on vapour adsorption and on immersion calorimetry[17,18] showed clearly that for monodisperse systems one should use the DA equation 3.1 with $n = 3$. It was also found that for true carbon molecular sieves with accessible widths of 0.5 and 0.7 nm, for example, E_0 was equal to 27.6 and 24.0 kJ mol^{-1}. It was further established that carbons such as CEP-35, following the DR equation over a wide range of relative pressures, present a certain degree of heterogeneity, confirmed by HRTEM[24] (Figure 3.3). This means that power $n = 2$ in the DA equation corresponds to an intermediate situation between monodisperse ($n = 3$) and strongly heterogeneous micropore systems ($n<2$). The DR equation may therefore be used to obtain information on average carbons.

On the basis of recent investigations described below, an empirical correlation can be postulated between the characteristic energy E_0 of the DR equation 3.3 and the average value of the micropore width $\bar{L}$,[25,26]

$$\bar{L}\,(\mathrm{nm}) = 10.8/(E_0 - 11.4) \quad (3.10)$$

It follows that the parameter $K = E_0L$, given in nm kJ mol^{-1}, can be expressed as a function of either the pore width L in general or E_0

$$K = 10.8 + 11.4L \quad (3.11)$$

and

$$K_0 = 10.8 + 123.1/(E_0 - 11.4) \quad (3.12)$$

where K_0 corresponds to the average micropore width $\bar{L}$ of the system.

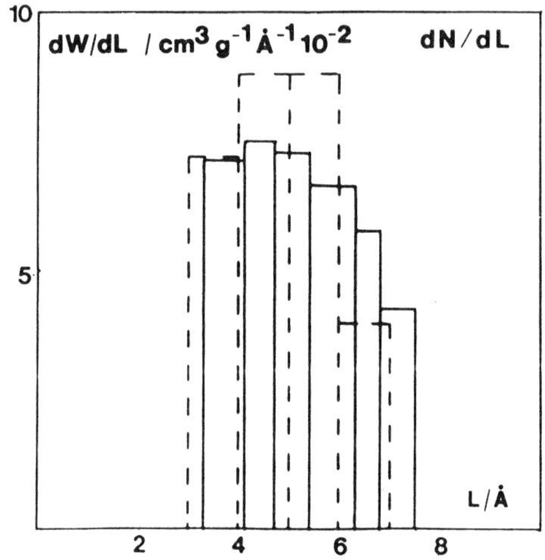

Fig. 3.3 Micropore distribution of carbon CEP-35, obtained from vapour adsorption and immersion calorimetry with molecules of different sizes [17,18] (histogram dW/dL), and from HRTEM[24] (frequency of widths dN/dL, dashed lines)

Equation 3.10 is based on moleculars sieve effects revealed by the adsorption of relatively light vapours ($L<0.8$ nm) and by immersion calorimetry with liquids of critical molecular dimension L up to 1.5 nm (Table 3.1). These experiments provide information on $W(L)$, the volume of the micropores larger than L, filled by the given adsorbate, and one obtains the histograms $\Delta W/\Delta L = f(L)$ shown in Figures 3.3–3.6. It is assumed, implicitly, that no 'gate' effects are present and that the larger micropores are directly accessible from the outside. This is not always the case, even for well-activated carbons and the adsorption of small molecules alone can be misleading.

If one postulates that micropores are slit-shaped, locally at least,[8] the following geometrical relationship should exist between the average pore width $\bar{L}$, the volume W_0 and the surface of their walls S_{mi},

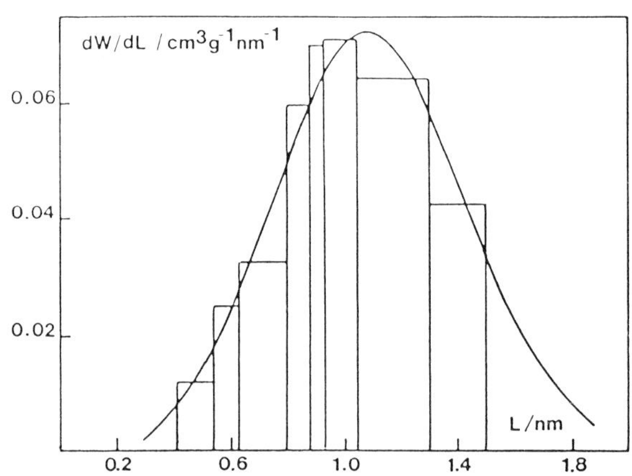

Fig. 3.4 Experimental micropore distribution for the microporous carbon black XC-72 (histogram) and the distribution f(L) corresponding to equation (3.17), calculated from the experimental adsorption isotherm

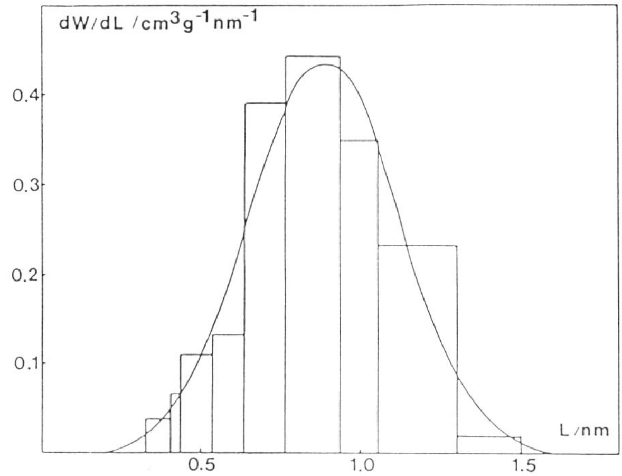

Fig. 3.5 Micropore distribution for active carbon CM. Same conditions as in Figure 3.4

$$\bar{L}\,(\text{nm}) = \frac{2 \times 10^3 [W_0 (\text{cm}^3\,\text{g}^{-1})]}{S_{\text{mi}} (\text{m}^2\,\text{g}^{-1})} \tag{3.13}$$

This area is related to the total surface area S_t of the carbon by

$$S_{\text{mi}} = S_t - S_e \tag{3.14}$$

and S_t can be obtained from the small-angle scattering of X-rays[27] (SAXS) or the adsorption of helium. It has been shown recently[26] that a technique based on the selective adsorption of caffeine from aqueous solutions, monitored calorimetrically, also provides a good estimate of the total area for pores wider than 0.6–0.7 nm (the limit is suggested by experiments with carbons such as CM, shown in Figure 3.5, which have a large proportion of smaller micropores).

It was first established that a simple linear relationship exists between the enthalpy of immersion of non-porous carbon blacks in aqueous solutions of caffeine and their BET surface area, the proportionality constant being -112 ± 15 mJ m^{-2}. As shown in Table 3.4, a similar relationship is found for selected microporous carbons when the enthalpies of immersion are compared with the total surface areas obtained from SAXS, helium adsorption and the cumulative surface area of the distribution to which S_e is added. In the case of carbon U-103 (Figure 3.6), where the histogram does not cover the entire micropore distribution, the data were fitted to a simple analytical curve.

It appears that for carbonaceous materials in general, the energy released by the interaction of the first layer of caffeine with the surface is higher than the enthalpy of disolution of caffeine into water. Although adsorption may not be limited to one layer, the caffeine isotherm shows a plateau in the range of relative concentrations c/c_0 of 0.1–0.6. Immersion calorimetry therefore gives a rapid estimate of the total surface area of carbons in general by using the approximate relationship

$$S_t(\text{m}^2/\text{g}) = -\Delta h_i / (112 \pm 15\ \text{mJ g}^{-1}) \tag{3.15}$$

The values of the average pore widths $\bar{L}$ obtained from molecular sieve

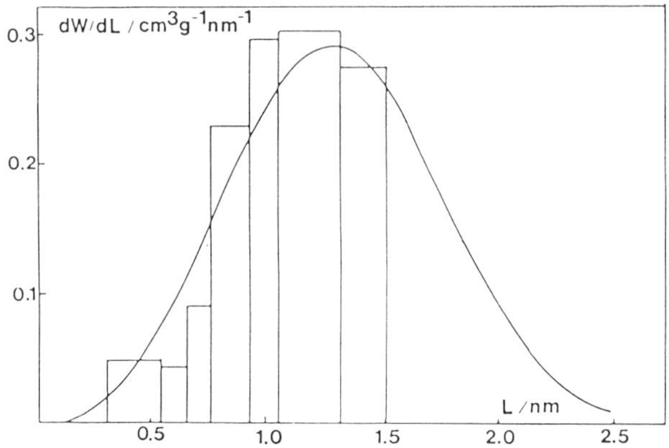

Fig. 3.6 Micropore distribution for carbon U-103. Same conditions as in Figure 3.4

experiments and from S_{mi} are in good agreement and lead to the overall correlation between $\bar{L}$ and E_0 expressed by equation 3.10.

In the case of unknown carbons, equations 3.10 and 3.13 can be used to estimate the average micropore width L and the corresponding area S_{mi} from parameters E_0 and W_0 of the DR equation 3.3. As an example, Figure 3.7 shows the evolution of L in different series of activated carbon,[28] with differences reflecting the history and the treatment of the material. Figure 3.8 shows the evolution of the micropore surface area S_{mi} in activation series CEP/87, calculated and partly cross-checked by the caffeine technique, and the corresponding value of S_{BET} derived from the same experimental adsorption isotherm. The two surfaces are equal for a burn-off near 50% where $L = 0.6$–0.7 nm (see Figure 3.7), and where the average micropore can accommodate approximately two layers of benzene. For larger micropore systems, S_{BET} corresponds practically to the monolayer equivalent of W_0.

From the observation that homogeneous systems of micropores are described by a DA equation with $n = 3$,[18] it is possible to re-examine the description of heterogeneous distribution. As shown recently,[29] a convenient approach is based on an integral transform of the type

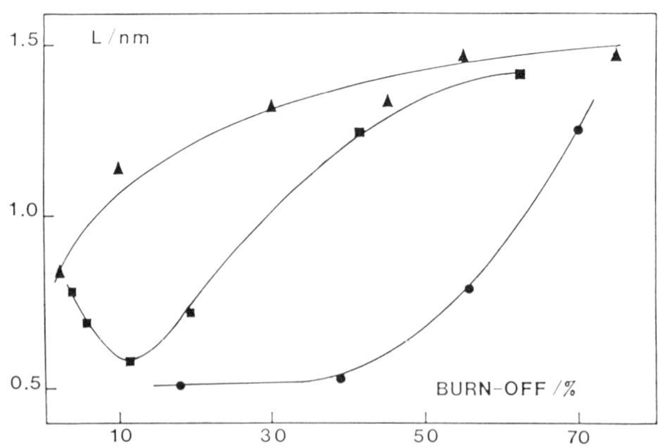

Fig. 3.7 Variation of the average micropore widths L, calculated by equation 3.10, as a function of the degree of burn-off in series D (▲) and B (■) and CEP 87 (●)

Table 3.4 Total surface areas, S_t, and enthalpies of immersion of selected carbons into aqueous solutions of caffeine at 293 K[a]

	Carbon black							
Parameter	V-3	M-100	M-400	M-800	L-100	L-400	L-800	Hoechst
$-\Delta h_i$ J g^{-1}	9.20	9.43	9.93	8.59	8.14	9.16	8.00	7.60
S_{BET} m^2 g^{-1}	71.3	89.0	77.0	76.0	75.4	83.3	81.6	73.1
$-h_i$ mJ m^{-2}	129	106	129	113	108	110	98	104

	XC-72	U-02	F-02	N-125	U-103
$-\Delta h_i$ J g^{-1}	31.0	76.0	84.5	92.5	62.0
S_t m^2 g^{-1}	227	684/620	752	715	630
Method used	$S_{cum} + S_e$	SAXS/He	SAXS	He	$S_{cum} + S_e$
$-h_i$ mJ m^2	136	111/123	112	129	98

[a] For carbons XC-72 and U-103, S_{cum} represents the cumulat ve surface area of the micropores derived from the histograms (Figures 3.4 and 3.6).

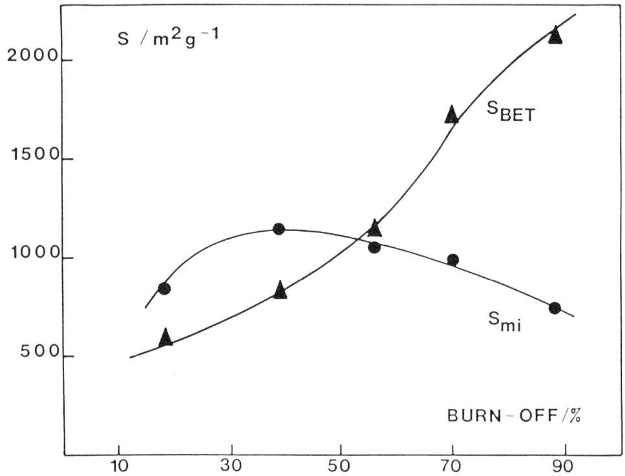

Fig. 3.8 Variation of the BET surface area (▲) and the true surface of the micropores S_{mi} (▲) in series CEP/87 (see also Figure 3.6). S_{mi} is calculated from L and W_0. For higher degrees of burn-off, it is in agreement with the value obtained from caffeine

$$W_t(A) = \int_0^\infty f(L)\exp[-(A/\beta E_0)^3]\,dL \tag{3.16}$$

and taking into account the fact that $E_0 = K/L$. Although K is a function of L, model calculations show that for a given micropore distribution $f(L)$ one may use a constant value K_0 given by equation 3.12 and corresponding to the characteristic energy E_0 of the carbon under investigation. This may be the result of internal averaging in the integral transform. The same is true for Dubinin's generalized isotherm, equations 3.8 and 3.9, with a constant value k_0 equal to 12 nm kJ mol^{-1} or better to $K_0/2$, given by equation 3.12.

The arbitrary but plausible choice of the normalized gamma function

$$f(L) = \frac{3L^{(3\nu-1)}a^\nu \exp(-aL^3)}{\Gamma(\nu)} \tag{3.17}$$

to represent the micropore distribution, and its combination with the DA core

$$g(A;L) = \exp[-(AL/\beta K_0)^3] \tag{3.18}$$

has the advantage that it leads to a known integral transform.[30] The normalized adsorption isotherm corresponding to this model is

$$\theta_t(A) = \left[\frac{a}{a + (A/\beta K_0)^3}\right]^\nu \tag{3.19}$$

Since equation 3.19 is still relatively complicated, it does not replace the DA equation 3.1 to describe the adsorption isotherm itself. It is a mathematical expression used to derive parameters a and ν, and consequently the approximate micropore distribution $f(L)$ of the system under investigation. From a practical point of view, W_0 and E_0 are first derived from a best fit to the classical DR equation 3.3, which leads to $\theta_t(A) = W/W_0$ and to K_0. The reduced adsorption data are fitted to equation 3.19, parameters and K_0 being known, and one obtains the values of a and ν. Inspection of equation 3.19 also shows that K_0 cannot be obtained independently from a. As shown in Figures 3.3–3.6, the distributions calculated from the adsorption data of simple gases are in good agreement with the histograms. The corresponding values of a and E_0 are given in Table 3.5. From the mathematical analysis of equation 3.17 it also appears that the quantity $a^{-1/3}$, expressed in nm, is close to the average width $\bar{L}$ of the distribution. The data for other carbons discussed in this chapter are given in Table 3.3.

If one uses a variable value for k_0 in equation 3.9, instead of 12 nm kJ mol^{-1} as suggested originally, the Gaussian and gamma-type distributions 3.7 and 3.17 are very similar. The latter is broader, which is not surprising since it is based on a DA core with $n = 3$.

3.5 Immersion calorimetry and characterization of active carbons

Dubinin's theory was originally derived for the adsorption of vapours and gases, and it is based on a simple relationship between the degree of micropore filling and the Gibbs free energy, $\Delta G = RT\ln(p/p^0)$, the reference state being the liquid. The DA and DR equations can therefore be developed within the framework of classical thermodynamics and interesting consequences are obtained.[7,31,32] One of them is an expression for the enthalpy of immersion of the micropore system into liquids whose vapours are adsorbed according to equations 3.1–3.3. It follows that the process of micropore filling can be examined by two different techniques and complementary information can be obtained.

By combining equation 3.1 with the definition of the isosteric heat of adsorption, q^{st},

$$\frac{\partial \ln(p)}{\partial T} = \frac{q^{st}}{RT^2} \tag{3.20}$$

one obtains q^{st} as a function of the degree of micropore filling, $\theta = a/a_0$,

$$q^{st}(T;\theta) = \beta E_0\left\{[\ln(a_0/a)]^{1/n} + \frac{\alpha T}{n}[\ln(a_0/a)]^{1/n-1}\right\} + \Delta H_{vap}(T;p) \tag{3.21}$$

where a and a_0 represent the amounts (mol g^{-1}) adsorbed at p/p^0 and at saturation, respectively, and α is the thermal expansion coefficient of the adsorbate.

From the definition of the net heat of adsorption

$$q^{net}(T;\theta) = q^{st} - \Delta H_{vap} \tag{3.22}$$

and knowing that the enthalpy of immersion of a microporous solid into a liquid is given by[5,33]

Table 3.5 Structural parameters for various microporous carbons of low and medium activation, shown in Figures 3.3–3.6 and in refs 18, 25 and 29

Parameter	Carbon					
	CEP-18	CEP-35	CM	CEP-59	X-72	U-103
W_0 cm^3 g^{-1}	0.251	0.347	0.252	0.48	0.059	0.330
S_e m^2 g^{-1}	29.0	27.0	28.0	31.0	107.0	60.0
E_0 kJ mol^{-1}	33.0	30.0	26.2	25.0	22.0	20.6
$\bar{L}$ nm	0.50	0.58	0.75	0.79	1.02	1.17
a nm^{-3}	6.09	5.27	2.27	2.99	0.886	0.39
$a^{-1/3}$ nm	0.55	0.57	0.76	0.69	1.04	1.18
ν	1.25	1.74	1.90	1.85	1.47	1.16

$$-\Delta h_i(T) = \int_0^1 q^{\text{net}}(T;\theta)\,d\theta \tag{3.23}$$

an exact expression can be obtained for Δh_i. The minus sign takes into account the fact that q^{st}, as defined by equation 3.20, is positive, whereas Δh_i is negative (energy is given up). In the case of the DR equation, taken as a useful working relationship for typical active carbons, one obtains the molar enthalpy of immersion for the filling of the micropores by 1 mol of liquid adsorbate at temperature T,

$$-\Delta h_i(\text{J mol}^{-1}) = \beta E_0 \pi^{0.5}(1+\alpha T)/2 \tag{3.24}$$

Since the carbon is the object of the study, equation 3.24 can be expressed in joules per gram of solid immersed into the liquid,

$$-\Delta h_i(\text{J g}^{-1}) = \beta E_0 W_0 \pi^{0.5}(1+\alpha T)/2V_m \tag{3.25}$$

As mentioned earlier, it is assumed that the molar volume V_m of the adsorbate in the micropores is close to that of the free liquid.

Since active carbons often present an external (non-microporous) surface area S_e, the experimental enthalpy of immersion will include an extra contribution, corresponding to the wetting of this surface, and

$$-\Delta h_{i\,\text{exp}}(\text{J g}^{-1}) = \frac{\beta E_0 W_0 \pi^{0.5}(1+\alpha T)}{2V_m} - h_i S_e \tag{3.26}$$

The quantity h_i, given in J m^{-2}, represents the specific enthalpy of wetting. Average values[34–36] obtained from experiments with non-porous carbon blacks are given in Table 3.1. Formally, h_i is negative, wetting being exothermic, and care should therefore be taken with the signs.

Following equation 3.26, immersion calorimetry can be used in different ways to complement the adsorption of vapours. In the course of a systematic characterization, starting with the adsorption of a simple vapour such as CH_2Cl_2 or C_6H_6 at 293 K, the DR equation 3.3 leads to the values of W_0 and E_0. The enthalpy of immersion into the corresponding liquid, also near room temperature, will lead to S_e, by taking the difference between the experimental and the calculated values of Δh_i. In the case of carbon N-125, for example, shown in Table 3.2 ($W_0 = 0.64$ cm^3 g^{-1} and $E_0 = 16.6$ kJ mol^{-1}), the experimental enthalpy of immersion into benzene is -160 J g^{-1}. The value of Δh_i calculated by equation 3.25 for the micropore system alone is -140.7 J g^{-1}. The difference of -19.3 J g^{-1} leads to $S_e = 19.3$ (J g^{-1})/0.114 (J m^{-2}) = 169 m^2 g^{-1}, with an error of approximately 10%. Experiments with carbon tetrachloride lead to S_e = 153 m^2 g^{-1}. As shown in Table 3.2, the values of S_e obtained by different techniques are in relatively good agreement. For the present carbon, the overall average is 157 ± 20 m^2 g^{-1}. In practice, the value of S_e obtained from immersion calorimetry can be checked against the value derived from the analysis of the upper part of the adsorption isotherm, which provides a test for self-consistency. The latter value is often less accurate, in particular when comparison plot[5] (t and α plots) are used to derive simultaneously the micropore volume and S_e. An alternative approach is therefore to use the micropore volume W_0 known from the DR plot and to determine only S_e, from the tangent to the curve.

Within the framework of Dubinin's theory, as summarized by equations 3.25 and 3.26, immersion calorimetry also provides a means to probe the structure of the micropore system. A first possibility is provided by the immersion of the carbon into liquids with molecules of variable dimensions, leading to the histograms shown in the previous section (Figures 3.3–3.6). From equation 3.26, it follows that for a molecule of critical dimension L, the total volume $W_0(L)$ of the pores larger than L should be given by

$$W_0(L) = -[\Delta h_i(\exp) - h_i S_e] \frac{2V_m}{\beta E_0 \pi^{0.5}(1 + \alpha T)} \tag{3.27}$$

The histograms shown in Figures 3.3–3.6 were obtained from the successive values of $W_0(L)$ of the various molecular probes. It is assumed, implicitly, that there are no gate effects and that E_0, corresponding to the value obtained for a small molecular probe such as CH_2Cl_2, for example, is a constant. The latter assumption is not completely true, since E_0 decreases slightly as the size of the molecular probe increases (see below and Tables 3.7–3.8). In the case of carbon CEP-35, for example (Figure 3.1), the adsorption of CCl_4 vapour shows a sharp decrease in W_0 from 0.34 to 0.16 $cm^3\ g^{-1}$. At the same time, the characteristic energy E_0 decreases from 30 to 27.5 kJ mol^{-1}. This means that the value of $W(L)$ for CCl_4 calculated from Δh_i by equation 3.27 is 8–10% smaller than the real value of 0.16 $cm^3\ g^{-1}$. Carbon CEP-35 is an unfavourable case and model cases show that the use of a constant value of E_0 in equation 3.27 leads in fact to a good approximation of the histogram. The latter can be improved, if necessary, by using modified values E'_0 for the characteristic energy of the successive molecular probe. These values are calculated by equation 3.10 from the average pore sizes L accessible to the probes, on the basis of the first distribution, and new values of $W'_0(L)$are recalculated from equation 3.27. This procedure leads to histograms of the type shown in Figures 3.3–3.6.

Unfortunately, the determination of histograms by immersion calorimetry is not always successful. Two possible reasons are gate effects, often found in carbons with low degrees of burn-off, and the total absence of the smaller micropores in strongly activated carbons. In this case, it is also likely that the structure looses its rigidity and the micropore system adapts itself to the liquid adsorbate. This possibility has still to be investigated, but experiments carried out with carbon U-02 before and after soaking with *tert*-butylbenzene (TBB)[37] and with dimethylmethylphosphonate (DMMP) clearly show structural changes. Table 3.6 gives the values of W_0 and E_0 obtained from vapour adsorption (C_6H_6; 273 K) before and after treating carbon U-02 with these bulky molecules. The

Table 3.6 Variation of W_0 and E_0 in carbon U-02, following immersion in dimethylmethylphosphonate (DMMP) and *tert*-butylbenzene (TBB) for several days[a]

Parameter	U-02 (initial)	U-02 (DMMP)	U-02 (TBB)
W_0 $cm^3\ g^{-1}$	0.43	0.47	0.50
E_0 kJ mol^{-1}	20.0	17.5	17.1
$\bar{L}$ nm (equation 3.10)	1.25	1.77	1.89
S_{mi} $m^2\ g^{-1}$ (equation 3.13)	545[b]	531	529

[a]The characterization was carried out by C_6H_6 adsorption at 293 K.
[b]Calculated from SAXS and He adsorption (see Table 3.4).

permanent change in E_0 clearly reflects an increase in the average micropore width, which can be estimated by equation 3.10. This structural modification can not be detected by immersion calorimetry alone, since the product of E_0 and W_0 is almost constant, the increase in W_0 being compensated by the decrease in E_0. It is also possible that during immersion the irreversible changes are more important than suggested by the permanent modifications given in Table 3.6. It is also interesting to point out the agreement between the values of S_{mi} recalculated from W_0 and L by using equation 3.13.

Another interesting application of immersion calorimetry to the study of active carbons is provided by its combination with the preadsorption of long-chain n-alkanes.[38] The preadsorption of n-nonane is a classical technique proposed by Gregg and co-workers[5,39] and frequently used to determine the external surface area of active carbons.[6] The micropore system is first saturated with the alkane and S_e is determined from the subsequent nitrogen adsorption isotherm at 78 K.

A second possibility to investigate microporous carbons by immersion calorimetry is based on the preadsorption of n-nonane and n-undecane to various degrees of micropore filling θ, followed by immersion in the corresponding liquids or into water (see the following section). This technique can provide useful information on the micropore system accessible to a second adsorbate. This approach is possible in view of the low desorption rate of the n-alkane, compared with the duration of an immersion experiment (30–60 mins). Control isotherms with N_2 at 78 K show that for typical active carbons the volume occupied by nonane does not exceed the calculated volume by more than 10–15%, provided that precautions are taken during preadsorption. Under these circumstances, the enthalpy of immersion into the liquid n-alkane is given by

$$-\Delta h_i(T) = \frac{\beta E'_0 W_0 (1-\theta)\pi^{0.5}(1+\alpha T)}{2V_m} - h_i S_e \tag{3.28}$$

Typical results obtained for carbon U-03 are shown in Figure 3.9 and the smoothed values of E'_0 are given in Table 3.7. They show, as expected, that the characteristic energy E_0 decreases as the prefilling θ increases. A similar pattern is observed for other carbons, such as U-103, for example (see Table 3.8). E'_0

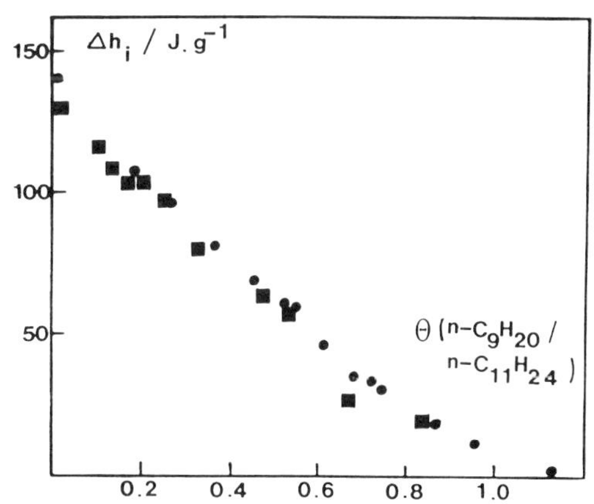

Fig. 3.9 Enthalpies of immersion Δh_i of carbon U-03 into n-nonane (•) and n-undecane (■) at 293 K as a function of the prefilling by the corresponding vapours, θ

Table 3.7 Smoothed values of the structural parameters of carbon U-03 after the preadsorption of *n*-nonane and *n*-undecane to various degrees of micropore filling θ.

Parameter	Prefilling θ					
	0	0.1	0.3	0.4	0.5	0.7
E'_0 kJ mol^{-1}	17.2	16.7	15.7[b]	15.0	14.0[b]	12.0
W'_0 cm^3 g^{-1}	0.52	0.47	0.36	0.31	0.26	0.15
S_{mi} m^2 g^{-1} [a]	570	458	338	268	213	88
L_w nm	1.8	2.0	2.1	2.3	2.4	3.4

[a] $S_{mi} = S_t - S_e$ (80 m^2 g^{-1}).
[b] The adsorption of N_2 (78 K) leads to E_0 = 15.6 and 13.2 kJ mol^{-1}.

decreases gradually and below 16 kJ mol^{-1} the calculated average micropore widths $\bar{L}$ become unrealistic. However, it is interesting that up to θ = 0.6–0.7, the quantity $L_w = 2 \times 10^3 W'_0/S_{mi}$ appears to be a reasonable width, but so far no direct evidence exists for these pore sizes. This suggests again some flexibility in the structure of supermicropores.

Since the preadsorbed alkanes do not mix with water, the technique based on the selective adsorption of caffeine from water can also be used to estimate the total surface area accessible to a second adsorbate. By applying equation 3.15 to the calorimetric data, one obtains the typical results shown in Figure 3.10 and in Tables 3.7 and 3.8. As expected, the accessible surface area decreases steadily as θ increases and near θ = 1 one is left with S_e. (In the case of carbon U-103, a small correction must be applied to Δh_i, to allow for the disolution of TEDA in water, -235 ± 10 J g^{-1}).

The combined preadsorption and immersion techniques can also be used to assess the distribution of an impregnating agent such as TEDA (triethylenediamine) within a microporous carbon, the agent being detected by a specific reaction. In the case of carbon U-103, impregnated with 1.65% by weight of TEDA, the variation of E'_0 with θ was first determined as described above, by using *n*-nonane (Table 3.8). In a second series of experiments, the amount of TEDA available at different degrees of prefilling was obtained from the enthalpy of immersion into CH_3I, by using equation 3.28 with a supplementary term taking into account the specific reaction of CH_3I with TEDA[40] (−1650 J g^{-1}),

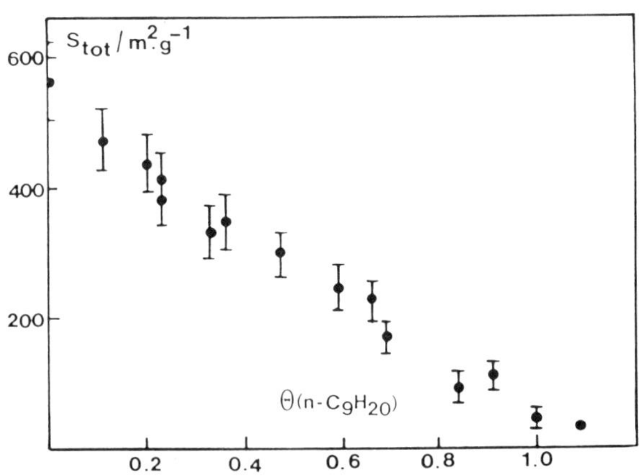

Fig. 3.10 Total surface area S_t of carbon U-02 as a function of *n*-nonane prefilling θ, calculated from the enthalpy of immersion into aqueous solutions of caffeine at 293 K

Table 3.8 Structural parameters of carbon U-103 and the accessibility of TEDA as a function of θ, the degree of micropore filling by *n*-nonane

	$\theta(n\text{-}C_6H_{20})$							
Parameter	0	0.1	0.2	0.3	0.4	0.5	0.6	0.7
E'_0 kJ mol^{-1}	20.6	18.9	18.0	17.6	17.1	16.7	15.8	13.2
W'_0 cm^3 g^{-1}	0.33	0.29	0.26	0.23	0.19	0.16	0.12	0.09
S_{mi} m^2 g^{-1}	570	410	370	314	257	201	157	105
L_w nm	1.2	1.4	1.4	1.5	1.5	1.6	1.5	1.7
TEDA mg g^{-1}	16.5	14.8	14.7	13.2	11.8	11.3	10.3	9.6

$$-\Delta h_i(\mathrm{J\ g^{-1}}) = \frac{\beta E'_0(1-\theta)W_0\pi^{0.5}(1+\alpha T)}{2V_m} - h_i S_e - 1650n \tag{3.29}$$

The value of n (grams of TEDA per gram of solid) is recalculated from this equation and one obtains its accessibility as a function of the prefilling θ of the micropores. As shown in Figure 3.11, the amount of TEDA detected by CH_3I decreases in a regular fashion. Near $\theta = 0.9$, approximately 35% of TEDA is still available, which suggests that it is located in supermicropores and on the external surface area.

This example shows the potential of the combined preadsorption and immersion techniques, which can be adapted to other systems in microporous carbons.

3.6 Interaction of water with active carbons

The interaction of water with active carbons does not follow Dubinin's theory and, in view of the hydrophobic nature of the untreated material, the adsorption isotherm is of Type V, with a hysteresis loop (Figure 3.12).[5,41] It is assumed that adsorption occurs through hydrogen bonding to hydrophilic centres, located on the edges of the aromatic sheets. Depending on the history of the material, a variety of oxygen-containing groups can act as primary centres.[7]

At present, the description of the water adsorption isotherm is still limited,

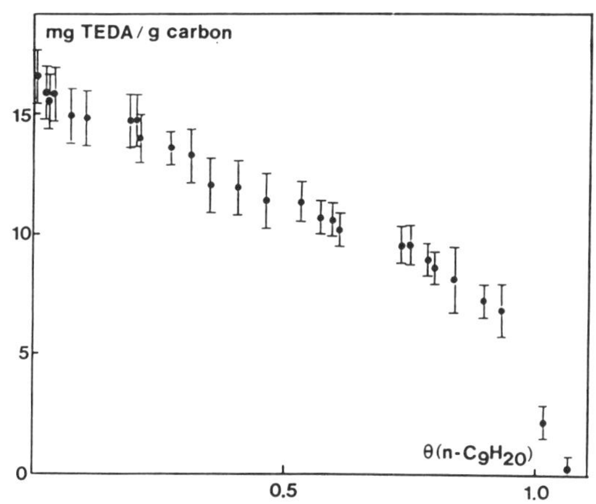

Fig. 3.11 Amount of TEDA accessible to CH_3I as a function of the prefilling θ of carbon U-103 by *n*-nonane. See also Table 3.8

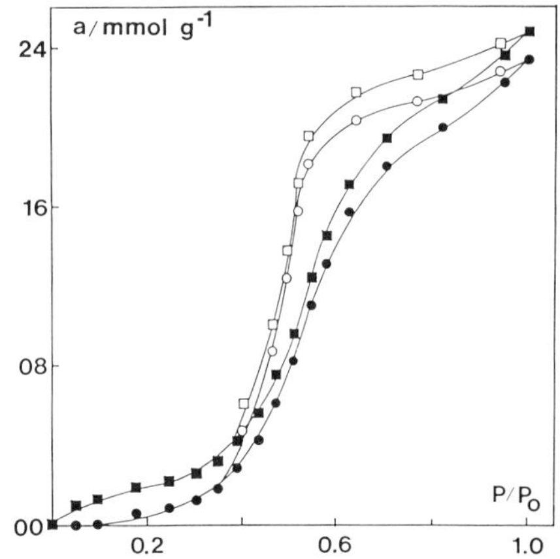

Fig. 3.12 Adsorption–desorption of water vapour at 293 K by an impregnated carbon of series U. Experimental isotherm (■/□) and Type V isotherm (●/○) obtained by subtracting the Langmuir contribution. Open symbols refer to desorption and at low pressures; not all points are shown. (Courtesy P. Rebstein, Neuchâtel)

but the model proposed by Dubinin and Serpinskii[42] can be used successfully in the range of relative pressures between approximately 0.35 and 0.80. However, no satisfactory model exists for the desorption branch of the isotherm. The basic equation is

$$p/p^0 = \frac{a}{c(a_0 + a)(1 - ka)} \tag{3.30}$$

where a represents the amount of water adsorbed at relative pressure p/p^0 and a_0 is the number of primary centres; c is the ratio between the rate constants of adsorption and desorption and k is a constant fixed by the condition that for $p/p^0 = 1$ the total amount adsorbed in the micropores is a_s. Implicitly, equation 3.30 is valid for a given type of primary centres and if different centres are present, parameters a_0 and c become apparent quantities only. The limitations of the Dubinin–Serpinskii (DS) equation have been discussed by Evans,[43] but so far no improved alternative has been proposed.

For many carbons, the experimental isotherm begins as Type I at low pressures and it can therefore be decomposed into a sum of contributions.[44] The first and limited domain is of the Langmuir type and corresponds to a distinct mechanism. The adsorption isotherm of water on an impregnated carbon of series U, shown in Figure 3.12, leads for example to $a_L = 1.4\ \text{mmol g}^{-1}$ for the sites of the Langmuir isotherm and to $a_0 = 0.80\ \text{mmol g}^{-1}$ for the DS contribution.

For pure active carbons outgassed near 450–500 °C,[45] the values of a_0 are typically in the range 0.5–1.5 mmol g^{-1} and c varies from 1.5 to 2.5 (see Table 3.9). The combination of adsorption and immersion data for this type of carbon reveals interesting correlations, probably due to their standard heat treatment. It was found that the enthalpies of immersion in water and the number of primary centres, probably of a uniform type, were related by the simple equation

$$\Delta h_i(\text{J g}^{-1}) = -25(\text{J mmol}^{-1}) \cdot a_0(\text{mmol g}^{-1}) - 0.6(\text{J mmol}^{-1}) \cdot (a_s - a_0) \tag{3.31}$$

Table 3.9 Values of a_0 and c for typical active carbons, obtained from the water adsorption isotherm near 293 K and by immersion calorimetry

Carbon	$-\Delta h_i$ (H_2O) J g^{-1}	a_0 mmol g^{-1} Isotherm	a_0 mmol g^{-1} Equation 3.31	c (isotherm)
U-02	27.9	0.35	0.50	1.79
F-02	47.0	1.49	1.12	1.62
N-125	31.6	0.49	0.52	1.66
AU-1	40.4	1.18	1.61	1.98
AU-2	49.4	0.95	1.16	1.80
U-03	34.5	0.56	0.62	1.63

In the case of carbons with an important external surface area, an extra term, $-S_e(m^2\,g^{-1}) \times 0.030$ ($J\,m^{-2}$), can also be added, to take into account the wetting of this area.

It was also verified that the constant c and the molar enthalpy of immersion, ΔH_{im} ($J\,mol^{-1}\,H_2O$) $= \Delta h_i/a_s$, were related by an expression of the type

$$c = c_0 \exp(-\Delta H_{im}/RT) \tag{3.32}$$

as implied by a thermodynamic treatment of the DS equation 3.30. The experimental value of c_0 is 1.00 ± 0.11.

A modified version of equation 3.31 has also been established for carbons containing carboxylic groups, with a prefactor of -55 J $mmol^{-1}$ instead of -25 J $mmol^{-1}$ primary centres. During outgassing these carbons were only subjected to a mild heat treatment (100–120°C), in order to avoid decarboxylation.

Equations 3.31 and 3.32 therefore provide useful correlations between immersion and adsorption data, at least for carbons having been subjected to standard physical and chemical treatments. Equation 3.31 has been used, for example, to assess the change in a_0 as a function of burn-off in different activation series. As illustrated by Figure 3.13, different patterns can be distinguished, depending on the origin of the initial carbon. (All the samples were washed with HCl/H_2O to remove inorganic salts.)

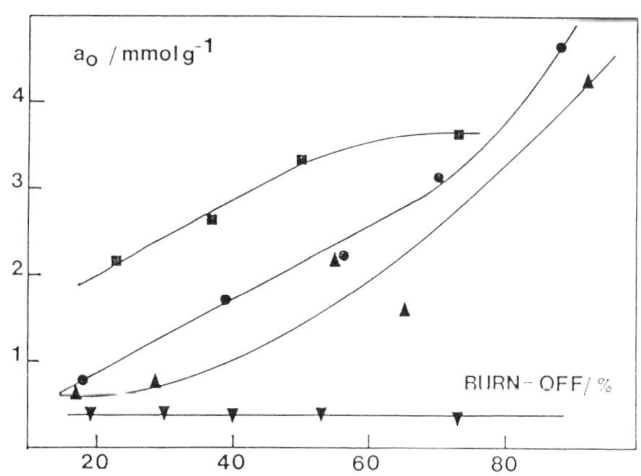

Fig. 3.13 Variation of the number of active sites a_0 as a function of burn-off in active carbons of series C-87 (●), B10 (■), CAF (▲) and TYRES (▼). (Courtesy L. Ballerini, Neuchâtel)

Table 3.10 Parameters a_0 and c for the adsorption of water at 293 K by carbon U-03, as a function of θ, the degree of prefilling by *n*-nonane

		Immersion calorimetry		Isotherm at 293 K	
θ (n-C_9H_{20})	$-\Delta h_i$ (H_2O) J g^{-1}	a_0 mmol g^{-1}	c	a_0 mmol g^{-1}	c
0	34.6	0.62	1.63	0.56	1.63
0.17	23.7	0.30	1.54	0.25	1.65
0.30	19.5	0.22	1.53	0.26	1.62
0.50	12.3	0.08	1.42	0.11	1.49
0.71	8.0	0.07	1.29	0.09	1.17
0.90	5.6	0.08	1.25	0.12	1.21

The problem of the location of the hydrophilic centres, and in particular their accessibility, can be dealt with by using the preadsorption technique described above and followed by immersion in pure water. As shown in Table 3.10, the enthalpy of adsorption of carbon U-03 in water decreases as the degree of prefilling with *n*-nonane and *n*-undecane increases. The original sample had been subjected to the standard heat treatment (450–500 °C *in vacuo*) and therefore equation 3.30 was used to recalculate a_0 from the enthalpy of immersion, as a function of θ. As shown in Table 3.10, a_0 decreases rapidly from 0.62 to 0.22–0.25 mmol g^{-1} between $\theta = 0$ and 0.3. This indicates that a large fraction of the sites are blocked off by the hydrophobic *n*-alkane and consequently they should be located within the smaller pores. Similar observations were reported by Zettlemoyer *et al.*[46] and Groszek.[47] The validity of the technique was verified also by measuring the water adsorption isotherms at 293 K, directly on the solid prefilled with various amounts of *n*-undecane (this adsorbate is virtually immobile, even over longer periods, and it does not allow water to penetrate into the structure). As shown in Table 3.10, the values of a_0 and c calculated from the enthalpies of immersion by equations 3.31–3.32 are in good agreement with those derived from the Dubinin–Serpinskii equation 3.30.

3.7 Concluding remarks

In this chapter we have shown that a reasonable description of the microporosity in carbons can now be obtained on the basis of systematic investigations carried out with adsorption and immersion techniques. The amount of information obtained on the structural and chemical properties of the carbon will depend on the degree of sophistication of the techniques used, but simple experiments, based on a single isotherm, can already provide much basic information. It is suggested that adsorption covers at least the region of $y = (T/\beta)^2 \log^2(p^0/p)$ up to 4×10^7 K^{-2}.

A classical adsorption isotherm, e.g. CH_2Cl_2 near room temperature, should provide a reasonable estimate of the micropore volume W_0 and its average width up to 1.5 nm at least, calculated on the basis of equation 3.10. The external surface area, on the other hand, is derived from the analysis of the upper part of the isotherm and it can be cross-checked by immersion calorimetry. This technique is less popular than the adsorption of vapours and gases, but its use can save time, as in the case of water adsorption, and provide more detailed information on the solid.

It is hoped that more information will become available on the size and shape of the supermicropores above 1.5 nm, which represent the transition between micropores and mesopores, with their specific adsorption mechanisms.

3.8 References

1. IUPAC. 1972: Manual of Symbols and Terminology, Appendix 2, Part 1, Colloid and Surface Chemistry, *Pure and Applied Chemistry* **31**, 578.
2. EVERETT, D.H. and POWL, J.C. 1976: *Journal of the Chemical Society, Faraday Transactions 1*, 619.
3. STOECKLI, H.F. 1974: *Helvetic Chimica Acta* **57**, 2105.
4. STOECKLI, H.F. and PERRET, A. 1975: *Helvetic Chimica Acta* **58**, 2318.
5. GREGG, S.J. and SING, K.S.W. 1982: *Adsorption, Surface Area and Porosity*. London: Academic Press, Chapters 2 and 4.
6. RODRÍGUEZ-REINOSO, F. 1989: *Chemistry and Physics of Carbons* **21**, 2.
7. BANSAL, R., DONNET, J.B. and STOECKLI, H.F. 1988: *Active Carbon*. New York: Marcel Dekker, Chapters 3 and 4.
8. STOECKLI, H.F. 1990: *Carbon* **28**, 1.
9. DUBININ, M.M. 1975: *Progress in Surface and Membrane Science* **9**, 1.
10. DUBININ, M.M. 1985 and 1989: *Carbon* **23**, 373; **27**, 457.
11. DUBININ, M.M., ZAVERINA, E.D. and RADUSHKEVICH, L.V. 1947: *Zhurnal Fizicheskoi Khimii*, 1351.
12. SING, K.S.W. 1989: *Carbon* **27**, 5.
13. DUBININ, M.M. and KADLEC, O. 1975: *Carbon* **13**, 263.
14. KRAEHENBUEHL, F. and STOECKLI, H.F. 1984: *Carbon* **22**, 296.
15. ALI, S. and McENANEY, B. 1985: *Journal of Colloid Interface Science* **107**, 355.
16. MARTIN-MARTINEZ, J.M., RODRÍGUEZ-REINOSO, F., MOLINA SABIO, M. and McENANEY, B. 1986: *Carbon* **24**, 255.
17. STOECKLI, H.F., KRAEHENBUEHL, F. LAVANCHY, A. and HUBER, U. 1984: *Journal de Chimie Physique* **81**, 785.
18. KRAEHENBUEHL, F., STOECKLI, H.F., ADDOUN, A., EHRBURGER, P. and DONNET, J.B. 1986: *Carbon* **24**, 483.
19. IZOTOVA, T.I. and DUBININ, M.M. 1965: *Zhurnal Fizicheskoi Khimii*, 2796.
20. DUBININ, M.M. and PLAVNIK, G.M. 1964 and 1968: *Carbon* **2**, 261; **6**, 183.
21. GUINIER, A. and FOURNIER, B. 1964: *Small-Angle Scattering of X-Rays*. New York: John Wiley, Chapter 2.
22. HUBER, U., STOECKLI, H.F. and HOURIET, J.PH. 1978: *Journal of Colloid Interface Science* **67**, 195.
23. DUBININ, M.M. and STOECKLI, H.F. 1980: *Journal of Colloid Interface Science* **75**, 34.
24. INNES, R.W., FRYER, J. and STOECKLI, H.F. 1989: *Carbon* **27**, 71.
25. STOECKLI, H.F., REBSTEIN, P. and BALLERINI, L. 1990: **28**, 907.
26. BALLERINI, L., HUGUENIN, D., REBSTEIN, P. and STOECKLI, H.F. 1990: *Journal de Chimie Physique* **87**, 1709.
27. JANOSI, A. and STOECKLI, H.F. 1979: *Carbon* **17**, 465.
28. STOECKLI, H.F., BALLERINI, L. and BERNARDINI, S.de. 1989: *Carbon* **27**, 501.
29. STOECKLI, H.F. 1990: *Carbon* **27**, 962.
30. GRADSHTEYN, I.S. and RYZHIK, I.M. 1980: *Tables of Integrals, Series and Products*. New York: Academic Press, p. 1143.
31. STOECKLI, H.F. 1981: *Izvestiya Akademii Nauk SSSR, Ser. Khim.*, 63.
32. STOECKLI, H.F. and KRAEHENBUEHL, F. 1981: *Carbon* **19**, 353.
33. CLINT, H.J. 1973: *Journal of the Chemical Society, Faraday Transactions 1*, 1320.

34. BARTELL, F.E. and SUGITT, R.M. 1954: *Journal of Physical Chemistry* **58**, 86.
35. ROBERT, L. 1967: *Bulletin. Societé Chimique de France*, 147.
36. WADE, W.H., DEVINEY, M.L., BROWN, W.A., HNOOSH, M.H. and WALLACE, D.R. 1972: *Rubber Chemistry and Technology* **45**, 117.
37. STOECKLI, H.F., PERRET, A. and MENA, Ph. 1980: *Carbon* **18**, 443.
38. STOECKLI, H.F., HUGUENIN, D. and REBSTEIN, P. 1991: *Journal of the Chemical Society, Faraday Transactions 1* **87**, 1233.
39. GREGG, S.J. and LANGFORD, J.F. 1969: *Transactions of the Faraday Society* **65**, 1394.
40. KRAEHENBUEHL, F., SCHMITTER, B. and STOECKLI, H.F. 1986: *Carbon* **24**, 649.
41. STOECKLI, H.F., KRAEHENBUEHL, F. and MOREL, D. 1983: *Carbon* **21**, 589.
42. DUBININ, M.M. and SERPINSKII, V.V. 1981: *Carbon* **19**, 402.
43. EVANS, M.J.B. 1987: *Carbon* **25**, 81.
44. DUBININ, M.M. and ISIRIKIAN, A.A. 1989: *Izvestiya Akademii Nauk SSSR, Ser. Khim.*, 2183.
45. KRAEHENBUEHL, F., QUEKET, C., SCHMITTER, B. and STOECKLI, H.F. 1986: *Journal of the Chemical Society, Faraday Transactions 1*, 3439.
46. ZETTLEMOYER, A.C., PENDLETON, P. and MICALE, F.J. 1981: in Ottewill, R.M. (ed.) *Adsorption from Solutions.* London: Academic Press, pp. 112–127.
47. GROSZEK, A.J. 1989: *Carbon* **27**, 33.

4 Characterization of Macropores in Carbons

Brian McEnaney and Timothy J. Mays
School of Materials Science
University of Bath
Claverton Down
Bath
Avon BA2 7AY
United Kingdom

4.1 Introduction

This chapter is concerned with macropores which are the largest of the classes of pores defined by IUPAC,[1] i.e. those of width, w, greater than 50 nm. The aim of this chapter is to review critically the three main experimental methods for quantitative characterization of macropores in carbons, viz. mercury porosimetry, fluid flow and quantitative microscopy. Note that the first two techniques are methods for characterizing open macropores whereas, in suitable conditions, the last method can be used to characterize both open and closed porosity.

The macropore/mesopore boundary

The boundaries between the classes of pores chosen by IUPAC are arbitrary, being based on adsorption criteria. Thus, the 50 nm boundary between mesopores and macropores is set by practical problems of measuring mesopore dimensions accurately using capillary condensation theories at high adsorptive relative pressures, p/p^0. This point can be illustrated by considering the adsorption of nitrogen at 77 K in cylindrical pores. From capillary condensation theory, the relationship between p/p^0 and the mean radius of curvature of the adsorbate meniscus in a cylindrical capillary, r_m, is given by the Kelvin equation,

$$r_m = \frac{2\gamma V_l}{RT \ln (p/p^0)} \tag{4.1}$$

where γ and V_l are the surface tension and molar volume, respectively, of the adsorbate, R is the gas constant and T the absolute temperature. Assuming that

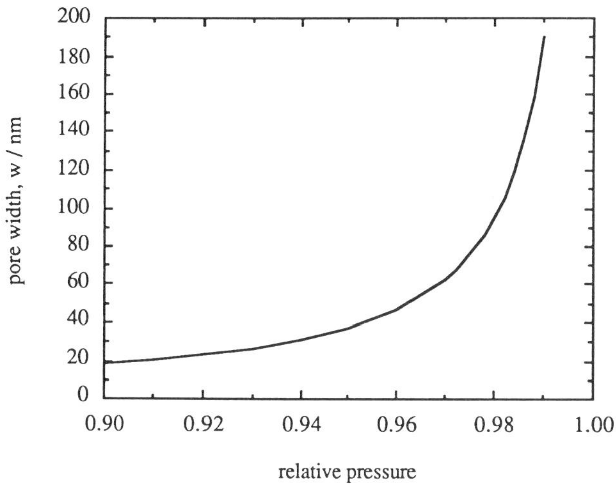

Fig. 4.1 Pore width, $w = 2r_m$, as a function of relative pressure for N_2 at 77 K, calculated from equation 4.1

the contact angle of the adsorbate meniscus with the wall of the capillary is zero, and neglecting the thickness of adsorbed surface layers, the pore width $w = 2r_m$. From equation 4.1 there is an exponential increase in w with increasing p/p^0; the relationship for adsorption of nitrogen at 77 K at high p/p^0 is shown in Figure 4.1. In this case the IUPAC upper limit for mesopores of 50 nm corresponds to $p/p^0 \approx 0.96$. At higher p/p^0 values it becomes increasingly difficult to measure w precisely.

The significance of macropores in carbon materials

Because macropores have low surface area-to-volume ratios, they make only a small contribution to the total surface area, but they are not without significance to the adsorptive properties of porous carbons, because they can act as conduits to mesopores and micropores. However, macropores have a dominating influence on many other properties of carbons, including mechanical, thermal and electrical properties. Thermal and electrical conductivities decrease with increasing porosity as a result of the decreasing cross-section of solid material to conduct heat and electrons. Clearly, the number of macropores in a solid has a proportionately greater effect on these properties than the number of meso- and micropores. The presence of pores in carbons and graphites adversely affects mechanical properties, such as elastic modulus and strength, since they reduce the volume of solid within which stresses are distributed. Also, pore shape has a strong influence on these properties, since there is a greater stress concentration around crack-like pores than around equi-axed pores.

When characterizing the macropore structure of a carbon, it is important to ensure that the experimental technique chosen is appropriate to the purpose. Thus, if the characterization is related to adsorption, percolation of fluids, or gas–carbon reactions, then techniques such as gas adsorption, mercury porosimetry and fluid flow, which measure open porosity, are appropriate. On the other hand, for thermal, electrical and mechanical properties, it is the total

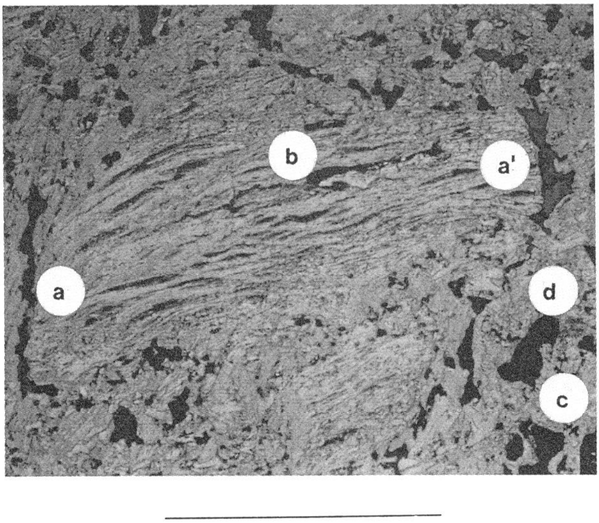

Fig. 4.2 Examples of macropores in an engineering graphite. b = Calcination cracks in the needle-coke filler particle, aa′; d = globular macropores in the binder phase, c

porosity which is important, and the use of these techniques can be misleading. In such cases, quantitative microscopy should be considered.

Also, macropores in carbons should not be regarded as an undifferentiated class of pores defined by their size. Often there are different sizes and shapes of macropores in a carbon which are introduced at different stages of manufacture. For example, several types of macropores and cracks in an engineering graphite[2] are shown in Figure 4.2. Pores marked b are volumetric shrinkage cracks formed during the calcination of the needle-coke filler particles, marked aa′. The cracks are about 100 μm wide and up to about 1 mm long. Graphitic basal planes lie parallel to these cracks, so that there is good thermal and electrical conductivity along the axis of the filler particles. The superior electrical conductivity parallel to the long axis of extruded graphite electrodes results from the orientation of needle-coke particles during the extrusion stage of manufacture. The shrinkage cracks in the filler particles also accommodate the anisotropic thermal expansion of the graphite during heating and so contribute to the low bulk thermal expansion of graphite electrodes.

The globular macropores marked d in Figure 4.2 are in the binder phase, marked c. These are formed during the baking stage of manufacture when the pitch binder is converted into carbon. The globular pores in the binder phase are often linked together by a network of fine cracks formed during shrinkage of the binder phase during baking or on cooling from graphitization temperatures. The macroporous network in the binder phase is usually open, whereas the shrinkage cracks in the filler particles are closed, probably because they are sealed with pitch during the mixing stage of manufacture.

This example illustrates the diversity of macropores that can be found in technical carbons. A full characterization of the macropore structure of such a carbon would require estimations of the proportions of open and closed pores, the sizes and shapes of the different macropore types and their spacial distribution in the material. Other examples of macropores in carbon materials have been described by McEnaney and Mays.[3]

4.2 Mercury porosimetry

Introduction

Mercury porosimetry is the most widely used technique for estimating the sizes of macropores in carbons and other porous solids. The technique is summarized in standard text books on characterization of porosity (e.g. by Gregg and Sing[4]) and it is treated in more detail by Lowell and Shields.[5] Early work on mercury porosimetry applied to carbon materials was reviewed by Scholten[6] and some applications to activated carbons have been illustrated by de Vooys.[7] Mercury porosimetry is applied to a wide range of pore sizes, which includes both mesopores and macropores; for convenience, the application of mercury porosimetry to mesopores is also considered in this chapter.

Mercury does not wet the surface of carbon, i.e. its contact angle, θ, with a carbon surface is greater than 90°. Consequently, mercury in a pore forms a meniscus which is convex to the vapour phase and a hydrostatic pressure p_1, greater than the vapour pressure above the meniscus, p_0, must be used to force mercury into the pores. The pressure difference, $\Delta p = p_1 - p_0$, required to force mercury into a cylindrical pore is related to the pore radius, r, by the Washburn equation[8]

$$r = -\frac{2\gamma\cos\theta}{\Delta p} \tag{4.2}$$

The Washburn equation is a special case of the Young–Laplace equation

$$\Delta p = -\gamma\left(\frac{1}{r_1} + \frac{1}{r_2}\right) \tag{4.3}$$

where, in this case, r_1 and r_2 are the principal radii of curvature of the mercury meniscus in the capillary. For a cylindrical capillary, $r_1 = r_2$, and it can be shown easily that the capillary radius $r = r_1\cos\theta$; combining this with equation 4.3 leads to the Washburn equation.

Measurement of mercury penetration

The experimental basis of mercury porosimetry was first demonstrated by Ritter and Drake.[9] Measurement of mercury penetration into a porous body is simple in principle, although modern commercial instruments are sophisticated as a result of continuous development over about 40 years. A diagram of the essential features of a mercury porosimetry is in Figure 4.3. In outline the procedure is as follows. The weighed, evacuated sample is placed in the sample

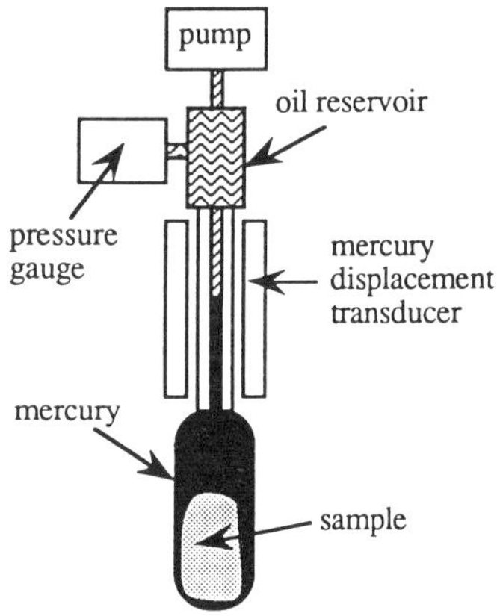

Fig. 4.3 Essential features of a mercury porosimeter

bulb, which is connected to a calibrated capillary stem and filled with mercury. The mercury in the sample assembly is pressurized and two measurements are made: the hydraulic pressure in the system and the displacement of the mercury in the calibrated capillary as it is intruded into the pores of the sample. Displacement can be measured by a contacting method, e.g. by measuring the electrical resistance of a platinum wire inserted coaxially inside the capillary, or by a non-contacting method, e.g. by measuring the capacitance between the mercury column and a coaxial plate outside the capillary.[5] The porosimeter can be operated in a stepwise or a dynamic mode. In the latter case the pressure is raised continuously at a uniform rate with frequent recording of the pressure and mercury intrusion data. As with dynamic adsorption methods, the effect of pressure scan rate on mercury intrusion should be investigated to ensure that the results are insensitive to this variable. In both modes the extrusion of mercury can be measured by progressive depressurization.

In modern instruments the pressure range is from atmospheric pressure (0.11 MPa) to about 414 MPa (15–60000 psia). From the Washburn equation, equation 4.2, assuming that $\gamma = 480$ mJ m^{-2} and $\theta = 140°$, this corresponds to a range of r from about 7 μm to 1.8 nm, i.e. from the macropore size range to the lower limit of the mesopore size range. However, the lower value of r must be regarded as nominal, since the applicability of the Washburn equation in pores of such small dimensions is open to question. The upper range of mercury porosimetry can be extended to about $r = 200$ μm by measuring mercury penetration at sub-atmospheric pressures. This is usually done as a first step using an ancillary vacuum apparatus, which also serves to pre-evacuate the sample before first contact with mercury. A standard experimental method for mercury porosimetry is described in ASTM C699.

For precise work it is necessary to make corrections for the compressibility of the hydraulic oil and of the mercury by carrying out blank runs. Corrections due to heating and compression effects are usually less than 1%. A more substantial correction may be required due to the compressibility of the sample. A linear increase in compression with applied pressure is usually assumed and at the highest pressures used (414 MPa) the correction ranges from 0.6 to 6.0%, depending on the compressibility of the solid.[5] Corrections to mercury porosimetry curves for coals are essential, owing to their high compressibilities ($1.3 \times 10^{-10} - 2.3 \times 10^{-10}$ Pa^{-1}).[10,11]

Results from mercury porosimetry

Densities

In principle, mercury penetration can be used to obtain various densities of the carbon material at stages in the pressurization cycle. For a particulate material, the specific volumes of the solid carbon, v_c, of open pores, v_p, and of interparticulate voids, v_v, can be defined. The bulk density, ρ_b, is defined as the mass per unit volume of particles, including open pores, plus the interparticulate voids, i.e.

$$\rho_b = \frac{1}{v_c + v_p + v_v} \tag{4.4}$$

The bulk density may be obtained from a measurement of the sample volume at a low pressure of mercury which encloses the sample but does not penetrate the interparticulate voids; a method is described in ASTM C699. The particle density $\rho_p = 1/(v_c + v_p)$ may be measured at a higher pressure where the mercury penetrates the interparticulate voids, but does not penetrate the largest macropores; in practice, this pressure may be difficult to define, particularly if the size ranges of the interparticulate voids and macropores overlap. The solid carbon density, ρ_c, may be defined from the specific volume of the sample at the highest pressure used: $\rho_c = 1/v_c$. Of course, if there is a significant amount of open microporosity or closed porosity of any size in the sample, then ρ_c is underestimated using this method, since these pore volumes are included in v_c. If the extent of open microporosity is significant, then an additional technique is required to estimate the open micropore volume, e.g. helium pycnometry or gas adsorption, although these techniques are not without difficulties. If open microporosity is negligible, then it is possible to obtain the open pore volume fraction from mercury porosimetry, $V_p = 1 - \rho_p/\rho_c$. If the volume of interparticulate voids cannot be separated from the macropore volume, then an effective open-pore volume fraction, V'_p may be obtained from $V'_p = 1 - \rho_b/\rho_c$, which includes the interparticulate voidage.

Pore-size distributions

Mercury porosimeter data are usually presented as a cumulative open-pore volume distribution, i.e. cumulative amount of mercury intruded, v cm^3 g^{-1}, versus open pore entrance radius, r nm, where r is obtained from the measured pressure, p, using the Washburn equation. Examples of cumulative pore-size distributions for several carbon materials are shown in Figure 4.4; the boundary between mesopores and macropores defined by IUPAC[1] is shown, assuming that pore width $w = 2r$. The cumulative intrusion curve for the carbon black suggests a broad unimodal pore-size distribution centred on $r \approx 20$ nm; the activated carbon has a wide range of pore sizes with a bimodal size distribution

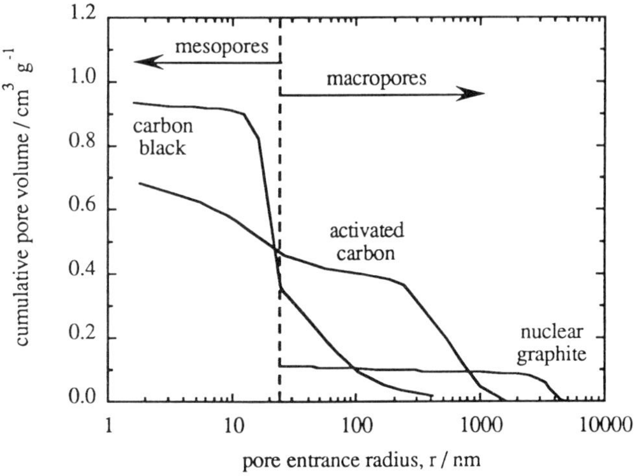

Fig. 4.4 Cumulative pore-size distributions for various carbon materials obtained using mercury porosimetry

of macropores and mesopores. The fine-grained graphite has a much smaller open porosity than the other two materials and a narrow, unimodal pore-size distribution of macropores centred on $r \approx 4\ \mu m$.

Differential pore-size distributions from mercury porosimetry are usually presented as a differentiation of the cumulative pore-size distribution, i.e. in the form $D(r) = -dv/dr$ versus r. An example for the activated carbon whose cumulative pore-size distribution is in Figure 4.4 is shown in Figure 4.5. Note that the distribution scales over five orders of magnitude; Lowell and Shields[5] proposed an alternative presentation in the form $D(\ln r) = -dv/d(\ln r)$ to compress the data.

An alternative way to present the pore-size distribution is as a probability density function, $f(r)$. If v_p is the total open-pore volume defined by mercury porosimetry, then the total intruded-pore volume, $v_p(1)$, at a pressure equivalent to a pore radius r_1 is given by

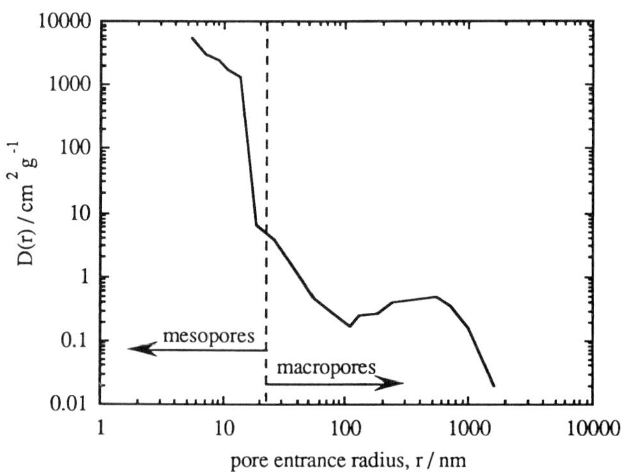

Fig. 4.5 Differential pore-size distribution for the activated carbon shown in Figure 4.4

$$\frac{v_p(1)}{v_p} = \int_{r_{max}}^{r_1} \left[\frac{v_p(r)}{v_p}\right] f(r)\,dr \tag{4.5}$$

where $v_p(r)$ is the volume of pores of radius r. In mercury porosimetry it is assumed that pores fill in serial order of their size, i.e. during intrusion, pores of radius $r+dr$ are completely filled before intrusion into pores of radius r commences. It follows from this that

$$f(r) = \frac{d}{dr}\left[\frac{v_p(1)}{v_p}\right] \tag{4.6}$$

An example of the pore-size distribution $f(r)$ versus r for the activated carbon shown in Figure 4.4 is given in Figure 4.6. An advantage of the function $f(r)$ is that it can be compared with similar functions obtained by other methods, e.g. fluid flow and quantitative image analysis.

Surface area distributions

Rootare and Prenzlow[12] showed that mercury porosimetry can also be used to obtain the cumulative open-pore surface area, A. The work done by intruding a differential volume of mercury, dv, into a pore at pressure p can be equated to the decrease in surface energy at the pore wall,

$$p\,dv = -(\gamma_s - \gamma_i)\,dA \tag{4.7}$$

where γ_s is the solid–vapour surface energy of the pore wall, γ_i is the interfacial surface energy between the pore wall and mercury and dA is the pore wall area

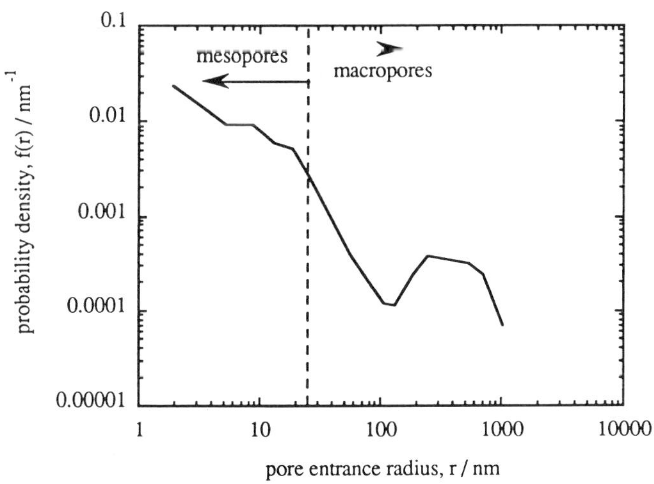

Fig. 4.6 Pore-size distribution for the activated carbon shown in Figures 4.4 and 4.5 presented as a probability density function, $f(r)$

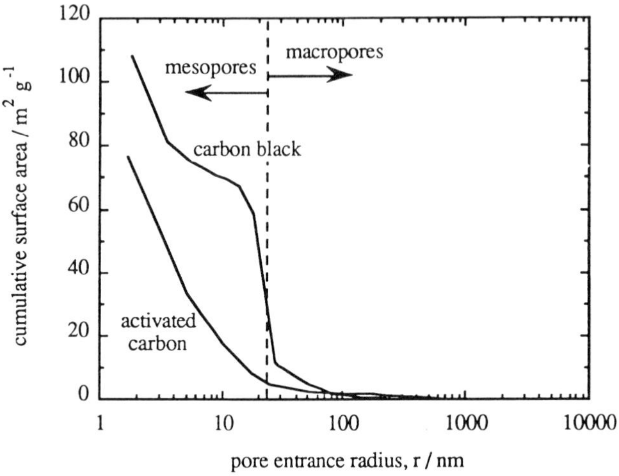

Fig. 4.7 Cumulative surface area distribution for the activated carbon and carbon black shown in Figure 4.4

covered by the intruding mercury. The relationship between γ_s, γ_i and the surface tension of mercury, γ, is given by the Young–Dupré equation

$$(\gamma_s - \gamma_i) = \gamma \cos\theta \tag{4.8}$$

Substitution of equation 4.8 into equation 4.7 and integration lead to

$$A(p) = -\frac{1}{\gamma\cos\theta}\int_{p_{min}}^{p} p\,dv \tag{4.9}$$

where $A(p)$ is the cumulative surface area up to pressure p. Equation 4.9 is obtained without making an assumption about pore shape, although it is assumed that γ and $\cos\theta$ are constant over the range of p. More usually a cylindrical pore shape is assumed so that p is transformed into pore radius r using the Washburn equation and the results are presented as $A(r)$ versus r.

Examples of cumulative surface areas obtained from the data in Figure 4.4 for a carbon black and an activated carbon are shown in Figure 4.7. From Figure 4.4, the mesopore volumes and macropore volumes for these carbons are comparable. However, from Figure 4.7, the mesopore surface areas are much greater than the macropore surface areas, because mesopores have higher surface area-to-volume ratios, S/V. It is easily shown that, for pores of a given shape, S/V is inversely proportional to pore width w. For example, a given total volume of mesopores of uniform shape and $w = 10$ nm has a surface area 1000 times greater than macropores with the same total volume and shape but with $w = 10$ μm.

Lowell and Shields[5] showed how it is possible to use equation 4.9 to obtain other distributions, including differential pore surface area distributions, pore length distributions and relative pore populations.

Fractal dimensions of pore walls

Pfeiffer and Avnir[13] have shown that the fractal dimension of a pore wall surface, D, is related to the pore size distribution, $\mathrm{d}v/\mathrm{d}r$, and the pore radius, r, by

$$-\frac{\mathrm{d}v}{\mathrm{d}r} \propto r^{(2-D)} \tag{4.10}$$

In a study of the fractal dimensions of coal particles, Friesen and Mikula[14] have shown that, if the pore-size distribution is determined by mercury porosimetry, then

$$\frac{\mathrm{d}v}{\mathrm{d}r} \propto p^{(D-4)} \tag{4.11}$$

A single value of D over the full range of p in equation 4.11 implies that $\mathrm{d}v/\mathrm{d}p$ is a continuously decreasing function of p, which in turn implies a unimodal pore-size distribution. It follows that a material which has a multimodal pore-size distribution cannot have a single-valued fractal dimension over the full range of p. This point is illustrated in Figure 4.8 for the activated carbon shown in Figure 4.4. The value of $D = 2.5$ at $p > 69$ MPa (10000 psia) is within the range $2 < D < 3$ expected for a surface fractal dimension. The apparent value of $D = 1.45$ at $p < 6.9$ MPa (1000 psia) lies outside this range and therefore must be treated with reserve. Friesen and Mikula[14] found a similar behaviour for coal char particles. Ehrburger–Dolle *et al.*[15] have also used this approach to estimate the fractal dimensions of oxidized resin-based carbons used as binders in coat-mix graphites.

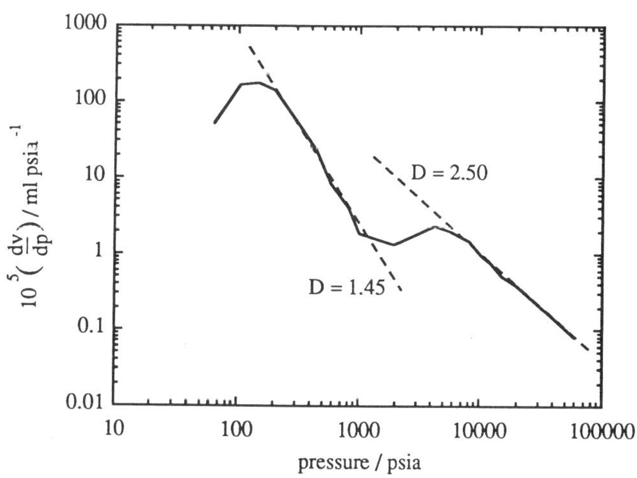

Fig. 4.8 Mercury porosimeter data for the activated carbon shown in Figure 4.4 replotted according to equation 4.11. D is the apparent fractal dimension. 1 psia = 6.895×10^3 Pa

Limitations of mercury porosimetry

Surface tension and contact angle

Although mercury porosimetry is one of the most widely used techniques to characterize porosity in carbon materials, it is subject to a number of serious limitations; these have been reviewed in standard texts.[4,5] First, there are the practical problems of selecting suitable values of γ and θ to use in the Washburn equation, equation 4.2. A value of $\gamma = 480\,\text{mJ m}^{-2}$ is usually assumed in mercury porosimetry, but mercury is particularly prone to contamination and even small amounts of impurities can change this value significantly. It is obvious from the Young–Dupré equation, equation 4.8, that the value of θ will depend on γ_s, and γ_i. However, in mercury porosimetry it is usual to assume a value of $\theta = 140°$ for mercury in contact with all solids. Measurements on a range of solids[5,9] suggest that θ can lie anywhere in the range 130–150°, giving rise to a possible error of up to about 16% in the determination of r. There is evidence that for some solids θ can lie outside this range. For example, Winslow and Diamond[16] measured values of θ = 105–117° for cement pastes. For mercury, as for other liquids, the value of the contact angle may also depend on the physical and chemical state of the solid surface and also on whether the liquid is advancing or retreating over it. This effect has been proposed as one mechanism to explain hysteresis in mercury porosimetry.

The cylindrical pore model

The Washburn equation, equation 4.2 assumes non-intersecting, cylindrical pores and different versions of the equation are obtained for other pore geometries. For example, in a parallel-plate model pore, which may be appropriate for slit- or crack-shaped pores, r_2 in the Young–Laplace equation is infinite and the Washburn equation becomes

$$r = \frac{\gamma \cos \theta}{\Delta p} \tag{4.12}$$

In such a case equation 4.2 would be in error by a factor of two. In the case of solids consisting of an agglomeration of packed spheres, the toroidal mercury meniscus around the point of contact of spheres has radii of curvature of opposite sign. There is a lower limit of r for applying the Young–Laplace equation, since the macroscopic concept of a meniscus will cease to be valid as pore widths approach molecular dimensions. Gregg and Sing[4] suggested a value of $r = 3.5$ nm. The assumption of non-intersecting pores also neglects pore network effects.

High-pressure damage to the sample

There is evidence that mercury intrusion at high pressure can damage the porous materials. The compressive strengths of porous carbon materials are generally less than 50 MPa and the maximum hydrostatic stress in mercury porosimetry is more than 400 MPa, so that in complex local stress fields generated, for example, by the presence of closed pores, fracture of the carbon structure can be easily envisaged. Existing open-pore walls may be prised apart and pore walls may be breached, allowing access to previously closed porosity. Conversely, compaction may occur leading to closure of previously opened pores. In graphites it has been shown[17] that mercury and helium pore volumes are in agreement provided that the mercury pressure is less than about 20 MPa, suggesting a lower limit of $r = 36$ nm. In a study of mercury intrusion into carbon black compacts,[18] there was evidence for distintegration of carbon particles at a pressure of 17.2 MPa, corresponding to $r = 43$ nm. Also, the cumulative surface area measured below this pressure corresponded to the BET surface area measured by nitrogen adsorption, whereas the BET surface area was less than the cumulative surface area measured at the highest pressure used, 413 MPa.

Pore network effects

Perhaps the most serious limitation of the Washburn equation is that the model of non-intersecting, cylindrical capillaries neglects pore network effects. In most porous solids pores of different sizes are interconnected and the manner of their connection can play an important role in mercury porosimetry. Most obviously, if the pore network is of the aperture-cavity type, then filling of pores by mercury will be controlled by the width of constrictions at the entrances to pores; it is for this reason that pore radii obtained from mercury porosimetry are usually referred to as pore entrance radii. There have been a number of theoretical and modelling studies of percolation effects in mercury porosimetry,[19–22] but these concepts appear not to have been applied to carbon materials.

More generally, intrusion and extrusion may be dominated by percolation of mercury through the porous network. Some of the consequences of pore networks on mercury porosimetry are illustrated using the model two-dimensional, aparture-cavity pore network in Figure 4.9. According to the Washburn equation the different size apertures $1>2>3$ should fill at pressures $p_1<p_2<p_3$. Figure 4.9a shows the extent of penetration after intrusion at p_1. Note that the hatched type 1 apertures in the centre of the pore network are

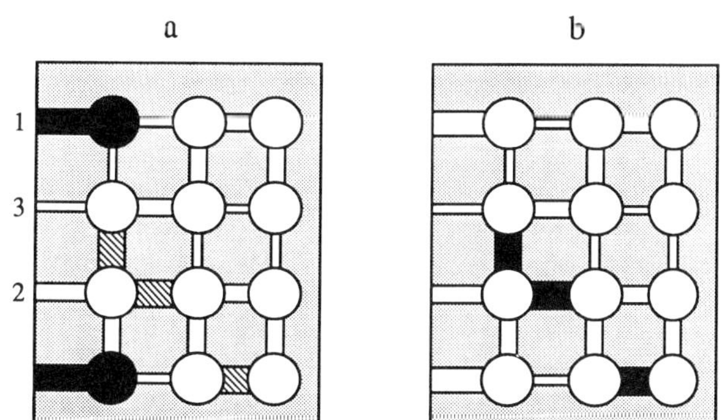

Fig. 4.9 Mercury intrusion into a two-dimensional model aperture-cavity network containing three types of aperture 1, 2 and 3, where $r_1>r_2>r_3$. (a) After intrusion at p_1, the pressure required to fill type 1 apertures (note that the hatched type 1 apertures are unfilled); (b) after extrusion at $p<p_1$ (note the retention of mercury in type 1 apertures at the centre of the network)

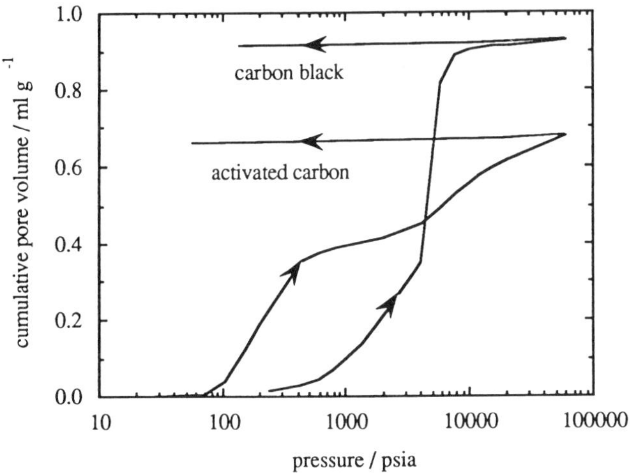

Fig. 4.10 Example of hysteresis during mercury intrusion and extrusion for an activated carbon and a carbon black. 1 psia = 6.895×10^3 Pa

unfilled, because they are shielded by narrower apertures which block access to the intruding mercury. In such a case the cumulative pore volume is influenced by the spatial arrangement of pores in the network and the pore-size distribution obtained using the non-intersecting cylindrical pore model is erroneous. Figure 4.9b shows the position after mercury extrusion at $p < p_1$. There is retention of mercury in the size 1 apertures in the centre of the network, because they have lost contact with the external surface during extrusion at $p < p_1$. This is relevant to hysteresis in mercury porosimetry.

Hysteresis effects

Examples of hysteresis effects during intrusion and extrusion of mercury into carbon materials are shown in Figure 4.10. These are characterized by retention of significant quantities of mercury at the end of the depressurization cycle; in the examples shown almost all of the mercury is retained. In other cases it has been shown[5] that, after an initial open hysteresis loop, reversible or near-reversible hysteresis loops can be obtained on second and subsequent intrusion–extrusion cycles.

There have been a number of explanations for hysteresis in mercury porosimetry. The difference between the contact angle of a liquid when it advances and retreats over a surface has been proposed as a cause of hysteresis. Lowell and Shields[23] showed that intrusion and extrusion curves could be brought into coincidence if a smaller contact angle was used for extrusion. However, Conner *et al.*[24] claimed that contact angle hysteresis cannot explain the phenomenon in mercury porosimetry and they favoured a mechanism based on percolation effects (see below).

Aperture-cavity effects, e.g. as modelled by an 'ink bottle pore', have been proposed to explain mercury porosimetry hysteresis[25–27]; similar mechanisms have been advanced to explain adsorption hysteresis. If an ink bottle pore has a neck of radius r_n and a body of radius r_b, then, from the Washburn equation, filling of the pore occurs at

$$\Delta p_n = -\frac{2\gamma\cos\theta}{r_n} \tag{4.13}$$

and emptying of the pore occurs at

$$\Delta p_b = -\frac{2\gamma\cos\theta}{r_b} \tag{4.14.}$$

since $r_n < r_b$ then $\Delta p_n > \Delta p_b$, and emptying of the pore on depressurization occurs at a lower pressure than filling of the pore during intrusion, giving rise to hysteresis. A hysteresis scanning technique for estimating the size distributions of apertures and cavities, based upon this model has been proposed by Reverberi *et al.*[28] The ink bottle model has also been criticized as an explanation for mercury porosimetry hysteresis.[29]

Hysteresis can also arise from the pore-network effects discussed previously. Androutsopoulos and Mann[19] showed that hysteresis curves similar to those found for a porous catalyst could be generated by a percolation network model. For the simple model in Figure 4.9b, there is retention of mercury in apertures of type 1 after extrusion at $p < p_1$, because these apertures have lost contact with the external surface of the solid. Finally, if intrusion of mercury causes irreversible rupture of the structure of the porous solid, as discussed previously, then hysteresis is to be expected.

Comparisons of mercury porosimetry with gas adsorption

The partial overlap in pore-size ranges between mercury porosimetry (~3.5 nm–200 μm) and pore-size distributions determined by gas adsorption (~2–50 nm) enables the two methods to be compared. Results have been variable, which is not surprising when the limitations of the two methods are considered. Usually it is necessary to employ a shifting factor to bring the pore-size distribution curves into coincidence, either by normalizing the volumes of mercury intruded and gas adsorbed at a reference point on the pore radius axis, or by shifting the latter axis by adjusting the contact angle used in the Washburn equation. In a study of an activated charcoal, Joyner *et al.*[30] found good agreement between the two methods by adjusting the contact angle. By normalizing pore volumes, Dubinin *et al.*[31] found good agreement for one activated carbon but poor agreement for another carbon.

A more obvious and direct comparison can be made between BET surface areas, $A_{(BET)}$, measured by gas adsorption, and surface areas obtained from mercury porosimetry, $A_{(Hg)}$. In the original study, Rootare and Prenzlow[12] found reasonable agreement between the two values for a range of non-porous materials with $A_{(BET)}$ values up to $110\ m^2\ g^{-1}$. Spitzer[11] compared $A_{(Hg)}$ and $A_{(BET)}$ for a series of coals with $A_{(BET)} < 10\ m^2\ g^{-1}$ and found $1 < R < 2$, where $R = A_{(Hg)}/A_{(BET)}$. A more complex picture emerges for porous materials with larger $A_{(BET)}$ values; R ranges from ~2 to ~0.01. A compilation of results from various workers made by Davis[32] for a range of oxides and recent results[18,33] for carbon black compacts and activated carbons are shown in Figure 4.11. The very low values of R for the activated carbons are undoubtedly due to the contribu-

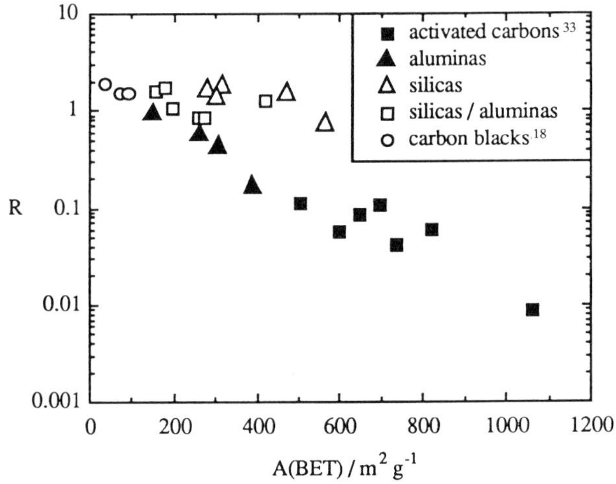

Fig. 4.11 Comparison of the ratio $R = A_{(Hg)}/A_{(BET)}$ as a function of $A_{(BET)}$ for a range of porous materials

tion to $A_{(BET)}$ from micropores which are not accessible to mercury. Values of R in the range $1<R<2$ which were found for carbon black compacts were attributed to disruption of the porous structure at high mercury pressures.[18]

4.3 Fluid flow methods

Introduction

In porosimetry, mercury penetrates but does not flow completely through the grains (or consolidated parts) of a carbon sample since the applied pressure, p, of the liquid that surrounds them is uniform. However, if a path in the open-pore structure extending from one external face of a carbon grain (say face 1) to another (face 2) is filled with mercury at constant temperature, and if there is a pressure difference in the liquid between these faces, for example if $p_1>p_2$, then, provided that the effects of external forces acting on the mercury (gravitational, electromagnetic, centrifugal, etc.) are negligible or absent, there will be a net flow through the grain via this path, in this case from face 1 to face 2. The flow rate of mercury or, more generally, of a component in any inert liquid or gas mixture, depends on external forces acting on the fluid, on gradients in fluid pressure, composition and temperature across a carbon grain (or across a consolidated piece of any size) and on the geometry of the open pores through which flow occurs. Standard texts on fluid flow in porous media include those by Collins,[34] Scheidegger,[35] Dullien[36] and Greenkorn.[37] In addition, there are more specialized monographs that consider only gas flow in porous media, e.g., by Carman,[38] Cunningham and Williams[39] and Mason and Malinauskas.[40] Concerning fluid flow specifically in carbons and graphites, there have been a number of reviews, e.g. by Hewitt and Morgan,[41] Hewitt[42] and Grove.[43]

Essentially, in fluid flow methods, the geometry of the open pore structure in a carbon sample is estimated from experimental measurements of flow that are made in controlled or known external fluid conditions. Usually in these methods, the temperature is constant, the effects of external forces acting on the fluid are negligible and components in the fluid do not interact significantly, either with each other or with the sample. Under these conditions, the flow of components is as a result only of fluid-phase pressure and composition gradients across a sample. As is shown below, fluid flow due to pressure and composition gradients is dominated, in general, by the presence of open macropores. Accordingly, fluid flow methods are of special use for the characterization of open macroporosity, in carbons in particular, but also in other porous media such as rocks, cement and ceramics.[34–37]

As was demonstrated earlier, the macropore structure in carbons is, in the main, far too complex and irregular to be described precisely. Therefore, the exact relationships between flow, fluid conditions and pore structure usually cannot be defined. This difficulty is normally overcome by assuming a model pore structure for carbons, and representing fluid flow in terms of parameters of this model.[41–43] Such an approach leads to explicit equations that relate flow, fluid conditions and pore structure. Subsequently, these equations may be used to analyse experimental flow data, and to obtain estimates of model parameters that characterize open macropore structure.

Almost exclusively, in fluid flow methods, as for the derivation of the Washburn equation, equation 4.2, in porosimetry, the model open-pore structure in carbons consists of an array of cylindrical capillaries extending throughout a sample, and connected to external surfaces.[41–43] An equation that relates flow rate to the radius and length of a single, straight capillary, and to the fluid conditions at its ends, is well established.[39,40] Thus, in principle, an equation for net flow through a sample can be derived by a summation or integration of this single capillary equation over the model array. The resulting equation for net flow will depend on fluid conditions external to the sample, and on functions that characterize the model pore structure, such as distributions of capillary radius, length and connectivity.

In summary, then, fluid flow methods involve: (i) the experimental measurement of flow rates through a carbon sample of components in a fluid mixture, under known or controlled external fluid conditions; (ii) the derivation of an equation that relates flow rates to an assumed model pore structure; and (iii) the estimation of functions that characterize porosity in the equation derived in (ii) from the data obtained in (i). The following three sections review the main aspects of fluid flow methods for characterizing the (macro)pore structure in carbons: the first considers models of pore structure and fluid flow, the next section describes the commonest methods for experimentally measuring fluid flow rates and the final section summarizes ways that model pore structures may be estimated from experimental flow data.

Models of pore structure and fluid flow in carbons

Models of pore structure

The model structure for the open porosity in carbons that forms the basis of established fluid flow methods consists of N_V, non-intersecting, straight, cylindrical capillaries per unit volume. This model, proposed by Wheeler[44] for porous catalysts, and later developed for similar materials by Johnson and Stewart,[45] was originally applied systematically to carbons and graphites by Hewitt.[42] The model assumes that porosity is isotropic, i.e. in a unit cube of carbon, equal numbers of capillaries extend between opposite external faces, so that the number of capillaries per unit surface area on each face is $N_V/3$. For anisotropic pore structures, e.g. as might be expected in extruded, synthetic graphites, the ratio of volume to surface capillary densities will, in general, be different from 1/3 in different directions. However, in the context of fluid flow, no detailed analyses of anisotropic pore structures in different carbons appear to have been attempted to date. In the model, the length, l, of each capillary in a sample is constant, but may be greater than or equal to the length of the sample across which flow occurs, L, by a factor $q \geq 1$, called the tortuosity, so that $l = qL$. The tortuosity is often considered as a factor which accounts for the irregular paths that real pores follow in porous media although, because of the simplistic nature of the capillary model, its physical significance is doubtful. To account for pores of different sizes (heteroporosity), the radii of the capillaries in the model are usually assumed to be distributed with a probability density function, $f(r)$. Homoporosity, i.e. where all the pores have the same size, e.g. $r = r_0$, may be considered as the special case when $f(r) = \delta(r - r_0)$, where $\delta()$ is the Dirac delta function.

In the model, the fractional open pore volume, V_p, and the surface area of open pores per unit volume, S_p, are given by

$$V_p = \pi N_V \langle r^2 \rangle q \tag{4.15}$$

and

$$S_p = 2\pi N_V \langle r \rangle q \tag{4.16}$$

where $\langle\,\rangle$ denotes a mean value.

In summary, the model open-pore structure in carbons is characterized by the two parameters N_V and q, and the probability density function $f(r)$. More realistic models of pore structure have been proposed[34–37] that account for anisotropy, interconnected and non-circular pores and pores of different lengths. However, none of these models is of much practical use, either for porous media in general or carbons in particular, on account of the often insurmountable difficulties encountered in obtaining from experimental data stable estimates of the large number of characteristic parameters and functions in the models.

Models of fluid flow

The two best established fluid flow methods for determining the pore structure in carbons, described below, are based on measurements of the isothermal, steady-state permeation of a pure gas and the isothermal, isobaric, steady-state diffusion of components in a binary gas mixture. Other fluid flow methods that have also been used in the study of pore structure in carbons include permeametry,[46] i.e. measurements of the steady-state flow of a pure liquid (a method related to porosimetry, see previously and below), and unsteady-state or transient gaseous diffusion (see Cunningham and Williams[39]).

The basis of both permeation and diffusion methods is the gas flow conductivity, C, defined as the molar flux through a sample per unit concentration gradient. For a component i in a gas mixture, the gas conductivity may be written as

$$C_i = \frac{J_i/A}{\Delta c_i/L} \tag{4.17}$$

where J_i is the molar flow rate through the sample of component i, A is the cross-sectional area of the sample perpendicular to the macroscopic direction of flow and Δc_i is the concentration difference for component i across the length, L, of the sample parallel to flow. The gas flow conductivity is a measure of the ease with which a component of a gas mixture flows through a material, and is analogous to the thermal conductivity for the flow of heat.

For the isothermal, steady-state flow of a pure, ideal gas, equation 4.17 may be written as

$$C_i = C = K = \frac{JRTL}{A\Delta p} \tag{4.18}$$

where K is the permeability, R is the gas constant and T is the absolute temperature.

For the isothermal, isobaric, steady-state diffusion of component i in an ideal, binary mixture of ideal gases, equation 4.17 becomes

$$C_i = Di = \frac{J_i RTL}{Ap\Delta y_i} \tag{4.19}$$

where D is the effective diffusivity of the material, p is pressure and y is mole fraction.

Both K and D depend on the flowing gas and the pore and surface structure of the carbon, in addition to the pressure and composition conditions. Experimentally, permeabilities and diffusivities are determined from equations 4.18 and 4.19 respectively, using measurements of molar flow rates, temperature and sample geometry, and of pressure and composition conditions at opposite ends of the sample. The required relationships that provide the means to estimate pore structure from experimental measurements of K and D are based on equations that describe flow through individual capillaries.

First, considering permeation, the isothermal molar flow rate, j, of a pure, ideal gas through a single, straight, cylindrical capillary of radius r and length $l \gg r$ may be written as[39,40]

$$j_i = \frac{\pi r^4 \langle p \rangle \Delta p}{8\eta RTl} + \frac{\pi r^2 D_K \Delta p}{RTl} \tag{4.20}$$

where the mean pressure $\langle p \rangle = (p_1 + p_2)/2$, the pressure difference $\Delta p = p_1 - p_2 > 0$ and η is the viscosity of the flowing gas. The first term in equation 4.20 is the Hagen–Poiseuille equation for a gas, and represents viscous or laminar flow through the capillary. The second term, which was in the past referred to as slip flow,[38] is now recognized as being due to Knudsen diffusion,[39,40] i.e. flow in the capillary at low pressures where gas-phase intermolecular collisions are rare compared with collisions between the gas molecules and the walls of the capillary. The factor D_K in this second term is the Knudsen diffusion coefficient, which, assuming completely diffuse reflection of molecules colliding with pore walls, is

$$D_K = \frac{2}{3} r \langle v \rangle \tag{4.21}$$

where $\langle v \rangle$ is the mean molecular speed of gas molecules of molecular weight M, which, from kinetic theory, is given by

$$\langle v \rangle = \left(\frac{8RT}{\pi M} \right)^{1/2} \tag{4.22}$$

Assuming a large number of capillaries in a sample, the total molar flow rate of gas, J, through the model, isotropic array of non-intersecting capillaries may be approximated by integrating equation 4.20 over the probability density function of pore radius, $f(r)$

$$J = \frac{N_V}{3} A \int_{r_{min}}^{r_{max}} j \cdot \mathrm{f}(r)\,\mathrm{d}r \tag{4.23}$$

where r_{min} and r_{max} are the lower and upper limits of $f(r)$. Combining equations 4.18, 4.20, 4.21, 4.22 and 4.23 leads to the Carman equation[34–37]

$$K = \frac{B_0}{\eta} \langle p \rangle + \frac{4}{3} K_0 \langle v \rangle \tag{4.24}$$

where the factors B_0 and K_0 are the viscous flow coefficient and the Knudsen flow coefficient, respectively. In terms of the model pore structure, these coefficients are given by

$$B_0 = \frac{\pi N_V \langle r^4 \rangle}{24q} \tag{4.25}$$

and

$$K_0 = \frac{\pi N_V \langle r^3 \rangle}{6q} \tag{4.26}$$

Clearly, then, if experimental permeability data, determined from equation 4.18, are plotted as a function of the mean pressure across the sample, $\langle p \rangle$, then, from the Carman equation, equation 4.24, a straight-line relationship should be observed, whose slope is related to B_0 using equation 4.25 and whose intercept is related to K_0 using equation 4.26. Accordingly, since both the viscous and Knudsen flow coefficients are related to parameters and functions of the capillary array, the desired relationship between gas permeation and pore structure is established.

Flow of a pure, incompressible liquid does not involve Knudsen diffusion, for in this case molecule–wall collisions are always rarer than molecule–molecule collisions in the fluid phase. The viscous flow coefficient for a pure, incompressible liquid is represented by equation 4.25. This coefficient may be determined from experimental measurements of volume flow rate, Q, rather than molar flow rate, J, as follows,[38]

$$B_0 = \frac{Q\eta L}{A\Delta p} \tag{4.27}$$

The basis of permeametry (or liquid permeability)[46] is the correlation of experimental measurements of B_0 determined using equation 4.27 with the model for the viscous flow coefficient, equation 4.25. However, there are complications in the modelling of permeametry data since, according to the Washburn equation, equation 4.2, at a given pressure, only pores larger than a certain radius are filled with liquid, and will therefore allow flow.

Considering diffusion, the isothermal, steady-state, molar flow rate, j_i, of a component i in an ideal, binary mixture of ideal gases through a single, straight, cylindrical capillary of radius r and length $l \gg r$ is given by[39,40]

$$j_i = \frac{\pi r^2 p D_{in}}{\alpha_{in} RTl} \ln \left(\frac{1 - \alpha_{in} y_{i_2} + \dfrac{D_{in}}{D_{\mathrm{K}_i}}}{1 - \alpha_{in} y_{i_1} + \dfrac{D_{in}}{D_{\mathrm{K}_i}}} \right) \tag{4.28}$$

where the flow rate ratio $\alpha_{in} = 1 + (j_n/j_i) = 1 - (M_i/M_n)^{1/2}$ and D_{in} is the normal or bulk binary diffusion coefficient for the mixture of components i and n ($i = 1$, $n = 2$ or $i = 2$, $n = 1$). In equation 4.28, the subsubscripts 1 and 2 for mole fraction, y, refer to the ends of the capillary, and the subsubscript i for the Knudsen diffusion coefficient, D_{K}, refers to equations 4.21 and 4.22 determined using the molecular weight, M, for component i.

It is important to note that, unlike the capillary flow rate for permeation, equation 4.20, the capillary flow rate equation for diffusion, equation 4.28, depends in a complex way on radius (since the Knudsen diffusion coefficient is proportional to r) and on pressure (since the bulk diffusion coefficient is inversely proportional to p).

The total molar diffusion rate of component i, obtained by assuming a large number of capillaries in the sample, and integrating equation 4.28 over the radius distribution, $f(r)$, is

$$J_i = \frac{N_{\mathrm{V}}}{3} A \int_{r_{\min}}^{r_{\max}} j_i f(r)\,\mathrm{d}r \tag{4.29}$$

This equation cannot be evaluated in terms of means values of powers of r, unlike equation 4.23 for permeation, on account of the complex equation for j_i, equation 4.28, referred to above.

Instead of working with the effective diffusivity, equation 4.19, it is convenient to define a new factor, the diffusivity ratio, λ, i.e.

$$\lambda_i = \frac{D_i}{D_{in}} = \frac{J_i RTL}{Ap\Delta y_i D_{in}} \tag{4.30}$$

The diffusivity ratio, which has values between zero and one, is a normalized measure of the ease with which a component of a gas mixture diffuses through a carbon. With increasing porosity, a carbon sample offers decreasing resistance to diffusion, and as the fractional porosity tends to one, then $D_i \rightarrow D_{in}$ and $\lambda_i \rightarrow 1$. An equation for the diffusivity ratio is obtained by combining equations 4.28, 4.29 and 4.30,

$$\lambda_i = \frac{\pi}{3\alpha_{in}\Delta y_i} \int_{r_{\min}}^{r_{\max}} r^2 \left(\frac{1 - \alpha_{in} y_{i_2} + \dfrac{D_{in}}{D_{K_i}}}{1 - \alpha_{in} y_{i_1} + \dfrac{D_{in}}{D_{K_i}}} \right) f(r)\,\mathrm{d}r \tag{4.31}$$

This generalized but complex relationship reduces to simpler forms under certain conditions. The Knudsen diffusivity ratio (for flow at low pressures) is

$$\lim_{p \rightarrow 0} \lambda_i = \frac{2\pi N_V \langle r^3 \rangle \langle v_i \rangle}{9qD_{in}} \tag{4.32}$$

and the equimolar, molecular diffusivity ratio (for the flow at high pressures of components with approximately equal molecular weights) is given by

$$\lim_{\alpha_{in} \rightarrow 0,\, p \rightarrow \infty} \lambda_i = \frac{\pi N_V \langle r^2 \rangle}{3q} \tag{4.33}$$

Equation 4.33 is often combined with the expression for the fractional open-pore volume, equation 4.15, to give

$$\lambda_i = \frac{V_P}{3q^2} \tag{4.34}$$

This equation is the one most often used in studies of pore structure and gaseous diffusion in carbons and other porous media. For $q = 1$, it demonstrates that the diffusivity at high pressures of components in a binary gas mixture, consisting of components with approximately equal molecular weights, is smaller than the normal, bulk diffusivity by factor equal to the fractional open porosity in the direction of flow.

The limiting forms of the generalized diffusivity equation, equations 4.32 and 4.33, indicate that at low pressures, where Knudsen diffusion predominates, the diffusivity ratio (which may be determined experimentally from equation 4.30) increases linearly with pressure (since D_{in} is inversely proportional to p), but at high pressures, where normal diffusion occurs in pores, the diffusivity ratio is a constant. These limiting equations also show that the diffusivity ratio depends on $(N_V/q)\langle r^3\rangle$ at low pressures (as does the Knudsen flow coefficient for permeation, equation 4.26), and on $(N_V/q)\langle r^2\rangle$ at high pressures.

It is important to note that permeability coefficients and limiting effective diffusivities (or diffusivity ratios) for the model pore structure assumed for carbons depend on mean values of powers of pore radius, r. This general conclusion that the values of flow coefficients and flow rates increase with increasing pore size is the same for more complex models of pore structure that involve connected pores and pores of non-circular cross-sections.[34–37] Hence flow in carbons, and in porous media in general, is dominated by large pores, and therefore fluid flow methods are of particular use for investigating macropores.

Experimental fluid flow methods

Gas permeation measurements

Experimentally, gas permeation measurements are based on the definition of the permeability, K,

$$K = \frac{JRTL}{A\Delta p} \tag{4.35}$$

where J is the molar flow rate of a gas at absolute temperature T through a sample of cross-sectional area A and length L, as a result of an applied pressure difference Δp. A schematic gas permeability apparatus is shown in Figure 4.12.

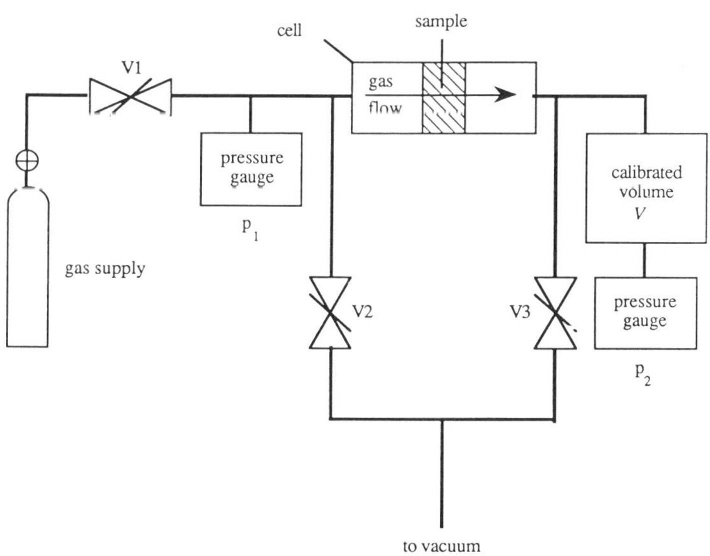

Fig. 4.12 Schematic diagram of gas permeability apparatus

Detailed descriptions of practical permeability systems have been given by Hewitt and Morgan.[41]

The principles of operation of the permeability apparatus in Figure 4.12 are as follows. The sample cell is evacuated to ~13.3 μPa. A pure gas (typically nitrogen) is introduced to side 1 of the sample cell, such that the pressure, p_1, is maintained constant, while side 2 of the cell is kept under vacuum. On isolating the cell from the vacuum pump, the rate of rise of pressure, dp_2/dt, in the previously evacuated, calibrated volume connected to side 2 of the cell may be used to determine the molar flow rate through the sample, J, for if the total volume of side 2 of the cell, V (the calibrated volume plus the volume of pipework), is known and sufficiently large that $p_1 \gg p_2$, then

$$J = \frac{V}{RT}\frac{dp_2}{dt} \qquad (4.36)$$

Accordingly, the permeability may be calculated from

$$K = \frac{V}{Ap_1L}\frac{dp_2}{dt} \qquad (4.37)$$

Essentially, then, the determination of permeability involves the accurate and precise measurement of the inlet and outlet pressures, p_1 and p_2. This is normally accomplished using sensitive manometers, capable of detecting high pressures on side 1 of the cell (around atmospheric pressure, ~100 kPa), and low pressures on side 2 of the cell (down to ~10 μPa).

The definition of permeability, equation 4.18, for the experimental conditions above, where the mean pressure $\langle p \rangle = p_1/2$, is

$$K = \frac{B_0}{2\eta}p_1 + \frac{4}{3}K_0\langle v \rangle \qquad (4.38)$$

Accordingly, the procedure for permeation analysis involves: (i) obtaining experimental measurements of permeability (from equation 4.37) for a selected gas and flow temperature (usually ambient), at different inlet pressures p_i; (ii) plotting these measurements as a function of p_1; and (iii) estimating the slope and intercept of a linear relationship assumed for these data (equation 4.38), from which the viscous and Knudsen flow coefficients, B_0 and K_0, and hence pore-structure factors from equations 4.25 and 4.26, may be calculated.

Gas diffusion measurements

The main element of most experimental systems to measure effective gas diffusivities or diffusivity ratios in carbons, and other porous media, is the Wicke–Kallenbach cell[47] (Figure 4.13). For a detailed description of a practical diffusivity system (that was used to investigate graphites, but is applicable to other porous carbons), see Hewitt and Morgan,[41] and for descriptions of diffusivity experiments in different carbons (mainly graphites) see, e.g., Evans *et al.*,[48] Weller and Watts,[49] Hawtin *et al.*[50] and Yang and Liu.[51]

In the Wicke–Kallenbach cell (Figure 4.13), pure gases, A and B, flow past opposite faces of a sample. When the total pressure is maintained constant,

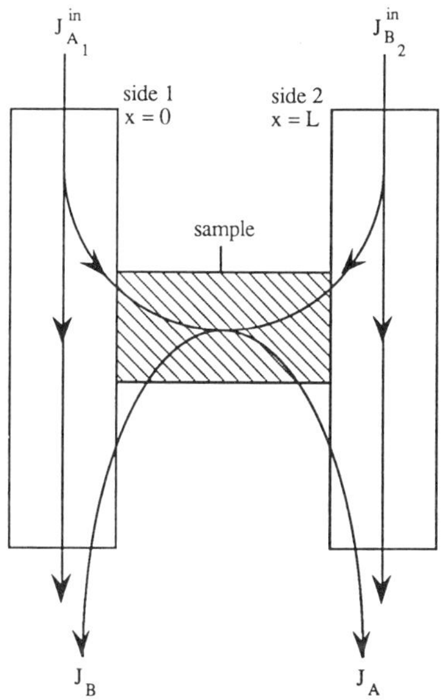

Fig. 4.13 Schematic diagram of Wicke–Kallenbach cell for the measurement of effective diffusivity. J denotes molar flow rate

countercurrent diffusion of the gases occurs through the sample, i.e. gas A diffuses in the positive x direction, (from side 1 of the sample to side 2, Figure 4.13), down its concentration gradient into the stream of gas B, and gas B diffuses in the negative x direction down its concentration gradient into the stream of gas A. In the steady state, the molar flow rates of the gases through the sample are given by a mass balance of flows into and out of the cell. The molar flow rate of gas A through the sample, J_A, is given by[39]

$$J_A = \frac{y_{A_1} y_{A_2} J^{in}_{B_2} - y_{A_2}(1 - y_{A_1}) J^{in}_{A_1}}{y_{A_1} - y_{A_2}} \tag{4.39}$$

where y is mole fraction, the superscript 'in' refers to the inlet molar flow rates and the sub-subscripts refer to sides 1 and 2 of the cell. A similar equation applies to gas B, except that subscript A becomes subscript B, and sub-subscript 1 becomes sub-subscript 2, and vice versa.

By measuring the inlet molar flow rates to the cell and the mole fractions in the outlet streams, the diffusion flow rates can be calculated by equation 4.39 (and its equivalent for gas B). These rates are those required to determine the diffusivity ratio, equation 4.30, rewritten here for component A as

$$\lambda_A = \frac{J_A RTL}{Ap(y_{A_1} - y_{A_2}) D_{AB}} \tag{4.40}$$

Experimentally, inlet and outlet molar flow rates may be measured using rotameters or mass flow meters, and the compositions of the outlet streams (which contain only small fractions of the minor components) may be determined by suitable methods such as infrared or mass spectrometry.

Most diffusivity measurements for carbons appear to have been made at around atmospheric pressure, where Knudsen diffusion is negligible, and using gas pairs of similar molecular weights such as CO_2 and Ar. Under these conditions, equation 4.33 or 4.34 is often used to equate to the experimental measurements obtained using equations 4.39 and 4.40. However, some diffusivity measurements in carbons have been made over ranges of pressures below atmospheric, where Knudsen diffusion is important, and using gases of different molecular weights, e.g. He and Ar (see Mason and Malinauskas[40] and Mays and McEnaney[52]).

Estimating pore structure from experimental flow data

Transport and blind porosity

In the steady state, fluid flows through only a fraction of the open porosity, called the transport porosity. The remaining open pores, which are connected to transport pores via single openings, are called blind pores. Pressure and concentration gradients and hence flow occur only in blind pores during unsteady-state flow. Accordingly, the fractional pore volumes referred to above in descriptions of steady-state fluid flow methods refer only to transport pores. It has been estimated from non-steady-state diffusion measurements[53–55] that the transport porosity in graphites is about 90% of the total open porosity. However, as is shown below, it is impossible using the cylindrical capillary model described above to estimate the exact value of transport porosity in carbons from steady-state fluid flow measurements.

Analysis of permeability data

For the purposes of illustration, Figure 4.14 is a plot of the permeability, K, of N_2 at 298 K in a nuclear graphite as a function of mean pressure, $\langle p \rangle = p_1/2$, measured using a pressure-rise apparatus described previously. Figure 4.14 shows that K increases linearly with increasing $\langle p \rangle$, as expected from the model.

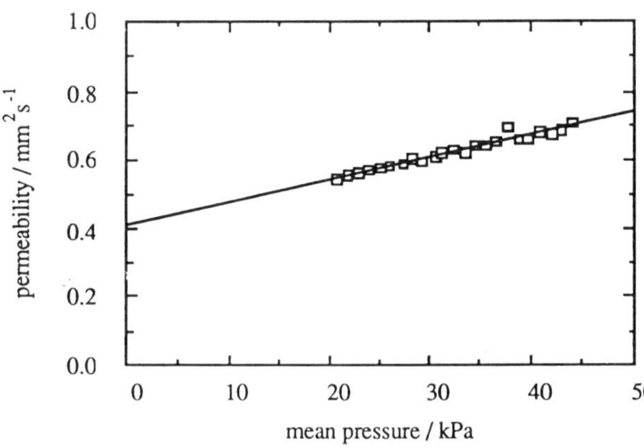

Fig. 4.14 Permeability of nitrogen at 298 K in a nuclear graphite as a function of mean pressure

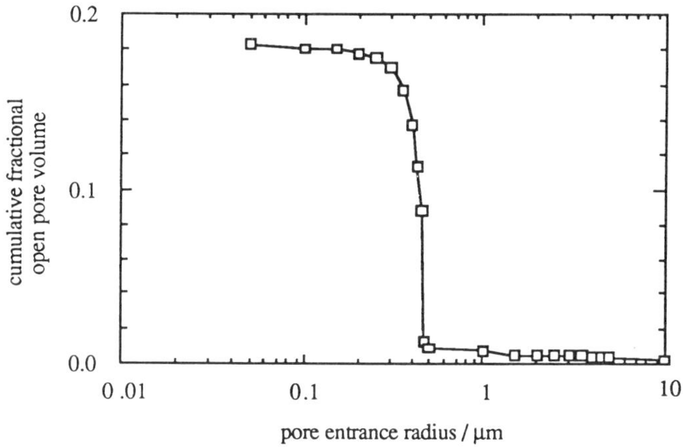

Fig. 4.15 Cumulative open pore volume as a function of pore entrance radius for a nuclear graphite, determined using mercury porosimetry

The estimated values of the viscous and Knudsen flow coefficients, B_0 and K_0, from the data, using the Carman equation, equation 4.24, are 230 nm^2 and 1.50 nm, respectively. Assuming homoporosity, i.e. that all pores are the same size, the (transport) pore radius for the graphite, determined from a combination of equations 4.25 and 4.26 for B_0 and K_0, is 0.46 μm. Further, assuming unit tortuosity ($q = 1$), the transport porosity determined from equations 4.15, 4.25 and 4.26 is determined as 2.7%, i.e. 16% of the open pore volume for the graphite measured using He pycnometry (17%). The value of tortuosity assuming that all open pores in the graphite were involved in transport is $q = 2.5$. As mentioned above, pore volume cannot be determined without assumptions concerning the value of tortuosity, and vice versa.

These data compare with the mercury pore-size distribution for the graphite (Figure 4.15), which indicates a mean pore-entrance radius of 0.3 μm and a total open-pore volume of 17% (in agreement with He pycnometry).

This example of an analysis of permeability data indicates that even using simplistic assumptions, values of characteristic pore-structure data obtained from a fluid flow method appear consistent within themselves, and compare favourably with information from a different experimental technique. That the pore size from permeability is close to the pore size from porosimetry also suggests that permeation in carbons is influenced by constrictions in the pore structure as is mercury intrusion.

Analysis of diffusion data

The equimolar diffusivity ratio for CO_2/Ar at atmospheric pressure and 298 K (where Knudsen diffusion is negligible, and where, therefore, equation 4.33 or 4.34 may be used for λ) for the same graphite considered previously was determined as 0.0108, using an experimental apparatus based on a Wicke–Kallenbach cell. From equation 4.34, assuming unit tortuosity, the transport porosity is therefore 3.2%, i.e. similar to the value from permeability (2.7%). Again, as for permeability, an assumption concerning the value of tortuosity is required to determine pore volume.

This correlation between pore-structure data obtained using different experimental methods supports the capillary model assumed for the pore structure, and illustrates the utility of steady-state flow methods for characterizing macroporosity. While the capillary model has been strongly criticized in the past on account of its simple representation of complex pore structures, for flow methods at least it is clear that it can provide useful data to characterize and compare different carbons.

Development of fluid flow methods

The simplifying assumptions of homoporosity and unit tortuosity in the analyses in previous sections illustrate that the basic assumptions in the cylindrical capillary model are surprisingly realistic, considering the well-known complexity of macroporisity in carbons and graphites. Nevertheless, the formal inclusion of pore-size distributions in the analyses of fluid flow data, along the lines of Johnson[56] and Mays and McEnaney,[52] is probably the way forward in the development of fluid flow methods applied to carbons. For example, the method summarized by Mays and McEnaney[52] and detailed by Mays[57] indicates that the generalized diffusivity ratio equation, equation 4.31, can be used to estimate the probability density function of pore radius, $f(r)$, in graphites from experimental diffusivity data obtained using a Wicke–Kallenbach cell system, provided that the mechanism of flow in samples is in the transition region between Knudsen and bulk diffusion. This method, which involves obtaining diffusivity ratio data for different pressures, still needs some refining, especially with regards to the statistical estimation of $f(r)$ (which is not a trivial problem), but it represents a positive, new development in fluid flow methods for characterizing macroporosity in carbons.

4.4 Quantitative microscopy

Introduction

A common feature of porosimetry and fluid flow methods is the assumption of a simple model pore structure whose parameters and functions can be correlated with experimental measurements. An alternative approach, which need not involve assuming a pore model, is to observe macropores directly using a microscope.

Using standard techniques derived from metallography, the commonest and most convenient way to prepare a carbon for microscopy[58] is to grind and polish flat a sample, usually held in a resin mount. This yields a planar surface that includes two-dimensional cross-sections through pores of all sizes, both open and closed. These pore sections can then be viewed either using an optical microscope, which can resolve macropores larger than ~1 μm, or using a scanning electron microscope (SEM), which can resolve the smallest macropores (width >0.05 μm). Hence, provided that there is sufficient contrast between sections through pores and the surrounding carbon matrix, microscopy can give, at the very least, qualitative information on macropore structure. This information may include rough estimates of geometric pore size and also qualitative data on pore shape, orientation and spatial distribution that are unobtainable using porosimetry and fluid flow.

Whereas qualitative information on macropore structure may be useful initially, it is quantitative data on pore sizes, shapes, etc., that are important in the characterization and comparison of different carbons. However, apart from experimental difficulties in preparing suitable images of polished sections, there are three additional problems with determining quantitative macropore structure using microscopy: (i) a macropore cross-section is, in general, a complex two-dimensional figure whose geometry is not easily defined; (ii) large numbers of macropores must be measured to ensure that results are statistically significant; and (iii) the estimation of three-dimensional macropore structure from measurements of two-dimensional sections poses special mathematical and statistical difficulties. Modern quantitative methods for analysing pore images are reviewed later. In particular, the uses of computer-based analysers that provide some solutions to the above-mentioned problems are discussed. However, first, a summary is given of standard microscopical techniques for investigating carbon macropores.

Microscopical techniques for investigating carbon macropores

Sample preparation

Polished sections, prepared as described above, may be viewed directly in an optical microscope or in an SEM if an electrically conducting path is provided between the polished face of the sample and its base. Important points to note in preparing samples for microscopy are that for anisotropic materials, the orientation of the section to be studied must be specified with respect to a reference axis or plane of symmetry, e.g. the extrusion direction for a synthetic graphite, or the forming plane for a hot-pressed carbon, and that non-optimum grinding and polishing may either open or close sections through pores,[59] Figure 4.16, so that apparent pore cross-sections are not strictly representative.

Optical and scanning electron microscopy of carbons

If a carbon section is viewed in an optical microscope using reflected white light, macropores appear black or dark grey, while the solid carbon matrix appears in much lighter shades (*see* Figure 4.2). Normally transport, blind and closed pores

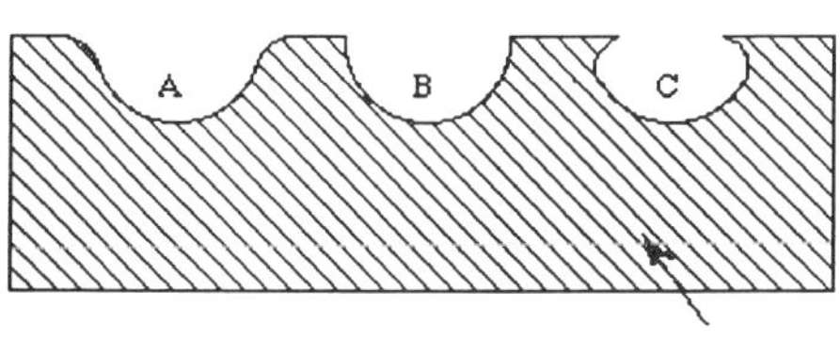

Fig. 4.16 Schematic diagram of polished sections through three spherical macropores. Section A indicates pore opening and section C pore closing. Section B indicates optimum polishing, with no pore opening or closing

are not differentiated, but open pores can be viewed separately from closed pores if they are filled with a material that enhances pore–carbon contrast, e.g. a metal[60] whose reflectivity in white light is much greater than carbon, or a resin that fluoresces when illuminated with ultra-violet radiation.[61]

The pore–carbon contrast in optical microscopy is often much greater than that in an SEM, as the depth of field in scanning electron microscopy is so large that carbon inside pores is usually imaged in addition to that on external, polished surfaces. Hence, despite the higher resolution of the SEM, optical microscopy provides the better option for characterizing macropores larger than ~1 μm. Accordingly, the remaining parts of this review are focused on optical methods.

Quantitative analysis of macropore images obtained using an optical microscope

General approaches to quantitative image analysis

Microscopical images of cross-sections through macropores in polished sections of a carbon are two-dimensional figures that, in general, have irregular perimeters, reflecting the complexity of the three-dimensional macropore structure. The perimeter of an image of a macropore cross-section is effectively a continuous, tortuous curve that usually cannot be defined in terms of simple parameters or functions. An additional problem is that usually there are a large number of pore sections distributed throughout a microscopic field, and it may be necessary to measure all the pores in a number of fields to obtain a representative sample. A final difficulty is to correlate structure in two dimensions with that in three dimensions. This last task is the subject of stereology, an aspect of quantitative microscopy that will be considered further below, after attention has been given to the basic problems involved in measuring two-dimensional structure.

A schematic macropore cross-section is illustrated in Figure 4.17. To characterize fully the macropores in the field of view of an optical microscope, the irregular curves that constitute the perimeters of each pore cross-section, e.g. Figure 4.17, could, in principle, be defined mathematically, for instance in terms of continuous functions of x–y Cartesian coordinates in the plane of the field.

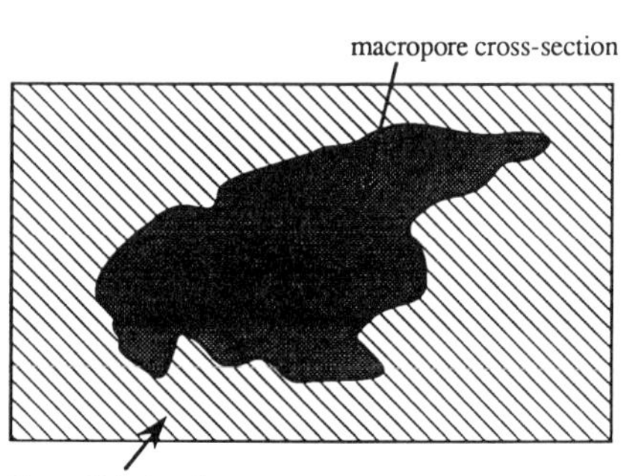

Fig. 4.17 Schematic diagram of cross-section of a macropore on a flat, polished carbon surface

However, in practice this is an impossible task, as it would require a potentially infinite sequence of approximations to the curves joining successively closer points along the perimeter of each pore section.

Early approaches to the problem of quantitative microscopy employed laborious manual counting techniques, where objects in two-dimensional fields, e.g. pores or grains, were characterized by the number of points they covered, or lines their perimeters intersected, in regular, superimposed meshes or grids of known point or line spacing. However, in recent years the development of computer-based image analysis systems has rendered these manual techniques obsolete. Modern methods of computer-based image analysis and quantitative microscopy, described below, together with their manual predecessors, are considered in detail in several monographs.[62–65] A number of papers concerned specifically with the use of computer-based quantitative microscopy for characterizing macroporosity in a range of different carbon materials have also been published.[60,61,66–77] These papers not only describe the application of image analysis methods to carbons, but also show how data from quantitative microscopy may be used in the understanding of properties that are related to pore structure, such as mechanical strength and chemical reactivity.

Computer-based quantitative image analysis

The essential principles of computer-based image analysis and quantitative optical microscopy are illustrated in Figure 4.18. The operation of the image analysis system illustrated is as follows. A flat, polished carbon sample is mounted in the microscope and viewed in reflected light. For unimpregnated samples illuminated with white light, pores appear black or dark grey while the

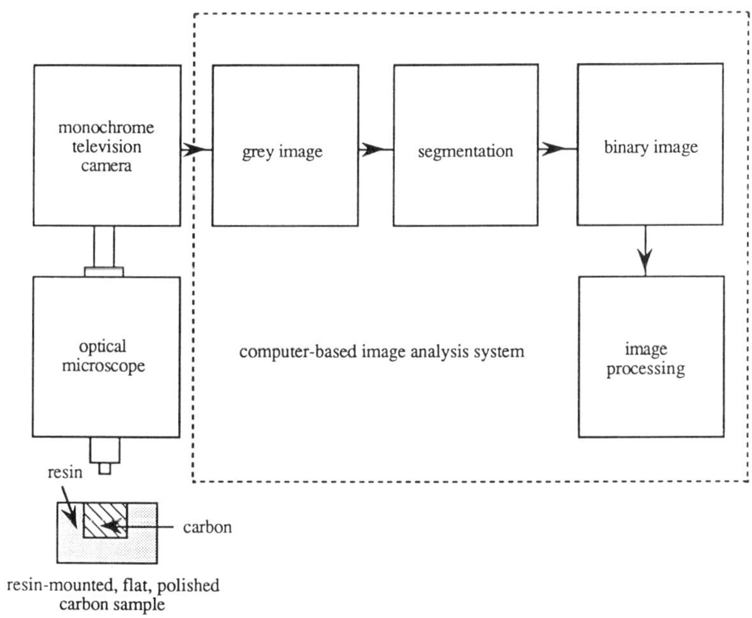

Fig. 4.18 Schematic diagram of computer-based image analysis system for quantitative optical microscopy

carbon matrix appears in brighter shades. The monochrome television camera mounted on top of the microscope converts the image from the microscopic field to an electrical signal that is directed to the computer for analysis.

The first stage of computer-based image analysis is digitization of the camera signal. This involves the division of the image into an array of square picture elements, or pixels, and the attribution of a number that correlates with the average brightness of the area of the image contained within each pixel in the array. In most commercial image analysis systems, the number of grey levels that can be resolved is an even power of two, e.g. $2^6 = 64$ or $2^8 = 256$. Normally pixels in dark regions of the image (pores) will be associated with low numbers and pixels in the lighter matrix regions with high numbers. In addition to the storage of grey levels, the computer also usually records the location of the geometric centre of each pixel, usually in x–y Cartesian coordinates that have been calibrated in convenient units, e.g. μm. This is in anticipation of measurements of the geometry of objects in the image that will be made at later stages in the analysis.

The second stage of computer analysis is the conversion of the grey image into a binary image by a process called segmentation. In this process, objects of interest in the image are identified by a narrow range of grey levels, e.g. a range of low grey levels if pores are being studied. A binary matrix is then created whose elements are either 1 (on) or 0 (off) according to whether a pixel is within or outside this range of grey levels, i.e. whether a pixel is in a pore or not. It is clear, then, that a pore in a binary image is a continuous array of 'on' pixels, where each pore is defined uniquely by the coordinates of the pixels along its boundary. These coordinates are the original data that are processed by the computer to yield geometric information on the pore cross-sections.

The coordinates of boundary pixels in a binary image may be considered as the coordinates of the vertices of irregular polygons that approximate to the continuous perimeters of pore sections. The final stage of computer-based image analysis is to measure and summarize the geometry of these polygons. The basic parameters that are commonly used to summarize pore geometry, illustrated for a polygonic approximation to the continuous pore section in Figure 4.19, are area, A, perimeter, P, and the maximum and minimum Feret diameters, d_{min} and d_{max}. From these simple parameters other factors may be derived for pore size, shape and orientation (Figure 4.20).

Quantitative microscopy of carbon macropores

Modern computer-based image analysis methods for quantitative microscopy have been used to characterize the macroporosity in a range of different carbons, as the following examples illustrate.

Example 1 Pore sizes and shapes obtained from image analysis were used in empirical models of strength–structure relationships in metallurgical cokes[71] and to demonstrate that pores of high aspect ratio are developed preferentially in the thermal oxidation of nuclear graphites.[69]

Example 2 The strengths of glassy carbon beams of use as catalyst supports were correlated with variations in macroporosity measured using image analysis.[61,76] These studies showed that macroporosity and flexural strength

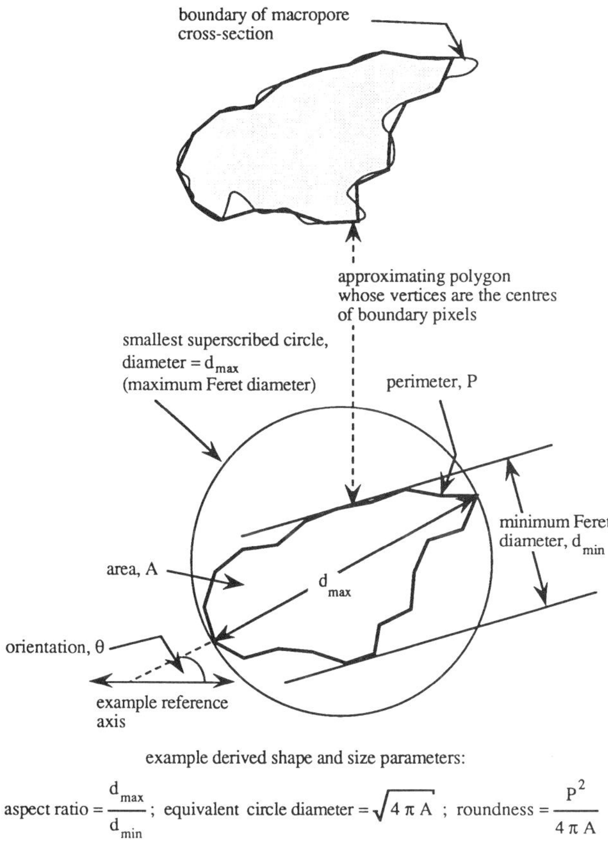

Fig. 4.19 Some parameters of the geometry of macropore cross-sections that may be measured using computer-based quantitative optical microscopy

were influenced by processing conditions in the manufacture of the glassy carbons and that the variation in the area fraction of macropores with distance across beams also influenced flexural strength. An example distribution of macropore area obtained from the image analysis of a glassy carbon used in these studies is shown in Figure 4.20.

Example 3 The sizes, shapes and locations of macropores in a carbon–carbon composite were determined using image analysis.[73] Also, data on the extent of cracking at fibre-bundle interfaces were obtained. These data were considered to be of potential use in understanding and predicting thermomechanical properties such as thermal expansivity.

Example 4 Image analysis measurements of the orientations of cracks in needle-coke particles were used to determine the orientation of filler grains with respect to the extrusion direction in an electrode graphite.[75] It was shown that orientation varied strongly with radial position in the electrode log, being greatest at the edge and least at the centre (Figure 4.21). This has implications for the radial variation of electrical conductivity in the log and hence its performance in service.

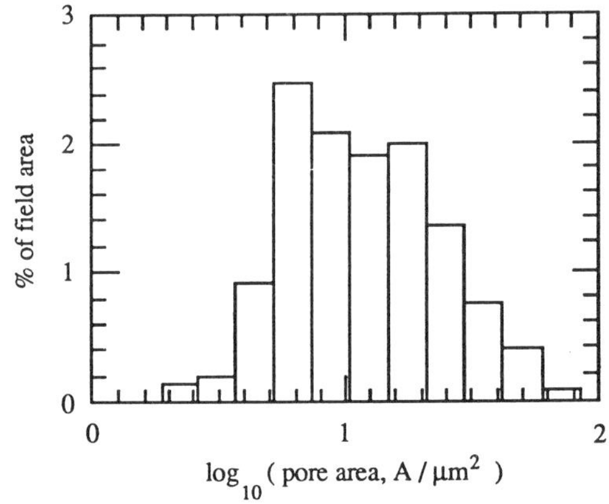

Fig. 4.20 Macropore area distribution for a field of a glassy carbon obtained using quantitative optical microscopy

Example 5 Cumulative open-pore size distributions for various graphites obtained from image analysis, using metallic and fluorescent impregnants for contrast enhancement, were compared with those obtained from mercury porosimetry.[77] In all cases mercury porosimetry underestimated the mean pore sizes when compared with those obtained from image analysis owing to the sensitivity of mercury porosimetry to constrictions in the pore network.

Concluding remarks

As has been mentioned already, the main advantages of quantitative microscopy for characterizing macropores in carbons compared with the methods of porosimetry and fluid flow are that a model pore structure need not be assumed initially and that, as the examples above illustrate, previously unobtainable data on pore shapes, orientations and locations can be determined. However, the principal limitation of quantitative microscopy is that information obtained from

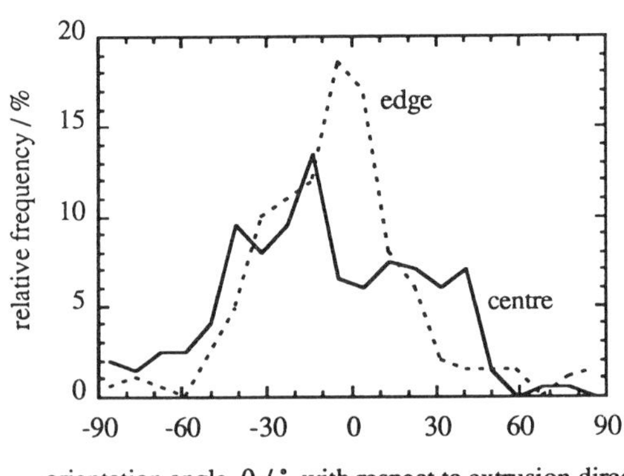

Fig. 4.21 Orientation of axial, macroporous cracks in the filler grains of an electrode graphite with respect to the extrusion direction of a cylindrical log (one representative crack per grain; 150 grains for each area of the log)

this method relates only to two-dimensional sections through macropores. One approach to obtaining three-dimensional structure from image analysis data is serial sectioning, i.e. by the analysis of a large number of adjacent sections through a sample. Alternatively, stereological techniques may be used, e.g. Underwood,[78] in which a model is assumed that relates two- to three-dimensional structure. Hence, although a model does not have to be assumed initially in image analysis studies, it is often necessary to do so eventually if a three-dimensional view of macropore structure is required.

Stereology usually involves difficult mathematical problems, for example in relating a pore-area distribution, e.g. Figure 4.20, to a pore-volume distribution. However, for some materials, such as glassy carbons, a random distribution of spherical pores is a realistic model of pore structure. This simple model allows three-dimensional structures to be computed easily. For example, the application of a fundamental theorem of stereology due to Delesse for this model shows that the pore-area fraction in a two-dimensional image is equal to the pore-volume fraction in three-dimensional space. Unfortunately, however, the assumption of simple model pore structures for most carbons yields erroneous, and possibly misleading, three-dimensional structures when applied to data from two-dimensional images.

Hence, it may be concluded finally that at present the main strength of quantitative microscopy lies in its ability to provide numerical information on pore structure that may be used to characterize and compare different carbons, and that may be incorporated into empirical or semi-empirical models of structure-related properties. Future work is required to develop practical experimental and analysis methods to relate two-dimensional data from image analysis to structures in three dimensions.

4.5 Acknowledgements

We thank Mr F.D. Botha and Dr B.H. Davis of the Center for Applied Energy Research (CAER), University of Kentucky, USA, for providing mercury porosimetry data. One of us (B.McE.) thanks Dr F.J. Derbyshire of the CAER for providing a Visiting Professorship during which part of this chapter was written.

4.6 References

1. SING, K.S.W., EVERETT, D.H., HAUL, R.A.W., MOSCOU, L. PIEROTTI, R.A., ROUQUEROL, J. A and SIEMIENEWSKA, T. 1985: *Pure and Applied Chemistry* **57**, 603.
2. YIN, Y., McENANEY, B. and MAYS, T.J. 1988: *Carbon* **27**, 133.
3. McENANEY, B. and MAYS, T.J. 1989: in Marsh, H. (ed.) *Introduction to Carbon Science*. London: Butterworth, p. 154.
4. GREGG, S.J. and SING, K.S.W. 1982: *Adsorption, Surface Area and Porosity*, 2nd edn. London: Academic Press.

5. LOWELL, S. and SHIELDS, J.E. 1984: *Powder Surface Area and Porosity*, 2nd edn. London: Chapman and Hall.
6. SCHOLTEN, J.J. 1967: in Bond, R.L. (ed.) *Porous Carbon Solids.* London: Academic Press, p. 225.
7. VOOYS, F.de. 1983: in Capelle, A. and Vooys, F.De (eds) *Activated Carbon—a Fascinating Material.* The Netherlands: Amersfoot, p. 13.
8. WASHBURN, E.W. 1921: *Proceedings National Academy of Sciences (USA)* **7**, 115.
9. RITTER, H.L. and DRAKE, C.L. 1945: *Industrial and Engineering Chemistry. Analytical Edition* **17**, 782.
10. DEBELAK, K.A. and SCHRODT, J.T. 1979: *Fuel* **58**, 732.
11. SPITZER, Z. 1981: *Powder Technology* **29**, 177.
12. ROOTARE, H.M. and PRENZLOW, C.F. 1967: *Journal of Physical Chemistry* **71**, 2733.
13. PFEIFFER, P. and AVNIR, D. 1983: *Journal of Physical Chemistry* **79**, 3558.
14. FRIESEN, W.I. and MIKULA, R.J. 1987: *Journal of Colloid Interface Science* **120**, 263.
15. EHRBURGER-DOLLE, F., FOX, T., LAHAYE, J. and LUHLEICH, H. 1990: in *Extended Abstracts of Carbone '90.* Groupe Français d'Etude des Carbones, Paris, p. 32.
16. WINSLOW, D.N. and DIAMOND, S. 1970: *Journal of Material Science* **5**, 564.
17. DICKINSON, J.M. and SHORE, J.W. 1968: *Carbon* **6**, 937.
18. MILBURN, D.R., ADKINS, B.D. and DAVIS, B.H. 1988: in Unger, K.K., Rouquerol, J., Sing, K.S.W. and Kral, H. (eds) *Characterisation of Porous Solids.* Amsterdam: Elsevier, p. 501.
19. ANDROUTSOPOULOS, G.P. and MANN, R. 1979: *Chemical Engineering Science* **34**, 1203.
20. WALL, G.C. and BROWN, R.J.C. 1981: *Journal of Colloid Interface Science* **82**, 141.
21. THOMPSON, A.H., KATZ, A.J. and RASCHKE, R.A. 1987: *Physical Review Letters* **58**, 29.
22. ZHDANOV, V.P. and FENELONOV, V.B. 1987: *Reaction Kinetics and Catalysis Letters* **33**, 377.
23. LOWELL, S. and SHIELDS, J.E. 1981: *Journal of Colloid Interface Science* **80**, 192, 273.
24. CONNER, W.C. Jr, LANE, A.M. and HOFFMAN, A.J. 1984: *Journal of Colloid Interface Science* **100**, 195.
25. FREVEL, L.K. and KRESSLEY, L.J. 1963: *Analytical Chemistry* **35**, 1492.
26. MAYER, R.P. and STOWE, R.A. 1965: *Journal of Colloid Interface Science* **20**, 893.
27. ORR, C. Jr, 1970: *Powder Technology* **3**, 117.
28. REVERBERI, A., FERAIOLO, G. and PELOSO, A. 1966: *Annali di Chimica (Rome)* **56**, 1552.
29. LOWELL, S. 1980: *Powder Technology* **25**, 37.
30. JOYNER, L.G., BARRETT, E.P. and SKOLD, R. 1951: *Journal of the American Chemical Society* **73**, 3158.
31. DUBININ, M.M., VISHNYAKOVA, E.G., ZHUKOVSKAYA, E.A., LEONTEV, E.A., LUKYANOVICH, V.M. and SARAKNOV, A.I. 1960: *Russian Journal of Physical Chemistry* **34**, 959.
32. DAVIS, B.H. 1984: *Applied Catalysis* **10**, 185.
33. BOTHA, F.D. and McENANEY, B., unpublished work.
34. COLLINS, R.E. 1961: *Flow of Fluids Through Porous Materials.* Tulsa: Penwell Books, OK.
35. SCHEIDEGGER, A.E. 1974: *The Physics of Flow Through Porous Media.* Toronto: University of Toronto Press, 3rd edn.
36. DULLIEN, F.A.L. 1979: *Porous Media: Fluid Transport and Pore Structure.* New York: Academic Press.

37. GREENKORN, R.A. 1983: *Flow Phenomena in Porous Media.* New York: Marcel Dekker.
38. CARMAN, P.C. 1956: *Flow of Gases in Porous Media.* London: Butterworth.
39. CUNNINGHAM, R.E. and WILLIAMS, R.J.J. 1980: *Diffusion in Gases and Porous Media.* New York: Plenum Press.
40. MASON, E.A. and MALINAUSKAS, A.P. 1983: *Gas Transport in Porous Media: the Dusty-Gas Model.* Amsterdam: Elsevier.
41. HEWITT, G.F. and MORGAN, J.R. 1964: *Progress in Applied Materials Research* **5**, 167.
42. HEWITT, G.F. 1965: *Chemistry and Physics of Carbon* **1**, 73.
43. GROVE, D.M. 1967: in Bond, R.L. (ed.) *Porous Carbon Solids.* London: Academic Press, p. 155.
44. WHEELER, A. 1951: *Advances in Catalysis* **3**, 249.
45. JOHNSON, M.F.L. and STEWART, W.E. 1965: *Journal of Catalysis* **4**, 252.
46. HEWITT, G.F. 1967: in Bond, R.L. (ed.) *Porous Carbon Solids.* London: Academic Press, p. 203.
47. WICKE, E. and KALLENBACH, R. 1941: *Kolloidnyi Zhurnal* **97**, 135.
48. EVANS, R.B. III, WATSON, G.M. and TRUITT, J. 1962: *Journal of Applied Physics* **33**, 2682.
49. WELLER, K.R. and WATTS, H. 1965: *Carbon* **2**, 337.
50. HAWTIN, P., DAWSON, R.W. and ROBERTS, J. 1969: *Transactions of the Institution of Chemical Engineers* **47**, T109.
51. YANG, R.T. and YIU, R.T. 1979: *Industrial Engineering Chemistry, Process Design Development* **18**, 245.
52. MAYS, T.J. and McENANEY, B. 1985: in *Proceedings of the 17th Biennial Conference on Carbon.* University of Kentucky, Lexington, MA: American Carbon Society, p. 104.
53. CLARK, J.D., GHANTHAN, C.S. and ROBINSON, P.J. 1979: *Journal of Material Science* **14**, 2937.
54. CLARK, J.D. and ROBINSON, P.J. 1972: *Journal of Material Science* **17**, 2649.
55. CLARK, J.D., McVEY, M.D., MURAD, M.D. and ROBINSON, P.J. 1979: *Journal of Material Science* **18**, 593.
56. JOHNSON, P.A.V. 1981: *Nuclear Energy* **20**, 231.
57. MAYS, T.J. 1988: *PhD Thesis.* University of Bath.
58. VANDER VOORT, G.F. 1984: *Metallography: Principles and Practice.* London: McGraw-Hill.
59. PIEKARCZYK, J. and PAMPUCH, R. 1976: *Ceramics International* **2**, 177.
60. BEST, J.V., STEPHEN, W.J. and WICKHAM, A.J. 1985: in *Proceedings of the 17th Biennial Conference on Carbon.* University of Kentucky, Lexington, MA: American Carbon Society, p. 27.
61. PICKUP, I.M., McENANEY, B. and COOKE, R.G. 1986: *Carbon* **24**, 535.
62. GONZALES, R.G. and WINTZ, P. 1977: *Digital Image Processing.* New York: Addison-Wesley.
63. PRATT, W.K. 1978: *Digital Image Processing.* New York: John Wiley.
64. ROSENFELD, A. and KAK, A.C. 1982: *Digital Image Processing.* London: Academic Press, Vols 1 and 2.
65. SERRA, J. 1982: *Image Analysis and Mathematical Morphometry.* London: Academic Press.
66. PAMPUCH, R. and PIEKARCZYK, J. 1976: *Science of Ceramics* **8**, 257.
67. DELLE, W., KOIZLIK, K., HOVEN, H., UHLENBRUCK, H. and WALLURA, E. 1978: in *Proceedings of the 5th London International Conference on Carbon and Graphite.* London: Society of Chemical Industry, p. 41.
68. EVANS, M. and MARSH, H. 1978: in *Proceedings of the 5th London International Conference on Carbon and Graphite.* London: Society of Chemical Industry, p. 32.
69. BURCHELL, T.D., PICKUP, I.M., McENANEY, B. and COOKE, R.G. 1986: *Carbon* **24**, 545.

70. GUPPY, R.M., MAYS, T.J., McENANEY, B., RODRÍGUEZ-REINOSO, F. and MOLINA-SABIO, M. 1986: in *Proceedings of the 4th International Carbon Conference, Baden-Baden.* Deutschen Keranishen Gesellshaft E.V., p. 329.
71. PATRICK, J.W. and WALKER, A. 1986: in *Proceedings of the 4th International Carbon Conference, Baden-Baden.* Deutschen Keranishen Gesellshaft E.V., p. 156.
72. PICKUP, I.M., MAYS, T.J. and McENANEY, B.1986: in *Proceedings of the 4th International Carbon Conference, Baden-Baden.* Deutschen Keranishen Gesellshaft E.V., p. 240.
73. MAYS, T.J. and McENANEY, B. 1987: in *Proceedings of the 18th Biennial Conference on Carbon.* Worcester, MA: Worcester Polytechnic Institute, American Carbon Society, p. 438.
74. PICKUP, I.M., McENANEY, B. and BODSWORTH, L. 1987: in *Proceedings of the 18th Biennial Conference on Carbon.* Worcester, MA: Worcester Polytechnic Institute, American Carbon Society, p. 484.
75. YIN, Y., McENANEY, B. and MAYS, T.J. 1988: *Carbon* **27**, 113.
76. McENANEY, B., PICKUP, I.M. and BODSWORTH, L. 1990: *Catalysis Today* **7**, 299.
77. BEST, J.V., STEPHEN, W.J. and WICKHAM, A.J. 1985: *Progress in Nuclear Energy* **16**, 127.
78. UNDERWOOD, E.E. 1970: *Quantitative Stereology.* Boston, Addison-Wesley.

5 Porosity and Thermal Reactivity

Heinz-Jürgen Mühlen and Karl H. van Heek
DMT—Geselleschaft für Forschung und Prüfung mbH
DMT—Institut für Kokserzeugung und Brennstofftechnik
Franz-Fischer-Weg 61
D-45307 Essen
Germany

5.1 Introduction

This chapter deals with the diverse relationships between the porosity of solid carbonaceous materials and their reaction behaviour with gases at elevated temperatures. Thus it deals with chemical gas–solid reactions, where the gases are such as O_2, H_2O, CO_2 and H_2 and the solids are materials such as graphite, 'carbons', coal, lignite, biomasses and their chars. This class of reactions is of high industrial importance being the basis for such processes as

- combustion and gasification for the production of heat and gas from coal and other solid fuels;
- activation of carbon-containing material to make adsorbents;
- regeneration of adsorbents used for the adsorption of organic substances;
- removal of carbon deposits from catalysts;
- corrosion of graphitic structural elements.

They are summarized under the term 'gasification reactions'. Reactions between gases and porous solids occur mainly at internal solid surfaces and generally in sequential steps: adsorption of the gasifying agent, reaction, desorption of the gaseous products. Hence it is evident that size, quality and accessibility of the surface area play important roles with regard to the course, products and rates of the reactions. For the gasification reactions under consideration it should be noted that the surface itself changes during the reactions by the consumption of carbon atoms. Hence an influence between 'porosity' and 'reactivity' is found in two ways: the internal structure affects the reaction rates on the one hand and the kind and the conditions of the reaction change the internal surface on the other.

The following sections first review the rates of conversion of coke and chars with gases and their dependence on solid properties and reaction conditions. Following this, the variation of porosity during the reactions is described. It follows a survey of the efforts to establish correlations between porosity and reactivity.

Concerning the experimental and theoretical methods to be used, the surface characterization is given in detail in other chapters. However, a certain basis has to be laid with respect to the gas–solid reactions, and this is done in the first part of the next section.

5.2 Reactivities of chars and cokes

Basic reactions and definitions

Solid carbonaceous material is mainly found as fossil fuels such as anthracite, coal, lignite and peat or as biomass. These materials are mixtures of organic and mineral matter. The organic matter undergoes thermal decomposition termed pyrolysis or devolatilization on heating at temperatures above about 300°C. Thereby condensable and non-condensable gaseous products may be formed; the former are called 'oil' and 'tar'. The solid may go through a plastic stage and solidify again. Generally the solid residue is called 'coke' when the precursor had become plastic or 'char' if not. In view of the fact that the gasification reactions occur at higher temperatures, pyrolysis is always a pre-step affecting to a high degree the char or coke properties with respect to the subsequent gas–solid reactions.

As the only or main constituent of the material under discussion here is carbon, the fundamental gasification reactions are as follows:

Steam gasification:	Heat consumption ($kJ\ mol^{-1}$)	
$C + H_2O \leftrightarrow H_2 + CO$	+118	(5.1)
$CO + H_2O \leftrightarrow H_2 + CO_2$	−42	(5.2)
$CO + 3H_2 \leftrightarrow CH_4 + H_2O$	−206	(5.3)
Hydrogasification:		
$C + 2H_2 \leftrightarrow CH_4$	−87.5	(5.4)
Combustion:		
$C + O_2 \leftrightarrow CO_2$	−406	(5.5)
$C + CO_2 \leftrightarrow 2CO$	+162	(5.6)
$C + \frac{1}{2}O_2 \leftrightarrow CO$	−123	(5.7)

The main reaction is the heterogeneous water–gas reaction (equation 5.1), in which the carbon reacts with steam to form hydrogen and carbon monoxide. It is a 'complex' reaction as the products can participate in two further reactions: the homogeneous water–gas reaction, also called shift-reaction (equation 5.2), and methanation (equation 5.3). Thus they react either with the gasifying agent or with one another, generating carbon dioxide and methane. The subsequent reactions (equations 5.2 and 5.3) are exothermic; however, the endothermic reaction 5.1 is dominant. Hence the overall reaction, which is the sum of the three included under steam gasification, is endothermic. To produce methane from coal hydrogasification (equation 5.4), the direct reaction of hydrogen with carbon can be used. The combustion reactions are the total oxidation (equation 5.5) and the partial oxidation (equation 5.7) reactions, whereby reaction 5.5 is always connected with the Boudouard reaction (equation 5.6) in which the product CO_2 reacts with the carbon.

The rates of the reactions are defined as the change in the mass of carbon as a function of time. As given in equations 5.8–5.10, it can be expressed by the degree of gasification X (equation 5.11), also called 'burn-off'.

$$r' = \frac{dX}{dt} \tag{5.8}$$

$$r'' = \frac{dX}{dt}(1-X)^{-1} \tag{5.9}$$

$$r_s = \frac{dX}{dt}(1-X)^{-2/3} \tag{5.10}$$

$$X = 1 - \frac{m_c(t)}{m_c(t=0)} \tag{5.11}$$

where r' is the rate related to the initial mass $m_c(t = 0)$. If r' is unchanged with burn-off the reaction is of zero order, and this is often found for catalytically influenced reactions; r'' and r_s relate the rate to the mass of carbon present in the reactor at time t. A constant value over the whole range of burn-off indicates a first-order reaction in the first case or an order of 2/3 in the second. The latter can be explained by the progress of the reaction at spherical surfaces. It has been found that r_s leads to a constant value over the whole range of burn-off in many cases.

For non-catalysed gasification reactions it has turned out that the reaction rate defined as r'' or r_s is nearly constant over the entire range of burn-off. Accordingly, from here on these two definitions will be used.

The term 'reactivity' describes a property of the solid fuel. It is expressed by the value of a reaction rate under well-defined conditions concerning gasifying agent, temperature and pressure.

A useful instrument for the investigation of the kinetics of gasification is the thermogravimetric apparatus.[1,2] It normally operates at temperatures up to 1100°C and pressures up to 10 MPa whereby the sample is treated isothermally or with a controlled temperature increase normally at a few K min^{-1}.

Comparison of different carbonaceous materials

Figure 5.1 compares the reactivities of lignites, coals, chars, metallurgical coke and pitch coke against steam at 4 MPa depending on temperature.[3] In all cases the rate of reaction increases exponentially with temperature. There are considerable differences between the various solid fuels tested; the reactivities of the lignites are highest, those of the coals and chars are found in a relatively narrow range regardless of coal rank and metallurgical coke and pitch coke show the lowest rates. The high reactivity of the lignites can be attributed to calcium, which is found finely dispersed in the samples as salts of humic acids. If the latter are removed by demineralization the samples are much less reactive and similar to chars or cokes from hard coal.

In the case of hard coal, the minerals have only a small or no effect owing to

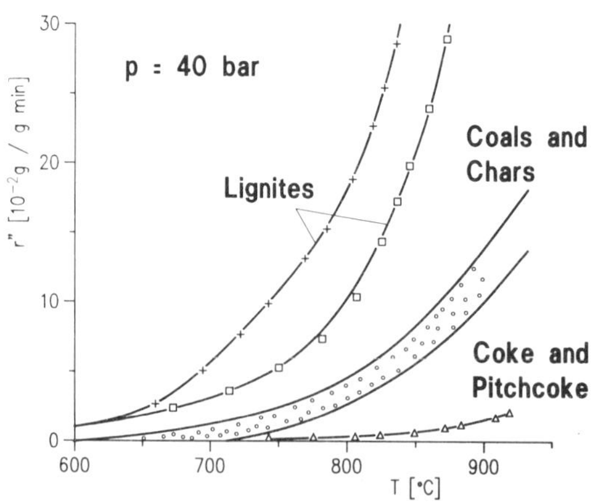

Fig. 5.1 Steam reactivity of different feedstocks vs temperature

less dispersion of the active elements or to their being bound in catalytically inactive types. For their chars as with cokes the physical structure of the solid and especially of its internal surface as influenced by pyrolysis conditions is of much greater importance.

Influence of pyrolysis conditions

Figure 5.1 shows that the cokes which have undergone treatment at high temperatures well in excess of 1000 °C exhibit low reactivities compared with the chars which were treated at considerably lower temperatures. This effect is understandable as heat treatment has a large influence on the physical structure. In particular, it destroys the active sites on the surfaces, i.e. edge atoms, dislocations and heteroatoms. The number of edge atoms is reduced by rearrangement, the dislocations disappear by annealing and the heteroatoms are removed by devolatilization of nitrogen-, sulphur- and oxygen-containing compounds. The effect of the final temperature (HTT) and the residence time at this final temperature during pyrolysis ('soak time') on the reactivity of the char produced will now be discussed in more detail.

Figure 5.2 shows the effect of temperature and residence time on the air reactivity of a lignite char as measured by Radovic *et al.*[4] The char was treated for 1 h at 700, 1000 or 1200 °C, and those processed at the higher temperatures proved to be considerably less reactive. Thus temperatures of 1000 and 1200 °C obviously lead to the same changes in physical structure of the char which are responsible for this drop in reactivity. For a further decrease in reactivity towards that of graphite, the final temperatures and the residence times have to be much higher or much longer, respectively. This behaviour corresponds to the course of the structural rearrangements in carbons depending on temperature as has been measured by Oberlin and co-workers.[5]

Figure 5.2 (right) illustrates the effect of residence time. Those changes in char properties which are responsible for the drop in reactivity depend on the soak time. At 1000 °C these processes are obviously fast because the reactivity

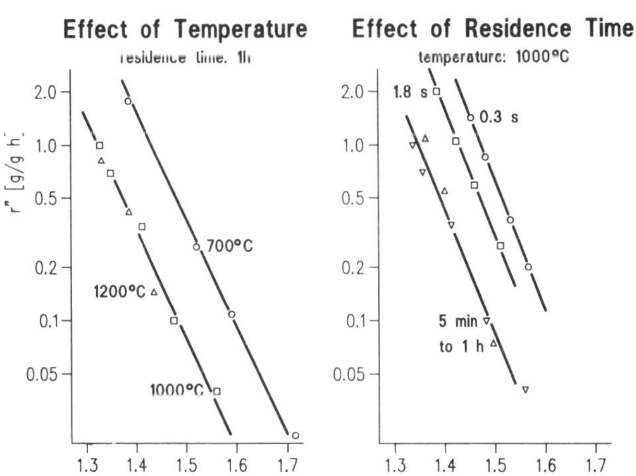

Fig. 5.2 Air reactivity of lignite char depending on (left) heat treatment temperature HTT and (right) residence time during pyrolysis (from ref. 4)

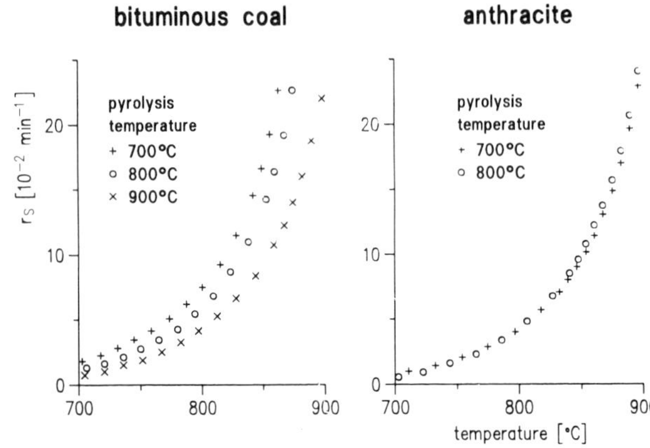

Fig. 5.3 Changes in reaction rates of (left) a highly volatile bituminous coal char and (right) an anthracite with HTT (from ref. 6)

drops considerably in the first few seconds and reaches a stable value after 5 min. These findings are important in practice whenever a pyrolysis process is used as a pre-step to gasification or activation. If the temperature goes beyond a certain limit, char can be quickly 'dead-burned'.

Figure 5.3 shows similar results for the reactivity of a coal and an anthracite in steam at 4 MPa.[6] For the char of the highly volatile bituminous (hvb) coal, the reactivity decreases with increasing HTT as expected. In contrast, the reactivity of the anthracite appears not to be influenced by the HTT, at least not between 700 and 900 °C. The differences in the behaviour of the chars compared with the two coals can be explained by their very different behaviour during pyrolysis: the hvb coal undergoes a drastic change in chemical and physical properties including melting and formation of a new solid phase, whereas the solid structure of the anthracite is hardly changed by carbonization.

Thermal treatment of the solid fuel also occurs during gasification or activation itself. Figure 5.4 gives the results of isothermal experiments with chars prepared in a fluidized bed at temperatures between 500 and 900 °C. Up to about 50 wt% conversion, the reactivities are of the order expected, i.e. highest for chars with a low HTT and lowest for those with high HTT. For higher burn-off,

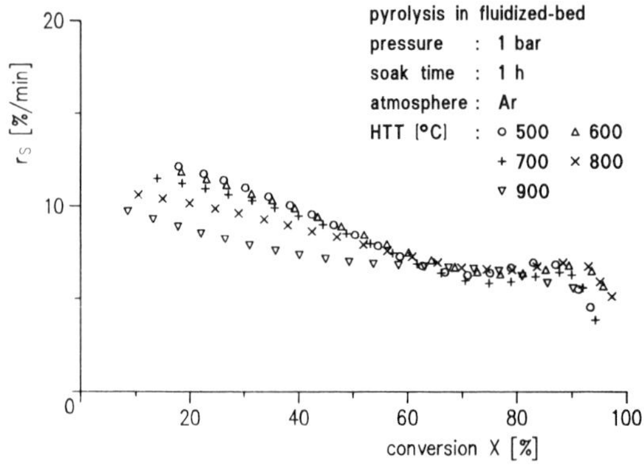

Fig. 5.4 Decrease in reactivity of an hvb coal char during gasification in 4 MPa steam at 900 °C prepared at different HTT (from ref. 7)

the different samples react with nearly the same rates. This suggests that after achieving 50 wt% burn-off the chars acquire the same quantity of active sites. That higher reactivity achieved during pyrolysis was finally compensated by the temperature treatment during gasification over a long enough residence time.[7]

Concerning the influence of pressure during pyrolysis, it had previously been found that it had no effect on the steam reactivity of chars or cokes, if the pyrolysis was carried out under inert conditions up to 4 MPa. The steam reactivity of a char from a hvb coal decreases considerably, however, with rising hydrogen pressure up to 10 MPa during pyrolysis.[8] A possible explanation is that at high pressures the pyrolysis products are very compressed and cannot open the structure of the forming semicoke. Also, it has to be taken into account, that hydrogasification takes place subsequently or simultaneously. Being a strongly exothermal reaction, this can cause an additional local heating of the particles. Hence the effect ascribed to the pressure could be an indirect temperature effect.

Reactivity changes with burn-off

In addition to the effects of thermal annealing, demonstrated in Figure 5.4, the rates can also be influenced by the development of porosity in different ranges of pore radii leading to a more or less pronounced change in the internal surface area. Moreover, the carbon atoms are bound with different strengths depending on their location in the solid structure. For that reason the more weakly bound sites will be consumed preferentially at the beginning of the reaction. This leads to an accumulation of more strongly bound carbon atoms and a decrease in reactivity. These changes are more pronounced the lower the activation energy of the reaction under consideration and the broader the range of bonding energies. Thus, in particular, gasification with hydrogen-containing agents exhibits a considerable decrease in reaction rate with increasing burn-off as shown in Figure 5.5. More recent interpretations of such findings propose the formation of very stable H—C bonds leading to a decrease or blocking of the active surface or the amount of active sites. This could explain why

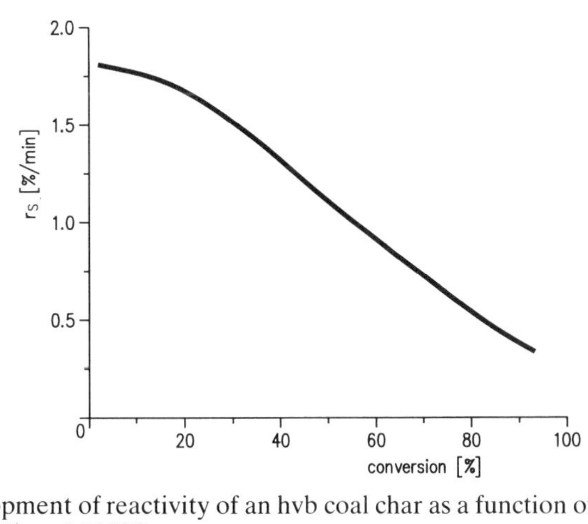

Fig. 5.5 Development of reactivity of an hvb coal char as a function of burn-off at 3 MPa ($H_2O:H_2 = 20:10$) and 900°C

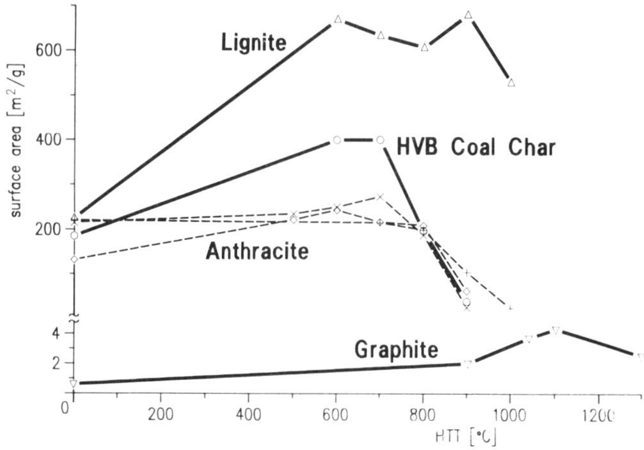

Fig. 5.6 Variation of CO_2 BET surface areas of a lignite, an hvb coal char, an anthracite and a graphite with HTT (from refs 9–11)

hydrogasification virtually ceases at temperatures around 800°C when medium degrees of burn-off have been reached.

5.3 Variation of internal surfaces and porosity during pyrolysis and gasification

Temperature, heating rate and pressure during pyrolysis

How and to what extent the processes occurring during pyrolysis affect the microstructure of the resulting char or coke is shown in Figures 5.6 and 5.7, where the specific internal surfaces are given as a function of the HTT. From the results in Figure 5.6 it can be derived that the internal surface areas pass through

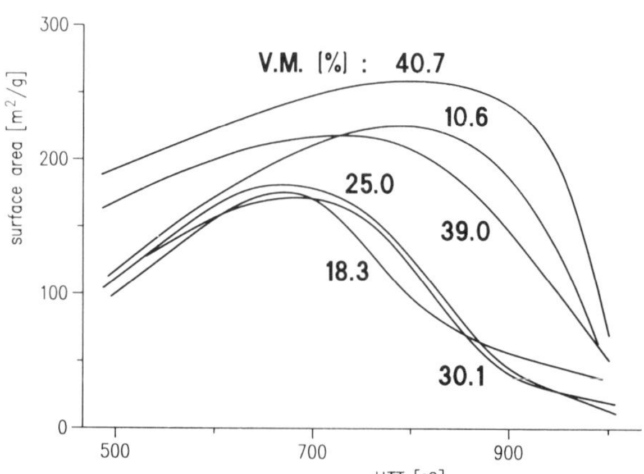

Fig. 5.7 Change of internal surface area measured with MeOH for chars from different coals, characterized by their volatile matter (V.M.), versus HTT (from ref. 12)

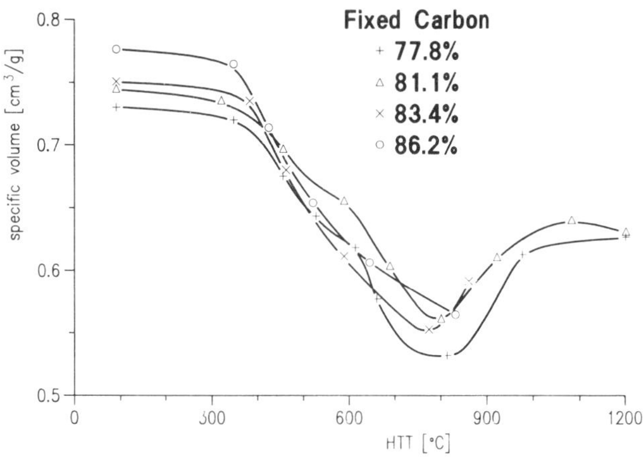

Fig. 5.8 Change in specific volume of different coal chars with HTT (from ref. 13)

a maximum between 600 and 1100°C depending on the type of carbon. The lignite investigated by Nandi and Walker[9] shows a broad maximum which already starts to develop below 600°C. At 1000°C there is a tendency for the surface area to decrease again. Bituminous coals behave differently. They reach the maximum surface at lower temperatures, the plateau is shorter and the shrinking process seems to be finished by 900°C. Moreover, the maximum values of the internal surfaces are considerably smaller than those of the lignites.

For anthracites Nandi *et al.*[10] found that the surface areas are even smaller and that the decline is shifted slightly to higher HTT. Also the graphite investigated by Walker *et al.*[11] changed its surface on heat treatment. However, the surface area reached only 4 $m^2 g^{-1}$ and declined beyond 1100°C.

More detailed results for hard coals have been reported by Chiche *et al.*[12] Figure 5.7 shows the surface areas for the char or coke samples chosen, made from coals ranging from 10 to 40 wt% volatile matter. In all cases the curves pass through a maximum with a steep decline at high temperatures. Finally, the surfaces reach values below those of the precursor materials.

Similar results are shown in Figure 5.8, which depicts the course of the specific pore volume with HTT for several chars. Toda[13] found that for all the coals he examined the specific volume remained constant up to temperatures of about 300–400°C. Beyond this it decreased drastically and passed through a minimum in the range 600–700°C. For the interpretation of these results it is necessary to take into account the fact that the changes in specific volume are mainly due to the changes in macroporosity, which are not or only slightly considered when CO_2 surfaces are measured. Nevertheless, the data presented complete the model developed above.

The extent to which the porosity, especially in the large macropore range, is affected by heating rate was investigated by Zygourakis and Glass,[14] who heated small particles of a bituminous coal at rates ranging from 0.1 to 1000 $K s^{-1}$. Examples of micrographs of the chars after devolatilization are given in Figure 5.9. In case (a), at the lowest heating rate, the particle does not swell. Obviously the pyrolysis products have enough time to escape from the particle during its plastic stage. With increasing heating rate the time of the plastic state is shortened and the gas production rate increases. This causes the formation of

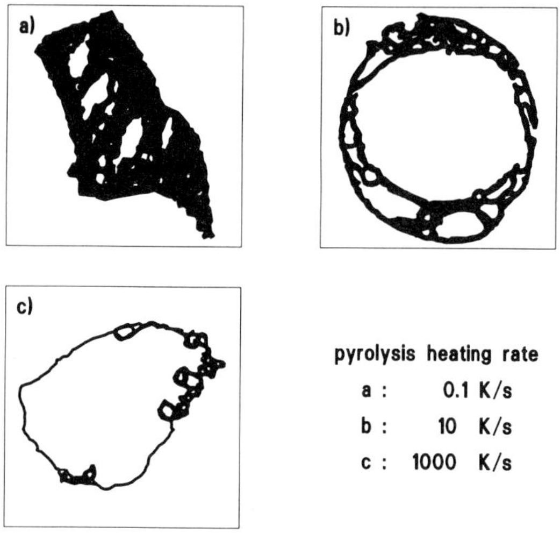

Fig. 5.9 Micrographs of hvb coal chars formed at different rates of heating (from ref. 14)

volatiles inside the particle. Consequently, it is enlarged while the grain is plastic. When the shell resolidifies before the volatiles have escaped, the resulting char particles are formed mainly as cenospheres (case b). We have obtained similar results.[16] At the highest heating rate and the shortest plastic state, these cenospheres are even more pronounced (case c). Finally, the thin balloon skin is no longer able to resist the internal pressure and more blow-holes are formed through which the gases can escape. This phenomenon leads to a restriction of swelling. Although this swelling is only illustrated by the macropores, one can imagine that the development of meso- and micropores is also influenced by similar mechanisms induced by the heating rate during pyrolysis.

In contrast to HTT and heating rate, the total gas pressure and kinds of gases used during pyrolysis play a minor role in the development of porosity and internal surfaces. As mentioned above, the gases produced during pyrolysis build up pressure inside the particle and try to escape through the already existing porous network. Compared with the outside pressure, the internal pressure may be considerably higher, especially if high heating rates are applied. Only if the external pressure is comparable to the internal pressure does it have any impact on the swelling or enlargement of all kinds of pores.

The outflowing pyrolysis products also hinder the diffusion of the surrounding reactive gases into the char particle, which explains the reported finding that the pyrolysis atmosphere has only a small influence on the porosity of the char so formed.[7] However, this holds true only during devolatilization, i.e. as long as there is a considerable production of volatile material. The effects of gasification which normally occur after pyrolysis will be dealt with in the next section.

Development of porous structure with burn-off

For technical purposes, as mentioned in the Introduction, it is important to know the porous structure of the material immediately after heat treatment in inert atmospheres and its changes in reactive environments. In 1989, Hoinkis

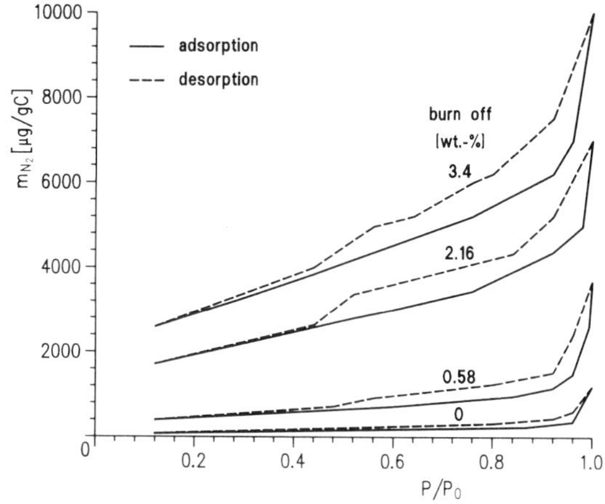

Fig. 5.10 Nitrogen adsorption and desorption isotherms at 77 K for an original and an oxidized graphite (from ref. 17)

and Robens[17] investigated the development of the porosity of a non-porous graphite during the early stages of gasification. Their results, shown in Figure 5.10, indicate that considerable amounts of either micro- or mesopores are already formed or made accessible at small degrees of gasification below 3.4 wt%. As can be seen, the slope of the isotherms in the range up to $p/p^0 = 0.4$, measured in N_2 at 77 K, increases steadily, indicating the enlargement of microporosity. Moreover, the hysteresis, a measure of mesoporosity, becomes increasingly pronounced with burn-off.

In 1953, Wicke and Hedden[18] published results concerning the surface changes of a pure carbon material which was gasified at 1035 °C in CO_2 up to total burn-off. For the determination of the resulting surface area they used butene as the adsorbate at 0 °C.

From Figure 5.11a it can be seen that the surface related to the mass of the non-activated carbon increases sharply at low degrees of gasification, passes

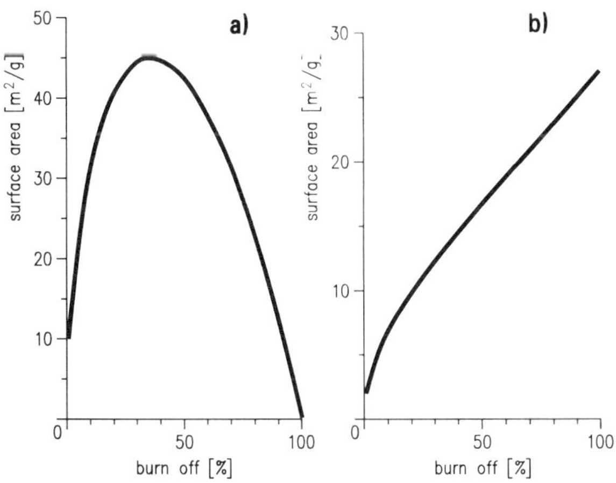

Fig. 5.11 Variation of the butene (0 °C) surface area of a 'spectral graphite' related to (a) the mass of ungasified material and (b) the actual mass of the material with burn-off in CO_2 at 1035 °C (from ref. 18)

through a maximum at 30 wt% burn-off and falls to zero at total conversion. The same results are also shown in Figure 5.11b, where the surface areas are related to the instantaneous amount of carbon. In this presentation the surface increases steadily with increasing burn-off and depicts a linear relationship in the range 20–100 wt%. Because nearly all of the carbon is gasified inside the particle on the internal surface, the outer shape of the particles remains more or less unchanged. Therefore, the porosity also correlates linearly with the surface.

Rate of reaction

The linear relationship between porosity, surface area and burn-off as described above cannot exist universally. The strength of the chemical bonding can differ depending on the location of the reacting carbon atom, as stated previously. In addition, the carbon atom must be accessible to the gasifying agent. The lower the bonding energy and the better the accessibility, the more probable it is that at a given place the carbon lattice is successfully attacked by the gasifying agent and the porosity enlarged. As the gas transport depends on pressure, temperature, rate and degree of conversion, it follows that at a given burn-off the growth of porosity will be less pronounced in the micropores if the rate of conversion controlled by the reaction temperature exceeds a certain value.

Figure 5.12 depicts this for an anthracite gasified in CO_2 at 850 and 950°C.[19] The graph for the higher temperature shows clearly that over the whole range of burn-off the porosity of the residue stays below the values of that prepared at the lower temperature. The same effect holds true for the internal surface which is higher at lower temperature, i.e. at lower reaction rate.

A similar behaviour for chars derived from xylitic brown coals was reported by Berger *et al.*,[20] who investigated the activation by CO_2, focusing on micro- and mesopores in the temperature range between 850 and 1000°C. Their results in Figure 5.13 support the above statement that the higher the temperatures and therefore the reaction rates, the slower the enhancement of microporosity. They found that the evolution of microporosity is inhibited by up to 30% if the gasification temperature is increased by 150°C, and that at the same time the volume of the mesopores is enhanced by a factor of approximately 2.

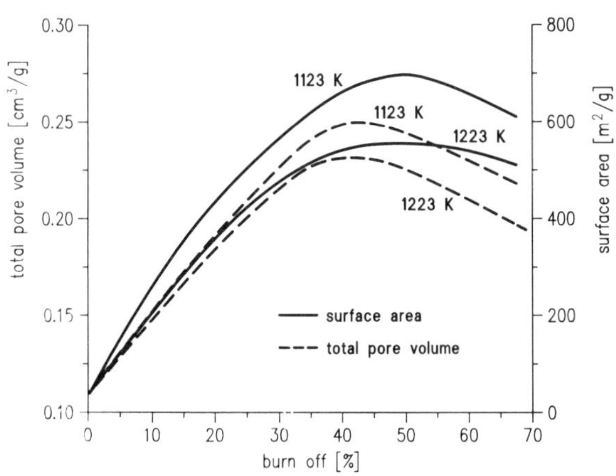

Fig. 5.12 Variation of the total pore volume and internal surface area with burn-off of an anthracite char during gasification with CO_2 (from ref. 19)

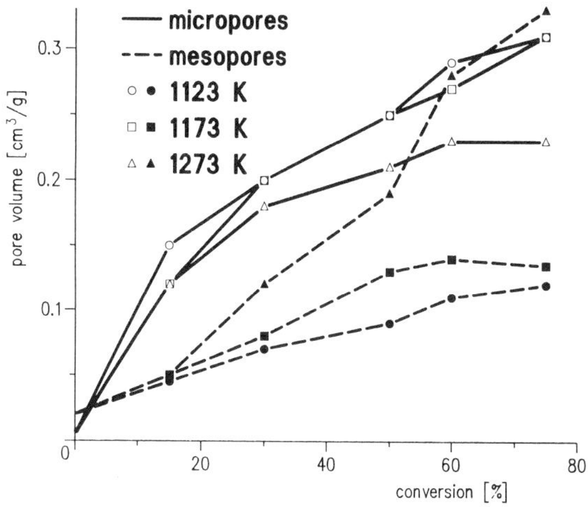

Fig. 5.13 Development of meso- and microporosity during CO_2 gasification at different temperatures (from ref. 20)

The same behaviour can also be observed if steam is used for the gasification of cokes from an hvb coal or a pitch coke as given in Figure 5.14. The two samples were treated at different temperatures and pressures up to a burn-off of 40 wt% in order not to mix the influences of burn-off and reaction rate. It can cleary be seen that with decreasing reaction rate the CO_2 BET surface areas vary by a factor of approximately 9. In Figure 5.15a–c the results of air gasification of a bituminous coal are demonstrated.[21] Here one has to take into consideration that before gasification can proceed, pyrolysis must have finished. Therefore, it is not surprising that at the lowest temperature of 300 °C the surface areas do not exceed 20 $m^2 g^{-1}$. However, between 400 and 500 °C the development of the internal surfaces turns out to follow the course described above. Figure 5.15d shows the results of CO_2 activation of the same coal carried out at 1000 °C. The development of surface area as a function of burn-off looks completely different. As an explanation for this effect, the thermal annealing of the coke due to the high temperature applied is proposed.

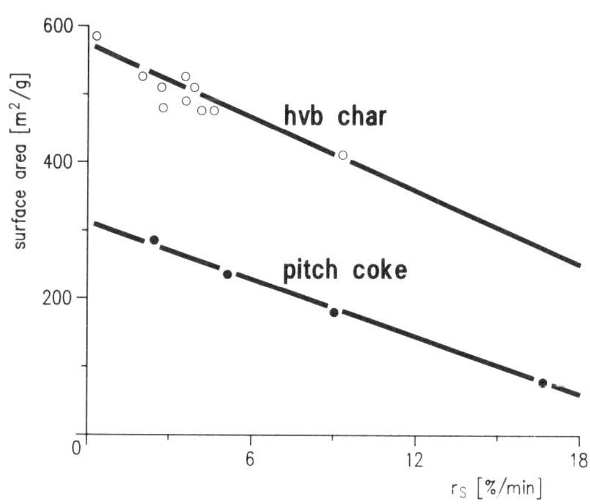

Fig. 5.14 Variation of CO_2 BET surface areas of an hvb coal char and a pitch coke gasified up to a burn-off of 40 wt% in steam vs reaction rate

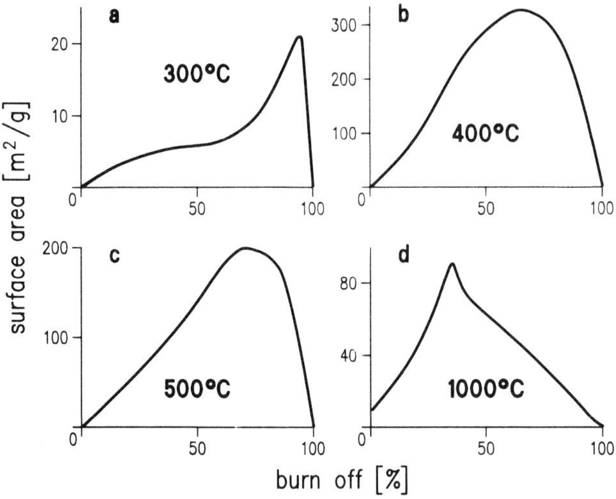

Fig. 5.15 Development of internal surface area during (a–c) air and (d) CO_2 gasification of a bituminous coal at different temperatures (from ref. 21)

Effect of gasifying agent and its pressure

In order to separate the influences of rate, temperature and pressure on the one hand, and of gasifying agent on the other, we carried out a series of experiments starting with the pitch coke and the char introduced in connection with Figure 5.14. Concerning the influence of kind of gasifying agent (Figure 5.16), the highest surfaces were yielded using steam followed by carbon dioxide and hydrogen. Although steam gasification proceeds with the highest reaction rate and although the reaction with hydrogen proceeds more than ten times more slowly, pure steam produces the highest surface areas, which were measured in this case applying CO_2 adsorption at room temperature in the pressure range 0.1–1.2 MPa.

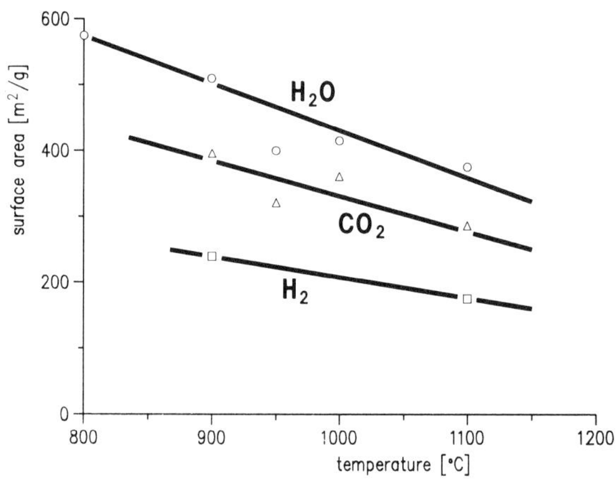

Fig. 5.16 Variation of CO_2 BET surface areas of an hvb coal char with temperature of gasification in different gasifying agents at the same burn-off of 40 wt%

Pressure ranging from 0.5 to 4 MPa during gasification had no marked influence on the development of the surface area. This is understandable, as for steam gasification the reaction rate as a function of pressure is nearly constant in this pressure range. For hydrogasification, the rates increase with increasing pressure, however. The lack of influence on the surface area developed may be due to the co-actions of two effects: the formation of stable CH groups and the resulting deactivation of special sites, and the increasing attack of the hydrogen.

Although oxygen is not commonly used as a gasifying agent, it is of interest to know how the exothermic reaction may influence the development of internal surfaces or porosity. Tomkow *et al.*[22] investigated O_2 activation at 350°C using a brown coal as starting material. They reported that during the first 25 wt% of burn-off the microporosity develops to a greater extent than during steam gasification at 840°C. However, at higher degrees of conversion the microporosity decreases considerably, being accompanied by a nearly constant volume of the mesopores. The explanation given is that micropores are produced at the beginning of the process by the opening of closed pores only, and these are subsequently blocked by chemisorbed oxygen. Hence the activation proceeds in larger pores only.

Rodríguez-Reinoso and co-workers[23–26] investigated extensively the changes of micro- and mesopores in chars produced from olive and almond stones occurring on activation in air (400°C) and CO_2 (825°C). For this purpose they first gasified the materials in CO_2 to burn-off levels of 35 and 32 wt%, respectively. Using these activated carbons as starting material they carried out a number of activation experiments in air at 350°C up to different degrees of burn-off. Their results confirmed those of Tomkow *et al.*,[22] showing an initial increase in microporosity followed by a drastic decrease in the range of higher burn-off. Regarding the mesopores, they have found that the activation in CO_2 causes a weaker decrease than in air and concluded that this effect is due to a continuous ablation of the particles in air.

5.4 Correlation of porosity and reactivity

Experimental results

Many researchers have tried to correlate the measured reaction rates with the internal surface area. Surprisingly, some of them succeeded in establishing a linear correlation between these two. An example is given in Figure 5.17, where Johnson[27] measured the rates of a coal char in steam–hydrogen mixtures at different degrees of burn-off and the corresponding CO_2 surface areas. Frequently it has been found that there is no correlation, however. In Figure 5.18 the results of Formella[28] demonstrate that although the surface area as a function of burn-off of the hvb coal char is approximately twice as high as that of the anthracite, the ratio of the reaction rates is just the opposite. The rates decrease from the start, reaching a constant value for the hvb coal char and passing through a minimum in the case of the anthracite, although the surface areas pass through a maximum in the range of 60 wt% burn-off.

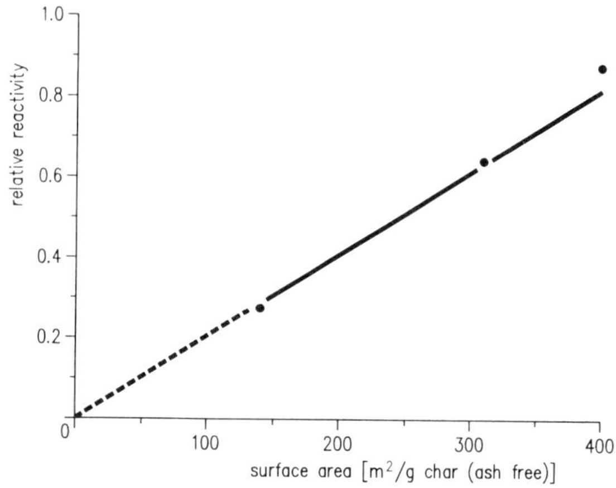

Fig. 5.17 Correlation of internal surface area and relative reactivity in H_2O–H_2 mixtures (from ref. 27)

Modelling of the development of porous structure

When a correlation was found between accessible surface area and reaction rate, some workers tried to develop an empirical model in order to predict the course of reaction on the basis of surface development. Bhatia and Perlmutter[29] did this for chemically controlled gasification up to 60 wt% burn-off, assuming the surface area to be a function of porosity, the latter depending exclusively on the degree of burn-off. Figure 5.19 shows such a development of surface area as a function of burn-off, and Figure 5.20 compares the reaction rates calculated on the basis of this model with the measured values. The quality of the fit is excellent.

However, the model of Bhatia and Perlmutter does not take into account that the development of the porosity and in particular the internal surface is dependent on the reaction conditions. To overcome these obstacles, Bliek[30] elaborated a model based on the 'dusty gas model'. It takes into consideration the initial pore-size distribution and the diffusion and flow into and out of the particles. The volume and pressure increases due to the heterogeneous water–

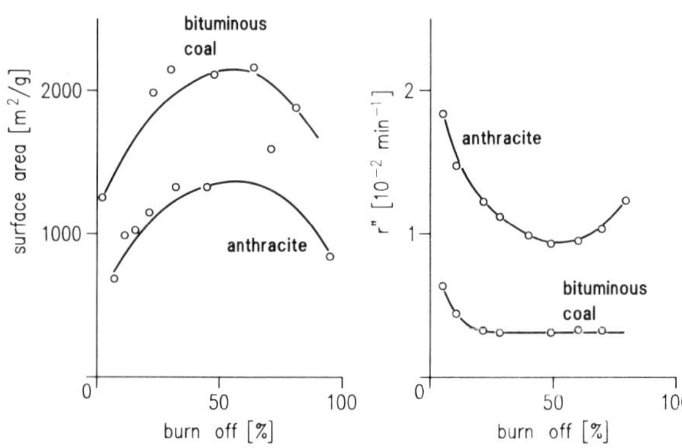

Fig. 5.18 Internal surface areas and reaction rates in 4 MPa steam at 700°C for an anthracite and an hvb coal (from ref. 28)

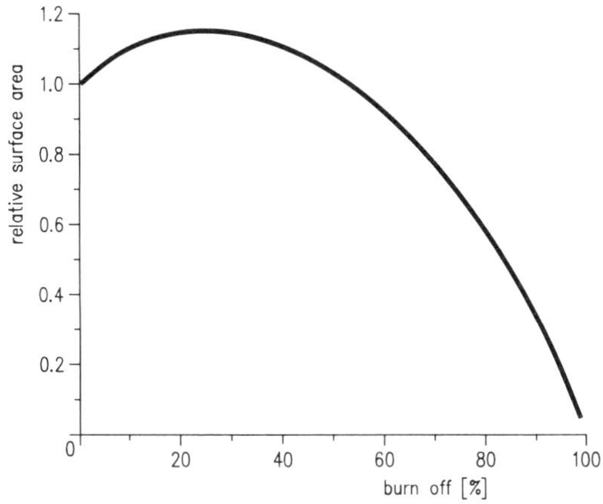

Fig. 5.19 Development of relative internal surface area as function of burn-off according to the model of Bhatia and Perlmutter[29]

gas reaction and the inhibition by the product gases according to the kinetic model are taken into account. Such a model should be able to predict the reaction conditions under which a special pore-size distribution and porosity can be generated. To what extent this will be possible has to be demonstrated by future work.

Concept of active sites

The above-mentioned difficulties in modelling the relationships between surface areas and reaction rates are not surprising as, following Walker, three aspects of internal surfaces have to be considered: the total specific surface area, its quality, i.e. the concentration of active sites, and its accessibility. The last factor might be restricted in the case of very large particles, in the case of pore diffusion-controlled reactions, especially at elevated temperatures, or if a part of the porosity is physically blocked.

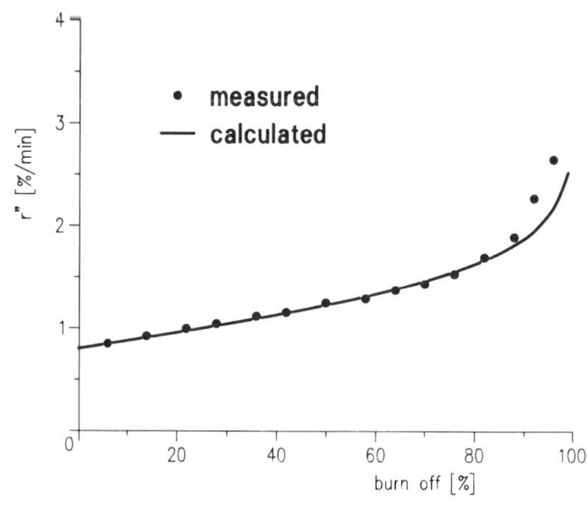

Fig. 5.20 Description of rate development at 800°C and 2 MPa H_2O applying the model of Bhatia and Perlmutter[29]

The first attempt to determine the quality of the internal surface, i.e. the active surface area (ASA) or the amount of active sites on the internal surface, was made by Vastola and co-workers[31,32] by measuring oxygen adsorption capacities of a graphitized carbon black (graphon) at 300°C which was oxidized to seven levels of burn-off between 0 and 35 wt%. They were able to show that these data correlate better with the reaction rates than the BET surface areas. They remarked on the fact that the ASA was markedly affected by the amount of prior burn-off which the graphon had undergone.

Since then, the concept of ASA for the characterization of carbonaceous materials has also been applied to feedstocks other than graphon with varying success. Some workers have tried to use it for coals or low-temperature chars and found that the oxygen adsorption capacities depend on the time allowed for chemisorption. Even after several days the weight increase had not finished, and there was no criterion to decide which chemisorbed amount of oxygen should be related to the reaction rate of gasification. Other research groups[33,34] investigated the chemisorption capacities as a function of the so-called cleaning procedure to remove surface oxide complexes, especially temperature and time, or as a function of chemisorption temperature, oxygen concentration or catalytic influences of mineral matter in the feedstock. They found that all these parameters affect the results of ASA, and up to now no procedure for a correct and unique ASA determination has been established. The reason could be that the gasification reactions do not run equally on all kinds of carbon sites and that oxygen chemisorption characterizes only a special group of them, not taking into account the energetic heterogenity of char or coke surfaces.

Recent investigations have focused on the *in situ* measurement of the active sites using temperature-programmed desorption (TPD) methods[35,36] and transient kinetics. A disadvantage of these methods is that the gasification reaction which should be characterized by the ASA has to be carried out during the same test. Nevertheless, these investigations help to understand what is happening on the surface of a carbonaceous material during gasification.

The most recent status of the ASA concept was reviewed and summarized by March and Radovic.[37]

5.5 References

1. MÜHLEN, H.-J. and SULIMMA, A. 1987: *Fuel Processing Technology* **15**, 145.
2. MÜHLEN, H.-J. and SULIMMA, A. 1986: *Thermochimica Acta* **103**, 163.
3. Van HEEK, K.H. and MÜHLEN, H.-J. 1985: *Fuel* **64**, 1405.
4. RADOVIC, L.R., WALKER, P.L. Jr and JENKINS, R.G. 1983: *Fuel* **62**, 849.
5. BOURRAT, X., OBERLIN, A. and ESCALIER, J.C. 1986: *Fuel* **66**, 1490.
6. CHITSORA, CH.T., MÜHLEN, H.-J., van HEEK, K.H. and JÜNTGEN, H. 1987: *Fuel Processing Technology* **15**, 17.
7. CHITSORA, Ch.T. 1986: *PhD Thesis*. Essen: GHS Universität.
8. MÜHLEN, H.-J., van HEEK, K.H. and JÜNTGEN, H. 1986: *Fuel* **65**, 591.
9. NANDI, S.P. and WALKER, P.L. Jr 1975: *Fuel* **54**, 169.
10. NANDI, S.P., RAMADASS, V. and WALKER, P.L. Jr 1964: *Carbon* **2**, 199.
11. WALKER, P.L. Jr, FORESTI, R.J. Jr and WRIGHT, C.C. 1953: *Industrial Engineering Chemistry* **45**, 1703.
12. CHICHE, P., DURIF, S. and PREGERMAIN, S. 1965: *Fuel* **44**, 5.

13. TODA, Y. 1973: *Fuel* **52**, 99.
14. ZYGOURAKIS, K. and GLASS, M.W. 1987: *Report of the American Institute of Chemical Engineers.*
15. Van HEEK, K.H. and MÜHLEN, H.-J. 1987: *Fuel Processing Technology* **15**, 113.
16. LÖWENTHAL, G., WANZL, W. and van HEEK, K.H. 1986: *Fuel* **65**, 346.
17. HOINKIS, E. and ROBENS, E. 1989: *Carbon* **27**, 157.
18. WICKE, E. and HEDDEN, K. 1953: *Zeitschrift für Elektrochemie* **57**, 636.
19. KAWAHATA, M. and WALKER, P.L. Jr 1963: in *Proceedings of the 5th Carbon Conference.* New York: Pergamon Press, Vol. 2, p. 251.
20. BERGER, J., SIEMIENIEWSKA, T. and TOMKOW, K. 1976: *Fuel* **55**, 9.
21. ADAMS, K.E., GLASSON, D.R. and JAYAWEERA, S.A.A. 1989: *Carbon* **27**, 95.
22. TOMKOW, T., SIEMIENIEWSKA, T., CZECHOWSKI, F. and JANKOWSKA, A. 1977: *Fuel* **56**, 121.
23. RODRÍGUEZ-REINOSO, F., LINARES-SOLANO, A., MARTÍN-MARTÍNEZ, J.M. and LÓPEZ-GONZÁLES, J.D. 1984: *Carbon* **22**, 123.
24. LINARES-SOLANO, A., LÓPEZ-GONZÁLES, J.D., MARTÍN-MARTÍNEZ, J.M. and RODRÍGUEZ-REINOSO, F. 1985: *Adsorpton Science and Technology* **1**, 123.
25. GARRIDO, J., LINARES-SOLANO, A., MARTÍN-MARTÍNEZ, J.M., MOLINA-SABIO, M., RODRÍGUEZ-REINOSO, F. and TORREGROSA, R. 1987: *Journal of the Chemical Society, Faraday Transactions 1* **3**, 1081.
26. GONZÁLES-VILCHEZ, P., LINARES-SOLANO, A., LÓPEZ-GONZÁLES, J.D. and RODRÍGUEZ-REINOSO, F. 1979: *Carbon* **17**, 441.
27. JOHNSON, J.L. 1975: *American Chemical Society, Division of Fuel Chemistry Preprints* **20** (4), 85.
28. FORMELLA, K. 1983: *Diplomathesis.* Münster.
29. BHATIA, S.K. and PERLMUTTER, D.D. 1989: *Journal of the American Institute of Chemical Engineers* **26**, 379.
30. BLIEK, A. 1984: *PhD Thesis.* Technische Hogeschool Twente.
31. LAINE, N.R., VASTOLA, F.J. and WALKER, P.L. Jr 1963: *Journal of Physical Chemistry* **67**, 2030.
32. LAINE, N.R., VASTOLA, F.J. and WALKER, P.L. Jr 1963: in *Proceedings of the 5th Carbon Conference.* New York: Pergamon Press, Vol. 2, p. 211.
33. MÜHLEN, H.-J., van HEEK, K.H. and JÜNTGEN, H. 1986: in *193rd ACS National Meeting Denver USA.* American Chemical Society, Division of Fuel Chemistry **31**, 286.
34. MÜHLEN, H.-J. 1990: *Fuel Processing Technology* **24**, 285.
35. HÜTTINGER, K.J. 1989: in *Proceedings of the 19th Biennial Conference on Carbon*, p. 592.
36. RADOVIC, L.R., LIZZIO, A.A. and JIANG, H. 1989: in *Proceedings of the 19th Biennial Conference on Carbon.* Penn State: University Park Campus, p. 596.
37. MARSH, H. and RADOVIC, L.R. 1991: in Lahaye, J. and Ehrburger, P. (eds) *Fundamental Issues in Control of Carbon Gasification Reactivity.* Dordrecht: Kluwer, p. 595.

6 Porosity and Radiation-induced Reactivity

Brain T. Kelly
Reactor Services
AEA Technology
NRL-S
Springfields
Nr Preston, Lancashire
United Kingdom

6.1 Introduction

The conventional oxidation of graphite in oxygen or carbon dioxide is a thermally-activated process with kinetics controlled by the supply of oxidant in different regimes.[1] This is well understood. The rate of the reaction may be substantially altered by the presence of irradiation damage and catalytic impurities.[2,3] In the presence of ionizing radiations the oxidation processes are found to proceed at temperatures where the thermal processes are negligible.

The radiation-induced reaction is of practical importance in nuclear reactors with a graphite moderator using carbon dioxide or air cooling where the weight losses due to the oxidation produce deleterious changes in the important graphite properties such as mechanical strength and thermal conductivity. The oxidation of graphite under these conditions, particularly with carbon dioxide gas, has been extensively researched because of its importance in the development of the gas-cooled nuclear power stations in the United Kingdom.[4,5] The content of this chapter is, for this reason, heavily biased towards the understanding of the reaction with carbon dioxide, but it should be remembered that similar considerations apply to other oxidants in the presence of radiation.

It is found experimentally that the radiolytic oxidation of graphite, as is known, is sensitive to the pore structure of the graphite and also to the presence of gas-phase inhibitors such as carbon monoxide or methane. However, it is not significantly altered by the presence of the usual oxidation catalysts.

In this chapter the generation of the reactive species in the gas phase is considered first, as it occurs in the open porosity of the graphite. The dependence of the oxidation on the size and shape of the pores in the graphite is then analysed, together with the effects of gas phase and surface inhibitors. The current theory is then compared with measurements of weight loss, gas-transport properties (diffusivity, permeability) and in-pore carbon deposition on a small number of graphites.

The observed changes in the important technical properties of graphite due to this oxidation are then presented, e.g. Young's modulus, strength, thermal conductivity and thermal expansion coefficient. Finally, the importance of these effects in the operational life of the Mk 1 (Magnox) and Mk 2 (advanced gas-cooled UK nuclear reactors) is briefly described.

6.2 Theory of radiolytic oxidation in graphite

Creation of oxidizing species

The presence of energetic radiation, capable of ionizing carbon dioxide (e.g. γ-rays, moving ions) creates reactive species in the gas phase which then diffuse to graphite surfaces and remove carbon atoms.[6] The ionizing radiation in the core of a nuclear reactor is a mixture of energetic photons and moving atoms produced by neutron–nucleus collisions. Early studies showed that to a good approximation the number of oxidizing species G produced for 100 eV of energy absorbed in the gas phase is constant. The ionization potential of the carbon dioxide molecule is 13.7 eV, hence G is expected to be about 10. The basic value of G, that is, the value in free gas without the pore walls, is referred to as G_0.

The value of E (watts per gram of carbon dioxide), the energy deposited in the gas phase, can be calculated using modern computer models of the radiation sources and attenuation[7] or measured using a suitable calorimeter. The calorimeters generally use graphite as the absorbing medium in order to produce a reasonable temperature rise for measurement. It is then assumed that absorption in carbon dioxide is identical with that in carbon per atom. This is not completely correct, since the elastic scattering cross-sections for neutrons as a function of energy can differ significantly between oxygen and carbon nuclei, and the electronic energy levels also differ, which affects the interaction with γ-rays. These corrections can be made in principle, but often are not. The proper value of G to use is further complicated because the oxidation of the graphite principally takes place in the open porosity and it is actually necessary to know the energy deposition in the gas in the pore. In this case it is essential to take account of the Bragg–Gray principle, which states that 'the energy absorbed in a gas-filled cavity is related to that in the surrounding solid material by the ratio of the stopping powers for secondary particles for gas and solid, provided that the size of the cavity is less than the range of the secondary particles in the gas.'

Linacre *et al.*[8] concluded that for the range of γ-ray energies found in the core of a nuclear reactor, the dose rate in carbon dioxide in the pores of graphite was within 1% of the ideal dose rate (that is, the rate which would be observed in a volume too small to cause significant attenuation and large compared with the range of the secondary radiation). The situation is not so simple for the recoil ions produced by neutron scattering. The range of a carbon ion of about 0.15 MeV in carbon dioxide at ~4 MPa (40 bar) pressure (as found in advanced gas-cooled reactors, for instance) is about 10 μm, whereas most recoils have lower energies. It must be concluded that some of the pores in graphite are too large to be considered a small cavity and too small to be considered a free gas volume.

In a free gas volume the dose rate would be 0.83 times that in carbon, whereas in a small cavity the value would be 0.92 times that in carbon. The correct value for an experiment therefore requires a knowledge of the open-pore spectrum of the graphite, and the values will change as pores grow due to weight loss.

In an advanced gas-cooled reactor the energy deposition in carbon dioxide is divided roughly into one third due to neutron scattering and two thirds due to energetic photons. It has become the custom to use the energy deposition rates measured in graphite in a calorimeter together with a value of $G_0 = 3/100$ eV independent of the ion/γ-ray photon energy ratio to estimate the rate of production of active species. However, the considerations above suggest that experimental values of G_0 may vary between different graphites and experimental facilities.

The oxidizing species produced lead to a reaction written most simply as:

$$CO_2 + \text{energy} \rightarrow CO_2^* \text{ (excited state)}$$

$$CO_2^* + C \text{ (graphite)} \rightarrow 2CO$$

but this simplicity conceals a complex chemistry which has occupied a substantial research effort. We shall assume that the rate of production of oxidizing species per unit volume of gas is

$$\frac{dN(\text{Ox})}{dt} = \rho_g E \frac{G_0}{100} \qquad (6.1)$$

where ρ_g is the gas density (g cm^{-3}) and E is the energy deposition rate (now in units of eV g^{-1} s^{-1}).

The first studies of the radiation chemistry date from 1908.[9] It is known that photolysis of carbon dioxide produces carbon monoxide and oxygen at low photon energies. In general, carbon dioxide appears to be remarkably stable under irradiation, one of the reasons for its selection as a nuclear reactor coolant. Dominey and Wickham[9] in 1971 studied the exchange of carbon atoms between carbon dioxide and carbon monoxide using isotopically labelled carbon monoxide in a pure γ-radiation field. The results correspond to a G value for the exchange of $G_{Ex} = 4.2$. The rates of ion-pair production in carbon dioxide have been measured for a variety of ionizing radiations.[10] The values obtained vary from about 1 to slightly greater than 3, increasing systematically with velocity of the ionizing particles. The apparent difference between G_0 and G_{Ex} has been the the subject of considerable speculation.[11]

The reaction of carbon dioxide with the graphite in a nuclear reactor core takes place at pressures much greater than those employed in the study of ion chemistry, between 0.9 and 4 MPa and at temperatures from 200 to 600°C. According to Sieck,[12] the positive ion chemistry of carbon dioxide and carbon monoxide–carbon dioxide mixtures deduced from the behaviour of thermal ions produced by photoionization is dominated by clustering reactions with carbon monoxide replacing carbon dioxide molecules successively, e.g.

$$CO_2 \rightarrow CO_2^+ + e$$

$$CO_2^+ + 2CO_2 \rightarrow (CO_2)_2^+ + CO_2$$

$$(CO_2)_2^+ + CO \rightarrow (CO_2 \cdot CO)^+ + CO_2$$

$$(CO_2 \cdot CO)^+ + CO \rightarrow (CO)_2^+ + CO_2$$

The $(CO_2)^+$ ion dominates for all carbon monoxide levels except those of less than a few hundred ppm. Best *et al.*[13] noted that at higher pressures further clustering is possible. The $(CO)_2^+$ ion reacts with carbon monoxide and carbon dioxide readily, e.g.

$$(CO)_2^+ + 2CO_2 \rightarrow (CO_2 \cdot CO \cdot CO)^+ + CO_2$$

$$(CO)_2^+ + 2CO \rightarrow (CO)_3^+ + CO$$

A cluster which is responsible for the overall reaction must contain at least one carbon dioxide molecule and it is observed that carbon monoxide significantly reduces the reaction rate with graphite. It is deduced that in this chemistry it is the dimer which must dominate the high-pressure reactions with graphite.[13] Experiments in pure carbon dioxide[13] permit estimates to be made of the cluster-size ratios present in reactor coolants.

The calculations predict that the dimers will dominate the chemistry except possibly at the upper end of the pressure and temperature range in an advanced gas-cooled reactor. There is no experimental evidence for carbon monoxide–carbon dioxide mixtures. However, the dimer $(CO)_2^+$ should not be able to oxidize graphite and the large observed effect of carbon monoxide is only to be expected when the monomer/dimer ratio is about unity.[13] Best *et al.*[13] considered the negative-ion chemistry to be represented by

$$e + CO_2 \rightarrow CO + O^-$$

$$O^- + 2CO_2 \rightarrow CO_3^- + CO_2$$

with the inhibition process represented by

$$CO_3^- + CO \rightarrow 2CO_2 + e$$

and to a lesser extent by

$$O^- + CO \rightarrow CO_2 + e$$

and concluded that a negative-ion chemistry is possible.

It is known that the products of the photolysis of carbon dioxide can oxidize graphite at rates comparable to those due to ionizing radiation.[14] The oxidation is due to the presence of oxygen species, but in carbon monoxide–carbon monoxide mixtures these would be rapidly neutralized by the carbon monoxide. This neutral species chemistry was rejected by Best *et al.*[13]

Ion chamber studies have been undertaken by Norfolk *et al.*[15] using graphite electrodes at reactor pressures and temperatures with careful control of the relative surface and in-pore oxidation rates. The results indicate that it is indeed the negative-ion chemistry which is dominant.

The preceding considerations are based on carbon dioxide or carbon monoxide–carbon dioxide mixtures. The practical cases of reactor coolants invariably concern the presence of appreciable quantities of hydrogen-containing compounds, usually molecular hydrogen, methane and water. Radiolysis of these mixtures produces a wide range of transient and stable organic compounds. In the earlier gas-cooled reactors (Mk 1, Magnox) hydrogen may be added deliberately or be present as a result of boiler or oil seal leaks. In the advanced gas-cooled reactors both carbon monoxide and methane are deliberately added to the coolant to control the graphite oxidation rate. One of the reaction products of the radiolysis is water, which is deliberately removed by desiccant beds, the net results being roughly equal concentrations of water and hydrogen present in the coolant at a few hundred parts per million by volume (vpm).

A number of studies[13] have shown important changes in positive-ion chemistry in the presence of hydrogen or water. Water can readily enter the positive-ion clusters from carbon dioxide to produce species such as $([CO_2]_m \cdot [H_2O]_n)^+$ and $(CO_2 \cdot H_3O)^+$. Consideration of the effects of water in the presence of carbon monoxide and hydrogen as well as carbon monoxide provide further evidence against the importance of positive-ion chemistry in graphite oxidation. Some information is available on positive-ion chemistry in the presence of methane. The reactions are complex leading to the formation of radical and ionized species which include CH_2CO^+, CH_3CO^*, HCO^+ and $HCO^* \cdot HCO^+$ apparently does not react with methane. There is little information on the effect of hydrogenous additions on negative-ion chemistry. The yield of CO_3^- is reduced owing to the competing reaction for O^-

$$H_2 + O^- \rightarrow H_2O + e.$$

The observations of Parkes[6] suggest that hydrogen will reduce the concentration of CO by $4[H_2]/[CO_2]$.

Best *et al.*[13] concluded that hydrogenous additives are unlikely to interfere with the basic negative-ion chemistry and also that the positive-ion chemistry is unimportant. Hence gas-phase effects of hydrogenous additives are expected to be small. The fate of all of the ions produced in irradiated carbon dioxide is neutralization either at the pore surface or in the gas phase. Gas-phase deactivation processes imply the existence of a range for the active species.

The value of this range may be larger or smaller than the size of the various open pores in graphite. Best *et al.*[13] quoted unpublished calculations of range which indicate that the number of oxidizing species reaching the walls may be substantially reduced in larger pores.

Surface chemistry of oxidation

The nature of the graphite crystal is such that the majority of atoms are bonded within basal planes, but there is also a population of atoms located at basal-plane edges or at lattice defects which are less strongly bonded. There is, in the presence of neutron irradiation, a population of more or less complicated lattice vacancies and edge-dislocation dipoles[17] which depends on the dose and temperature of irradiation.

Studies carried out using γ or UV radiation to avoid the creation of new lattice defects showed that the oxidizing species migrated over the surface of the basal planes to edge sites to produce gasification.[18–21] In particular Feates and Fryer[20] showed, using a stack of graphite sheets cleaved from a single crystal, with very small gaps between the sheets (0.01 cm) and a gas pressure of 8 kPa, that a G value of 2.9 could be obtained of which the basal plane surfaces contributed only $\sim 10^{-8}$, the remainder being due to the edges. Hence the mobility of the oxidizing species over the basal planes to the edges was very important.

Transient studies carried out in the BEPO reactor at Harwell using nuclear graphite labelled with ^{14}C substitutionally and both CO_2 and helium gases circulating through the irradiation capsules showed that:

1. There was a significant delay in the appearance of ^{14}C after irradiation in CO_2 commenced, although carbon monoxide appeared immediately. This was interpreted as the formation of surface oxide

   $$CO_2 + C \rightarrow \underset{\text{(surface oxide)}}{C(O)} + CO$$

   It is only when the equilibrium coverage of surface oxide is attained that newly arriving oxidizing species produce equivalent oxidation. Production of ^{14}C ceases with irradiation.

2. When the gas is changed to helium the net carbon monoxide yield decreased more rapidly than that of ^{14}C, that is, decomposition of the surface oxide to CO can be achieved in the presence of helium and irradiation, whereas the reverse occurs when carbon dioxide is substituted for helium.

Best *et al.*[22] have shown also that the existence of CO_2 and CO in the gas post-irradiation is not compatible with the apparent simplicity of the reaction scheme. The same work, in line with the BEPO transient studies, did not show, as would be expected, a doubling of the carbon monoxide yield once equilibrium surface coverage had been attained. It may be that exchange of ^{14}C between carbon monoxide and carbon dioxide is an explanation of these results, but the studies of Feates and Walker[23] indicate that it is unlikely.

Best *et al.*[13] wrote the oxidation process as

$$CO_2 + \text{energy} \rightarrow \text{Ox} + \text{products (Inc. CO)}$$

$$\text{Ox} + C \rightarrow C(O)_{\text{basal plane}}$$

$$C(O)_{\text{basal plane}} \rightarrow C(O)_{\text{edge}}$$

$$C(O)_{\text{edge}} + C(O)_{\text{edge}} \rightarrow CO_2 + C$$

$$C(O)_{\text{edge}} \rightarrow CO$$

where the carbon atoms, in all except the first equation and where contained in CO or CO_2, are part of the graphite structure. The first and second equations cannot be studied separately. Studies of the desorption of carbon monoxide and carbon dioxide showed discrete activation energies,[24] but whilst these values were found for pure carbon dioxide there was only one with carbon monoxide also present.

The presence of a surface carbon dioxide molecule leads to difficulties with observations of $G_0 = 3$ since it should lead to a situation in which the net corrosion is less than the supply of oxidizing species. The surface (carbon monoxide/carbon dioxide) ratio is found to vary with temperature from 0.25 at 498 K to 0.50 at 648 K. A number of alternative explanations were considered by Best *et al.*[13]

The presence of hydrogen in the reactor coolant is important practically. The compounds of major importance are hydrogen, water and methane, but other higher hydrocarbons are generated radiolytically. The Magnox reactors contain a relatively low hydrogen content and the gas-phase chemistry is dominated by a hydrogen–methane equilibrium with hydrogen absorbed on the graphite, which is presumed to react with the surface oxide. In the Magnox reactors, for hydrogen levels of below 0.004%, there is little methane or water, and reductions in radiolytic oxidation of about a factor of three must be due to surface effects.

The presence of water does not have any effect in the absence of carbon monoxide, but is significant in its presence, the effect saturating at ~0.03% concentration. Transient studies suggest that it is due to the formation of a sacrificial surface layer on the graphite, one carbon monoxide molecule plus one water molecule protecting two carbon atoms.

The effect of methane on the graphite corrosion is much more powerful, and cannot be explained by reaction of methane with the oxidizing species in the gas phase alone. This reaction forms a carbon-depositing species which gives rise to a sacrificial protective layer on the graphite surface.

Figure 6.1 compares the inhibiting effects of carbon monoxide, methane and hydrogen. However, the detailed knowledge of the oxidizing species, gas-phase and surface reactions leave much to be desired and in the following, very simplified schemes will be used, which in fact fit experiment very well.

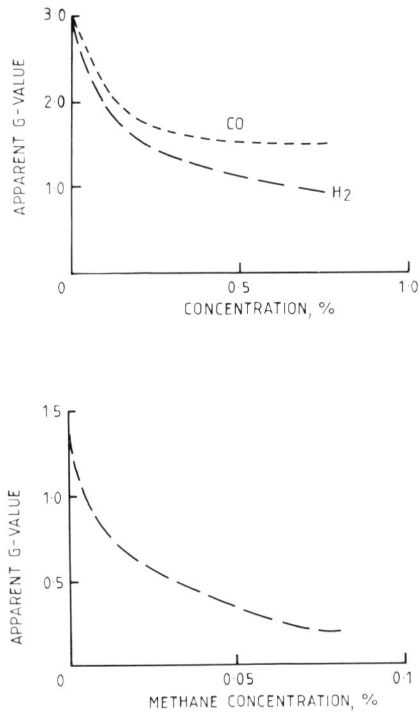

Fig. 6.1 Comparison of the inhibiting effects of carbon monoxide, methane and hydrogen on the radiolytic oxidation of Pile Grade A graphite. (a) Gas-phase inhibiting effect of hydrogen compared with carbon monoxide; (b) inhibiting effect of methane added to 1.2% CO–CO_2 gas composition (after Best *et al.*[13])

Development of porosity by radiolytic oxidation

It is assumed that the porosity of a graphite sample, either in its virgin condition or after oxidation, can be described simply, neglecting any surface roughness effects (e.g. no fractal considerations). The initial bulk density of the graphite is denoted by ρ_B g cm^{-3}, the crystal density by ρ_c g cm^{-3} and the helium density by ρ_{He} g cm^{-3}. It is assumed, without identification, that G_0 oxidizing species are produced for each 100 eV deposited in the gas phase in the graphite porosity. These species can be deactivated in the gas phase by reaction with carbon monoxide, methane, etc., and it is also assumed that the reaction with methane produces a depositing species containing a single carbon atom.[25]

There are three conditions of interest:

1. The oxidation of graphite in 'pure' carbon dioxide. It is not possible to achieve this condition exactly, because the carbon monoxide reaction product of the oxidation is capable of gas-phase deactivation of the oxidizing species. However, it is possible by the choice of experimental conditions to keep the in-pore gas concentration of carbon monoxide so low that deactivation is not important.

2. The oxidation of graphite in carbon dioxide containing one or more compounds which can deactivate the oxidizing species and thus define a range or mean free path L for the oxidizing species which is less than characteristic pore dimensions.
3. The oxidation of graphite in carbon dioxide-containing compounds which lead to carbonaceous deposition on pore surfaces. This carbon deposition gives surface protection for the graphite since it provides carbon for sacrificial oxidation.

These cases are now considered in turn.

If the open pore volume of a graphite is V_p in unit volume, then the number of oxidizing species produced is proportional to the amount of gas in the pores $\rho_g V_p$, where ρ_g is the gas density, and the energy deposition E eV $g^{-1}(CO_2)\,s^{-1}$. Each of these oxidizing species reaches the pore wall and removes a carbon atom of mass m_c. The rate of weight loss referred to unit volume of graphite and unit dose (dose $d = Et$) is

$$\frac{dx}{dd} = \frac{m_c}{\rho_B}\frac{G_0}{100}\rho_g V_p. \tag{6.2}$$

The rate of change of pore volume, assuming no breakthrough into closed porosity, depends on whether the graphite at the pore walls has true density ρ_c g cm^{-3} or contains small pores compared with that under study, which reduces its effective density to ρ_{He}. The latter is apparently a good assumption for the graphite that we shall consider most and so this assumption is made in all of the following except where indicated, but it should be checked when comparison with experimental data is being made. This assumption leads to

$$\frac{dV_p}{dd} = \left(\frac{\rho_B}{\rho_{He}}\right)\frac{dx}{dd}. \tag{6.3}$$

Equation 6.3 can be used to integrate equation 6.2 for the porosity as a function of dose or the weight loss. The results are

$$V_p = V_0 \exp(d/d_0) \tag{6.4}$$

and

$$x = V_0\left(\frac{\rho_{He}}{\rho_B}\right)[\exp(d/d_0) - 1] \tag{6.5}$$

where V_o is the initial pore volume and

$$d_0 = \frac{\rho_{He}}{m_c}\frac{100}{\rho_g G_0} \tag{6.6}$$

is a standard dose constant which will appear frequently in the modelling of oxidation.

The three standard pore shapes which are generally analysed theoretically for their behaviour are a slab of width W, a cylinder of radius R_c and a sphere of radius R_s. In order to conform to equation 6.4, these characteristic dimensions must obey

$$R_\nu = R_{0\nu} \exp(d/\nu d_0) \tag{6.7}$$

where $\nu = 1$ for a slab-shaped pore, $\nu = 2$ for a cylindrical pore and $\nu = 3$ for a spherical pore. In each case the pore shape is maintained, and although no proof has been given, this is likely to be true for other shapes also when all oxidizing species reach the pore wall.

The first calculations of the flux of oxidizing species to pore walls when their range was restricted to a length L were made by Tait,[26] who also considered the effect of a reflection coefficient or albedo for the oxidizing species at the graphite surface. The experimental evidence suggests that the reflection coefficient is effectively zero at least in polycrystalline graphite, and we consider this case only. Similar calculations have been made by Best and Wood[27] and Johnson.[25] For the three interesting cases of slab, cylinder and spherical pores, the probability that an oxidizing species will reach a pore wall is given by

(a) Slab

$$P(W) = \tanh(W/2L)/(W/2L)$$

(b) Cylinder

$$P(R_c) = \frac{I_1(R_c/L)}{I_0(R_c/L)} \left(\frac{2L}{R_c}\right) \tag{6.8}$$

(c) Sphere

$$P(R_s) = \left[\coth\left(\frac{R_s}{L}\right) - \frac{L}{R_s}\right]\left(\frac{3L}{R_s}\right)$$

where I_1 and I_0 are the appropriate Bessel functions.

Johnson[25,28] has shown that the characteristic pore dimension $R\nu$ is given by

$$\frac{dR\nu}{dd} = \frac{R\nu}{\nu d_0}\left[\frac{2L}{R\nu} y(R\nu)\right] \tag{6.9}$$

where

$$\left.\begin{aligned} y_1(R_1) &= \tanh(R_1/2L) \text{ (for a slab)} \\ y_2(R_2) &= I_1(R_2/L)/I_0(R_2/L) \text{ (for a cylinder)} \\ y_3(R_3) &= [\coth(R_3/L) - (L/R_3)] \text{ (for a sphere)} \end{aligned}\right\} \tag{6.10}$$

Equation 6.9 can be integrated numerically to give $R\nu$ as a function of dose d. If $R\nu/L \geqslant 2$ the three functions tend to very similar behaviour and the equations have simple solutions which will be used later. For the slab and cylinder these are

(a) Slab

$\tanh(W/2L) \rightarrow 1$ with increasing W

and then

$$W = W_0 + (2L/d_0)d \tag{6.11}$$

(b) Cylinder

$I_1(R_c/L)/I_0(R_c/L) \rightarrow 1$ with increasing R_c

and then

$$R_c = R_0 + (L/d_0)d \tag{6.12}$$

In virtually all of the practical situations there is a significant neutron flux which produces dimensional changes in the graphite crystallites and in consequence the bulk dimensions, which modify the open and closed porosity,[10] generally decreasing both.

In most modern reactor systems the graphite shrinks roughly linearly with dose to some maximum dose and thereafter expands. In early low-temperature reactors the graphite always expanded.[29] It has been suggested that the equation for pore width or radius as a function of dose is modified by the addition of a term $-W_0/d^*$ for the slab or $-R_0/d^*$ for the cylinder. The dose d^* is that at which the pore would apparently disappear in the absence of oxidation. Harper[30] has suggested other forms for this term particularly appropriate for cylindrical pores.

The presence of such a term modifies the pore size for a given dose d compared with that in its absence. In order to calculate the weight loss from a pore, it is now necessary to calculate the pore size as a function of dose, and then the number of oxidizing species reaching the pore walls.

Consider first the case of a slab-shaped pore of a size initially much smaller than the range of oxidizing species. The width is given by

$$\frac{dW}{dd} = \frac{W}{d_0} - \frac{W}{d^*} \tag{6.13}$$

with the solution

$$W/W_0 = [1 - (d_0 d^*)]\exp(d/d_0) + (d_0/d^*) \tag{6.14}$$

The pore expands if $(d_0/d^*) < 1$ and contracts if $(d_0 d^*) > 1$. The weight loss is now given by

$$x = V_0(\rho_{He}/\rho_B)\left\{\left(\frac{d}{d^*}\right) + \left\langle 1 - \left(\frac{d}{d^*}\right)\right\rangle [\exp(d/d_0) - 1]\right\} \tag{6.15}$$

If the relative values of d_0 and d^* are such that the pore closes, this occurs at a dose

$$d' = d_0 \ln\left[\frac{d_0/d^*}{(d_0/d^*)-1}\right] \tag{6.16}$$

with the corresponding weight loss given by equation 6.15 with $d = d'$.

If the range L is significantly less than the pore width, then equation 6.9 becomes

$$\frac{\mathrm{d}W}{\mathrm{d}d} = \frac{W}{d_0}\left[\frac{\tanh(W/2L)}{W/2L}\right] - \frac{W_0}{d^*} \tag{6.17}$$

and given $W(d)$, the weight loss can be calculated from

$$\frac{\mathrm{d}x}{\mathrm{d}d} = \frac{W(d)}{d_0}\left(\frac{\rho_{\mathrm{He}}}{\rho_{\mathrm{B}}}\right)\left(\frac{V_0}{W_0}\right)\left[\frac{\tanh(W/2L)}{W/2L}\right] \tag{6.18}$$

A similar treatment can be given for cylindrical or spherical pores but has not been used practically. It is easy to see for the slab that once the pore width is greater than the range L, the rate of weight loss becomes constant at

$$\frac{\mathrm{d}x}{\mathrm{d}d} = \left(\frac{\rho_{\mathrm{He}}}{\rho_{\mathrm{B}}}\right)\left(\frac{V_0}{W_0}\right)\frac{2L}{d_0} \tag{6.19}$$

The most important practical case is that in which methane is added to the carbon dioxide gas in addition to carbon monoxide. It is assumed that the presence of the methane produces two distinct inhibiting processes, the first of which is a deactivating reaction similar to that of carbon monoxide. The fraction of oxidizing species deactivated by methane is assumed to be given by

$$f = \frac{a[\mathrm{CH_4}]}{[\mathrm{CO}] + a[\mathrm{CH_4}]} \tag{6.20}$$

where the brackets denote concentrations,

$$a = k_2/k_1 \approx 10 \text{ at } 673\ \mathrm{K}$$

and k_2 and k_1 are the respective rate constants for reaction of the oxidizing species with methane and carbon monoxide (see later).

The reaction between methane and the oxidizing species is assumed to produce a species which deposits a single carbon atom on a graphite surface. This atom is then available for sacrificial oxidation. It is readily shown that the equation for the rate of pore growth is now

$$\frac{\mathrm{d}R_\nu}{\mathrm{d}d} = \frac{R_\nu}{\nu d_0}\left[\frac{2L}{R_\nu}\, y_\nu(R_\nu)(1+f) - f\right] \tag{6.21}$$

The reaction mean free path L is given by

$$L = \left(\frac{D}{k_1[\mathrm{CO}] + k_2[\mathrm{CH_4}]} \right)^{1/2} \tag{6.22}$$

Equation 6.2 is interesting because the pore can either grow or shrink depending on the sign of the bracketed term.

The effect of irradiation damage can again be included by adding a term $-R_0/\!/d^*$ to the right-hand side. Equation 6.21 also predicts that all pores will tend to the same dimension R_M, larger pores contracting by deposition of carbon and smaller pores expanding by oxidation.

It is possible to integrate equation 6.21 when the pores are initially larger than the range L. Two useful solutions for this case are

(a) Slab

$$W = W_\mathrm{M} + (W_0 - W_\mathrm{M}) \exp\left(-\frac{fd}{d_0} \right) \tag{6.23}$$

(b) Cylinder

$$R = R_\mathrm{M} + (R_0 - R_\mathrm{M}) \exp\left(-\frac{fd}{2d_0} \right) \tag{6.24}$$

where

$$W_\mathrm{M} = 2L\left(\frac{1+f}{f} \right) - \frac{W_0}{f} \left(\frac{d_0}{d^*} \right) \tag{6.25}$$

and

$$R_\mathrm{M} = 2L\left(\frac{1+f}{f} \right) - \frac{2}{f} R_0 \left(\frac{d_0}{d^*} \right) \tag{6.26}$$

In the absence of a neutron effect ($d^* \to \infty$), oxidation ceases when $R_\nu = R_\mathrm{M}$ because the oxidizing and depositing fluxes to the surface are equal. However, if the pore is shrinking because of the neutron effect, this corresponds to an oxidation rate because the pore is always smaller than the equilibrium size.

It is of some interest to consider an elliptical pore in the case when the range is less than the pore dimensions (and neglecting any neutron effects). It is easy to see that the pore will tend to a circular cross-section by considering the oxidizing/depositing fluxes along the principal axes of the ellipse. Thus a slab remains a slab, but elliptical pores tend to circular cross-sections with radius R_M. Circular pores tend to radius R_M also, either by deposition or oxidation.

This completes the required results on the weight loss and changes in dimensions due to individual pores of simple geometric shapes.

6.3 Comparison of theory and experiment

The experimental data gainst which these relatively simple models can be tested have been generated almost exclusively by the Berkeley Nuclear Laboratories of the Central Electricity Generating Board (UK), the United Kingdom Atomic Energy Authority (now called AEA Technology) and the French nuclear programme. The first of these has a relatively low-energy deposition rate γ-ray source well suited to measuring initial oxidation rates as a function of gas composition whereas the others have used nuclear reactors, including materials testing reactors, in which weight loss can be measured as a function of dose. Data have also been obtained from monitoring the graphite in power reactors. The high-dose experiments have permitted measurements of the changes in graphite properties due to weight loss. The important technical properties are diffusion and permeation of gases, Young's modulus, Poisson's ratio, thermal expansion coefficient, strength and thermal conductivity.

The construction of graphite-moderated gas-cooled reactors in the UK has been based on the use of two very different graphites, each manufactured by the two major UK graphite manufacturers, Anglo-Great Lakes at Newcastle upon Tyne and Union Carbide (previously British Acheson Electrodes) at Sheffield. The Calder Hall reactors, the Mk I (Magnox) gas-cooled reactors and the prototype Windscale advanced gas-cooled reactor (WAGR) used a graphite known as Pile Grade A for the moderator. This graphite was manufactured from a needle-coke with a pitch binder and impregnant, extruded in square sections ($\sim 20 \times 20$ cm $\times$ 100 cm long). The manufacturing process resulted in an anisotropic graphite with the properties given in Table 6.1. The properties of Pile Grade A graphite were not suitable for the more onerous conditions of energy deposition rate and neutron damage in the Mk II advanced gas-cooled reactors[31] and special high thermal expansion isotropic graphites were produced with a reduced pore volume and higher strength.[32] The final choice of candidate materials was based on coke from the naturally occurring Gilsonite pitch found in Utah.[33] This coke contains characteristic spherical particles with a readily recognizable internal structure. The properties of Gilso-carbon graphite are also summarised in Table 6.1, and Figure 6.2 shows sensitive tint pictures of the microstructure of Pile Grade A and Gilso-carbon graphite. The products of the two manufacturers were very similar. Experimental data have also been obtained on the Pechiney graphite used in French gas-cooled graphite reactors and on a number of other graphites used as fuel sleeves (or possible replacements) in advanced gas-cooled reactors, and these will be referred to as appropriate.

The earliest UK studies used Pile Grade A, sometimes with a variety of impregnant carbons and deliberately varied porosity to establish the dependence of oxidation rate on open-pore volume and gas pressure in carbon dioxide; the dependence was found to be linear in each case.[34] Experimental measurements were made on Pile Grade A graphite in carbon dioxide with low concentrations of carbon monoxide generated by the reaction. Experiments carried out in the Harwell Materials Testing Reactor, PLUTO, showed that the apparent *G* value for the gas in the pores of Pile Grade A graphite, for a carbon monoxide concentration of 0.01%, was 2.35. Extension of the studies to significant weight loss showed that after a 2% initial weight loss, almost half of the initially closed

Table 6.1 Properties of Pile Grade A and Gilso-carbon reactor moderator graphites

Property		Pile Grade A graphite (extruded)[a]	Gilso-carbon graphite (moulded)
Isotropy		Anisotropic	Isotropic
Density g cm^{-3}		1.72	1.815 ± 0.03
Open pore volume $cm^3\ cm^{-3}$		0.198	0.109
Closed pore volume $cm^3\ cm^{-3}$		0.01	0.086
Electrical resistivity $\mu\Omega\ cm^{-1}$	Parallel to extrusion	616	900 ± 150
	Perpendicular to extrusion	1081	
Thermal conductivity $W\ m^{-1}\ K^{-1}$	Parallel to extrusion	200	121 ± 21
	Perpendicular to extrusion	109	
Thermal expansion coefficient K^{-1} (20–120 K)	Parallel to extrusion	0.9×10^{-6}	$4.6 \pm 0.21 \times 10^{-6}$
	Perpendicular to extrusion	2.8×10^{-6}	
Poisson's ratio		0.07	0.21
Tensile strength $MN\ m^{-2}$	Parallel to extrusion	16.8	17.5 ± 3.5
	Perpendicular to extrusion	11.2	
Bend strength $MN\ m^{-2}$	Parallel to extrusion	18.8	23.0 ± 3.5
	Perpendicular to extrusion	12.3	
Compressive strength $MN\ m^{-2}$	Parallel to extrusion	27	70 ± 7.0
	Perpendicular to extrusion	27	
Diffusivity, λ_0		13.6×10^{-3}	3.0×10^{-3}
Permeability:			
Viscous flow coefficient, B_0 cm^2	Parallel to extrusion	712×10^{-16}	6.5×10^{-15}
	Perpendicular to extrusion	157×10^{-16}	
Slip flow coefficient, K_0 m	Parallel to extrusion	108×10^{-9}	5.8×10^{-9}
	Perpendicular to extrusion	21×10^{-9}	

[a]The values for Pile Grade A graphite can only be considered as representative; no comprehensive survey has been made.

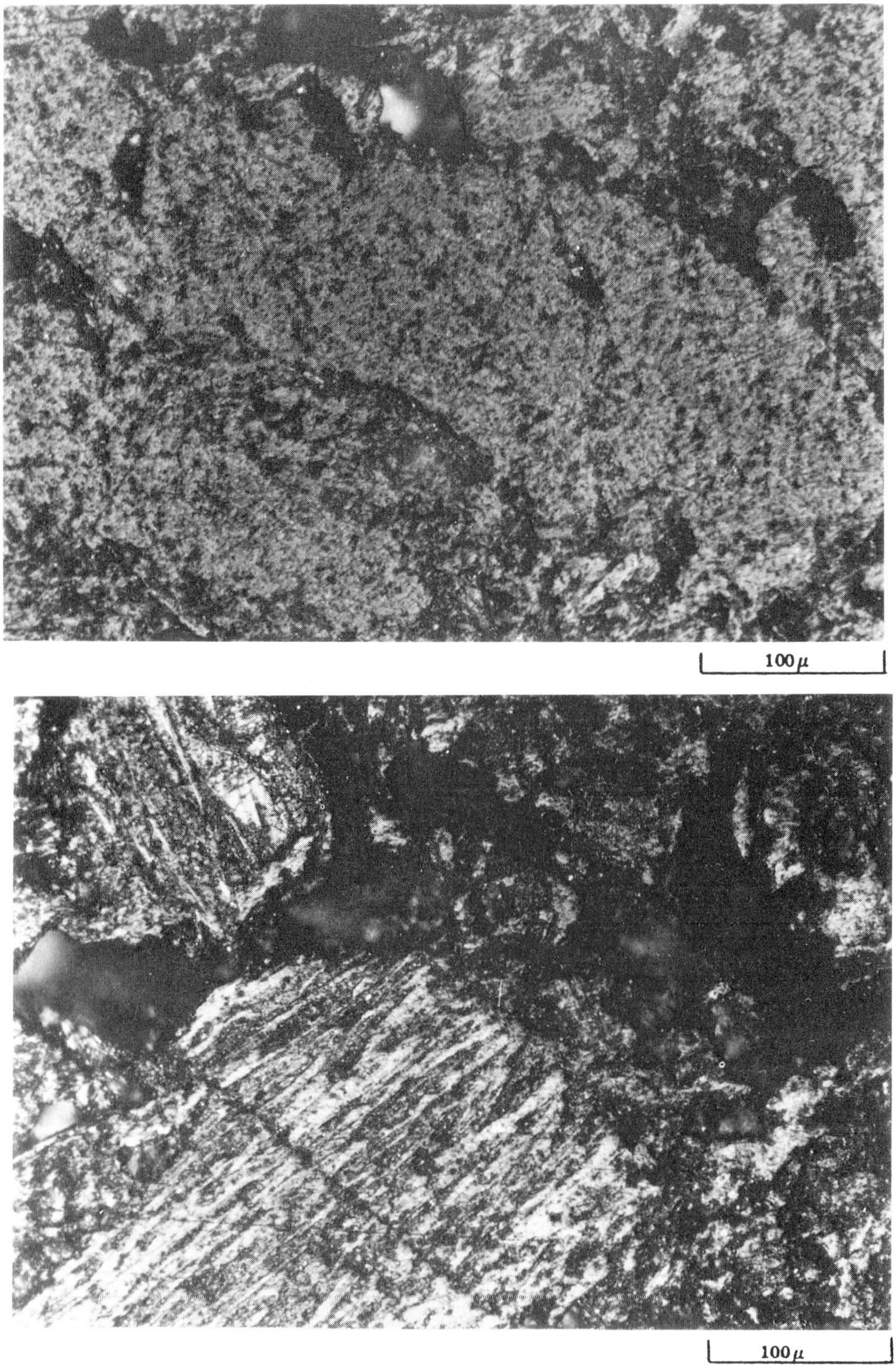

Fig. 6.2 Structures of (left) Pile Grade A and (right) Gilso-carbon moderator graphites

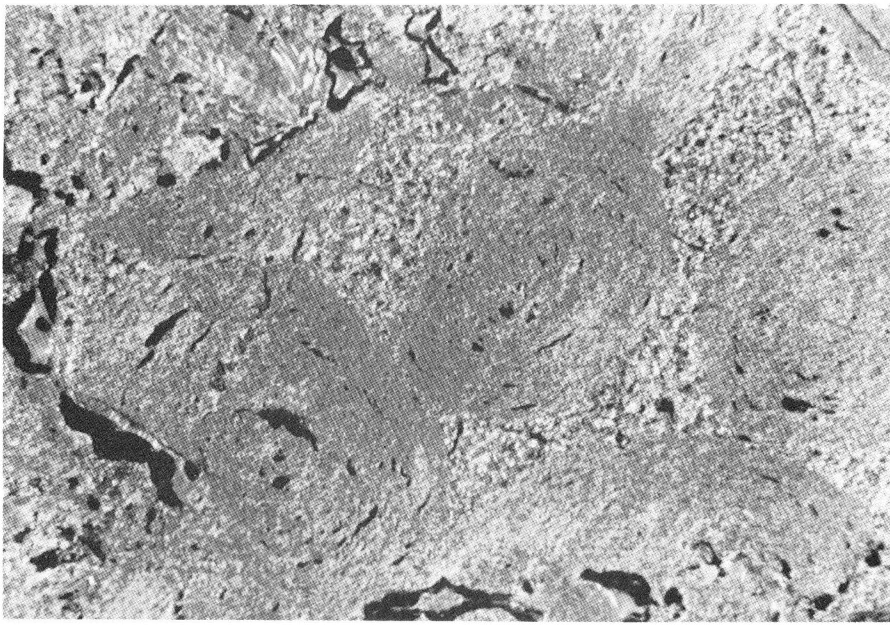

100μ

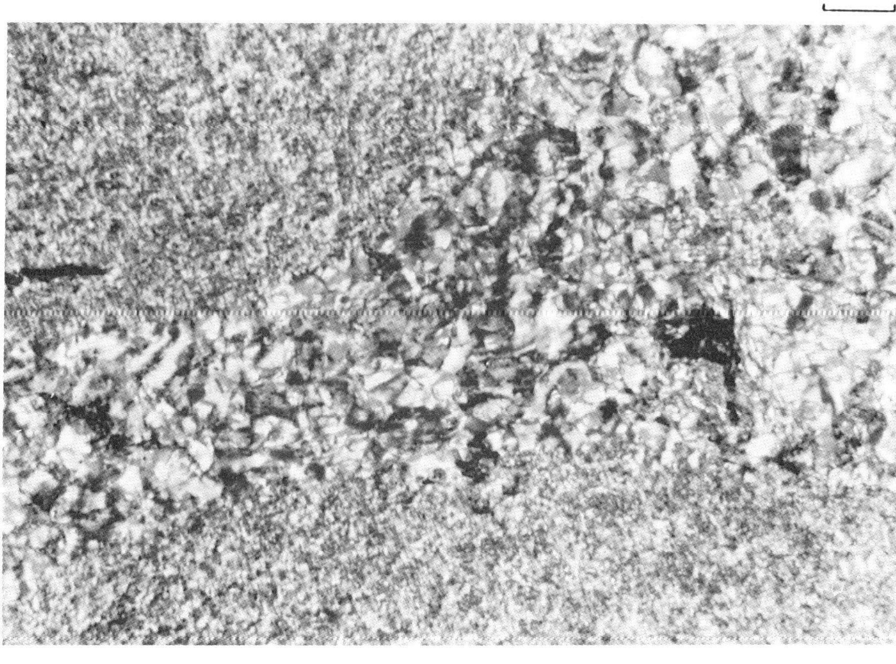

20μ

porosity became accessible.[34] This work also demonstrated again the proportionality between oxidation rate and accessible pore volume except for highly oxidized specimens. This was the first hint that not all of the oxidized species produced in large pores reached the pore walls and oxidation was inhomogeneous. Studies of the effect of carbon monoxide concentration on the oxidation rate on the same graphite were made in the range up to 1.2%. The apparent G value was found to fall to about unity at the highest carbon monoxide level.

The first attempt to correlate oxidation rate with a measured pore-size distribution was made by Labaton *et al.*[35] using Pile Grade A graphite and specially prepared graphite with deliberately modified pore-size distributions. The pore-size distributions were measured before and after oxidation using a high pressure mercury porosimeter. The pore-entrance diameter was assumed to be given by

$$D = \frac{4T\cos(180-\phi)}{P} \tag{6.27}$$

where T is the surface tension (480 dyn cm^{-1}), ϕ is the contact angle (140°) and P the applied pressure. The pressure range covered corresponded to pore entrance diameters of between 60 and 0.014 μm. The gas pressures used were ~5 MPa and the temperatures were in the range 150–270 °C. The dose rates to the gas varied from 0.14 to 0.7 W g^{-1}. The carbon monoxide concentrations were either 0.05 or 0.10%, controlled by the flow rate of gas through the rig. Initially it was attempted to correlate the oxidation rates with the mean accessible porosity during the oxidation for graphites with a range of porosites from 0.05 to 0.20 cm^3 g^{-1}.

A reasonably linear relationship with a correlation coefficient of 0.898 was obtained, but this gave an intercept of zero oxidation rate at about 0.025 cm^3 g^{-1}. A similar plot derived from many successive weight loss measurements on a single Pile Grade A graphite sample taken to high weight loss passed through a maximum at about 0.28 cm^3 g^{-1}, although agreeing reasonably well at lower values with the other data. The effect of a limited range for the diffusion of reactive species was then examined, assuming that this would define an 'operative' open or accessible pore volume. The pores were assumed to be cylindrical, but similar results were found for spherical pores. Two values were assumed for the range of the oxidizing species, 5 and 2.5 μm. The fraction of a pore of radius R_c, length l was taken to be

$$F = \frac{\pi(R_c^2 - L^2)l}{\pi R_c^2 l} = 1 - \left(\frac{L}{R_c}\right)^2 \tag{6.28}$$

ineffective in oxidation. The effect of an arbitrary cut-off dimension below which the pores were completely ineffective was also assessed. The correlation coefficients were improved by all these assumptions, but they differed little. In the same paper,[35] Tait's[26,36] theory was used to calculate the survival probability P of reactive species as a function of

$$x = a/L \tag{6.29}$$

where a is the ratio of pore volume to pore surface area. Expressed in this form, P is very similar for slab, cylindrical and spherical pores, and presumably also similar for any non-re-entrant shape. The apparent G value for the oxidation was taken to be given by

$$G = A \sum p(a)F(a) \tag{6.30}$$

where $p(a)$ was taken from the equation for slab-shaped pores, $F(a)$ is the probability that the ratio of pore volume to pore surface is a and A is a constant independent of pressure and temperature. In the case when $1/L$ is large, $p(a) \approx L/a$ and

$$G = A \sum \frac{L}{a} F(a). \tag{6.31}$$

Dominey and Morley[37] had reported experimental results on the variation of the radiolytic oxidation rate of Pile Grade A graphite with carbon monoxide concentration. The results could be fitted, in the range 0.1–1.0% carbon monoxide, by

$$G = \alpha + \frac{\beta}{1 + \Delta[\mathrm{CO}]} \tag{6.32}$$

with $\alpha + \beta \approx 2.35$ and $\Delta \approx 10$ (this form suggests the presence of *two* oxidizing species, only one of which reacts with carbon monoxide in the gas phase). The results were fitted using two adjustable parameters, A and the relative cross-section for deactivation of species between carbon monoxide and carbon dioxide. The results gave $A \approx 9/100$ eV and a cross-section ratio of ~180. In the experiments taken to high burn-up this implied $a/L \gg 1$, for which equation 6.31 would be valid, and it proved possible to fit the the decreasing value of G with weight loss up to ~20% of weight loss ([CO]≈0.1%). Some conclusions were drawn as to the nature of the two oxidizing species.

A major attempt to correlate the initial oxidation rates of a range of graphites with their pore structure was made by Best and Wood.[27] The experiments were carried out in a gamma irradiation facility using ^{14}C-labelled graphites with a range of very different pore structures. Two of the graphites were AGR moderator materials (Gilso-carbon based), Pile Grade A Magnox reactor graphite and a number of materials with open porosity in a narrow size range. The pore-size distributions were determined by mercury porosimetry. Measurements were made of the production rate of carbon monoxide and the rate of gasification of ^{14}C. Experiments carried out in carbon dioxide containing less than 0.02% carbon monoxide gave G values of ~2.5 for all of the materials, but the G values were different for each at significant levels of carbon monoxide. Results were obtained in three gas mixtures:

Mixture A 1% CO–0.03% H_2O–CO_2

Mixture B 1% CO–0.03% CH_4–0.03% H_2O–CO_2

Mixture C 2% CO–0.015% CH_4–0.03% H_2O–CO_2

An initial correlation for the gas composition with 1% carbon monoxide was obtained between the apparent G value and the pore volume <2 μm entrance diameter. The flux of oxidizing species at a pore surface was calculated as a function of pore size for cylindrically shaped pores for different carbon monoxide concentrations at 673 K and 2.7 MPa pressure. The rate constant K for reaction of the oxidizing species and carbon monoxide was taken to be 2×10^{-12} and the diffusion coefficient D for the species was calculated assuming that it was $(CO_2)_2^+$, using

$$L = (2D/K[CO])^{1/2} \tag{6.33}$$

A value of 0.62 μm was obtained for 1% carbon monoxide concentration. Two features immediately became clear for the experimental conditions explored, first that nearly all of the graphites possessed pore-size distributions in the range where the value of L modified the flux of oxidizing species, and second that the sensitivity to the value of the rate constant K is not large. The effect of methane on the range L is included by assuming its reaction cross-section with the oxidizing species to be eight times that with carbon monoxide, this value being derived from measurements of the rate of destruction of methane in carbon dioxide in a radiation field. This gave values of L for the two methane-containing experiments,

0.03% CH_4–1% CO–0.03% $H_2O \approx 0.54$ μm

0.015% CH_4–2% CO–0.03% $H_2O \approx 0.42$ μm

The estimates of range are ~50–100 times greater than the mean free path of a carbon dioxide molecule under these conditions, and to a good approximation each collision of an oxidizing species with a carbon monoxide molecule leads to deactivation. Defining a total probability of oxidizing species reaching a surface as

$$P = \sum_i P_i V_i \tag{6.34}$$

where V_i is the volume with a particular characteristic dimension corresponding to P_i. The probability p that such a species reacts with the graphite surface rather than a surface protective layer is introduced in

$$G = G_0 P p \tag{6.35}$$

Obviously p decreases from unity in a methane-free, dry gas.

The model can be tested by plotting the observed value of G as the calculated F for all of the graphites in each coolant composition. In each case a linear relationship passing through the origin was obtained (Figure 6.3), but one of the graphites with a narrow pore-size distribution at ~8 μm consistently was much too high and the scatter about the line was greater at high p values. The values of p for the three coolants were 0.61, 0.38 and 0.30, respectively, assuming $G_0 = 2.5$.

A calculation of the effect of gas pressure on the apparent G value for a Gilso-carbon graphite in 1% CO–0.03% H_2O coolant between 1.4 and 4 MPa

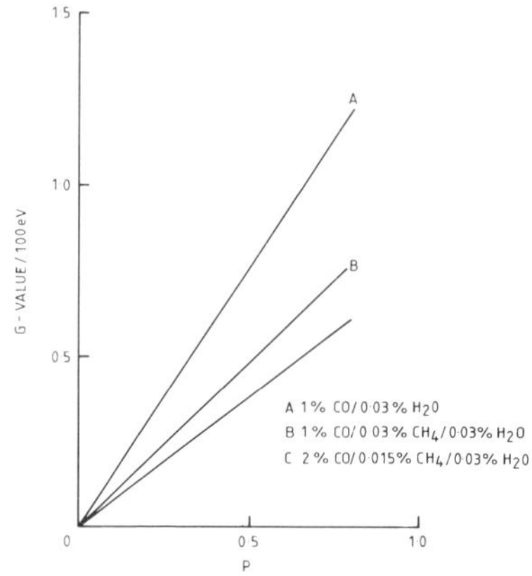

Fig. 6.3 Variation of G value as a function of the pore structure parameter p for three gas compositions (after Best and Wood[27])

gave too small a decrease with increasing pressure compared with experiment, unless a correction was made for a decrease in p due to the effect of partial pressure of carbon monoxide on the water–carbon monoxide surface protection. It was noted, however, that this would not be correct for all graphites, some of which (including Pile Grade A) showed very little or no variation of G with pressure.

This work is clearly a major improvement on previous attempts to correlate with pore structure, but it is of course subject to well-known uncertainties between the true pore-size distribution and that measured by mercury porosimetry.[38]

It is possible to study the changes in dimensions of particular porosity due to radiolytic oxidation by studying the changes in flow of gas through graphite under a variety of conditions. Extensive studies of the flow of gas through graphite using both concentration and pressure differences had demonstrated that the parameters describing these flows were related to the characteristic dimensions of the pores which completely penetrated the graphite.[39] These pores will be referred to as 'transport' pores.

The first parameter with which we are concerned is the diffusivity λ of the graphite evaluated for gas pressures such that the collision mean free paths of the gas molecules are much less than those of any characteristic pore dimension. The diffusive flow of A in B through unit thickness of graphite for unit concentration gradient is related to the free gas diffusion coefficient of A in B, D_{AB}.

$$D = \lambda D_{AB} \tag{6.36}$$

where λ is less than unity. Obviously λ is a measure of the effective cross-sectional area of the transport pores.

The flow of gas through graphite under a pressure drop ΔP is, in the laminar flow regime, described by the Carman equation[39]:

$$\text{Permeability } K = \frac{B_0 \Delta P}{\eta} + \frac{4}{3} K_0 \left(\frac{8RT}{\pi M} \right)^{1/2} \tag{6.37}$$

where η is the gas viscosity, M is the molecular weight, T the temperature (K), and R the gas constant, B_0 the viscous flow coefficient and K_0 the slip flow coefficient. It is found experimentally that for a very wide range of graphites $B_0 \propto K_0^{4/3}$.[39]

These dependences strongly suggest a model of non-interacting cylindrical transport capillaries of density N_T cm^{-3}, radius r_0 and tortuosity q which may be distributed with a probability function $P_T(r_0)$. In such a model,[25]

$$\left.\begin{aligned}
\lambda_0 &= \frac{\pi N_T}{3q^2}\int_0^\infty r_0^2 P_T(r_0)\,dr_0 = \frac{\pi N_T}{3q^2}\langle R^2\rangle \\
K_0 &= \frac{\pi^2 N_T}{32q}\int_0^\infty r_0^3 P_T(r_0)\,dr_0 = \frac{\pi^2 N_T}{32q}\langle R^3\rangle \\
B_0 &= \frac{\pi N_T}{24q}\int_0^\infty r_0^4 P_T(r_0)\,dr_0 = \frac{\pi N_T}{24q}\langle R^4\rangle
\end{aligned}\right\} \tag{6.38}$$

where $\langle R^i\rangle$ is the ith moment of the pore-size distribution, e.g.

$$\langle R^i\rangle = \int_0^\infty r_0^i P_T(r_0)\,dr_0 \tag{6.39}$$

If the gas composition in each pore of this type is uniform and the pores oxidize independently, then after a significant exposure these parameters will be[25]:

$$\left.\begin{aligned}
\lambda &= \frac{\pi N_T}{3q^2}\int_0^\infty r^2 P_T(r_0)\,dr_0 = \frac{\lambda_0}{R_2}\int_0^\infty r^2 P_T(r_0)\,dr_0 \\
K &= \frac{\pi^2 N_T}{32q}\int_0^\infty r^3 P_T(r_0)\,dr_0 = \frac{K_0}{R_3}\int_0^\infty r^3 P_T(r_0)\,dr_0 \\
B &= \frac{\pi N_T}{24q}\int_0^\infty r^4 P_T(r_0)\,dr_0 = \frac{B_0}{R_4}\int_0^\infty r^4 P_T(r_0)\,dr_0
\end{aligned}\right\} \tag{6.40}$$

The first attempt to test these relationships was made by Kelly *et al.*,[40] using experimental data on AGR moderator graphites, obtained in carbon dioxide with $\leqslant$0.1% carbon monoxide in the BR-2 reactor, which included measurements of λ, K and B as a function of energy deposition dose d W h g^{-1}. In equation 6.40, for this case, the results of equation 6.7 are relevant and they reduce to

$$\left.\begin{aligned}
\lambda &= \lambda_0 \exp(d/d_0) \\
K &= K_0 \exp(3d/2d_0) \\
B &= B_0 \exp(2d/d_0)
\end{aligned}\right\} \tag{6.41}$$

These relationships were well-obeyed and gave a value for $d_0 = 3.625 \times 10^3$ W h g^{-1} for a temperature of 673 K and 4 MPa pressure, corresponding to a value of $G_0 = 3.8/100$ eV, remarkably close to the fundamental value of 3. Blanchard[41] carried out a similar analysis of the diffusivity data alone and found that $G_0 \approx 3.2/100$ eV. Considering the assumptions made and the complexity of the pore structure of graphite, this is a remarkably good result.

The model was extended to coolants containing carbon monoxide and methane by Johnson[25] using the pore growth given by equations 6.21 and 6.22. Experimental data were available using the same AGR moderator graphites based on Gilso-carbon graphite in a coolant composition

0.02% CH_4–2% CO–0.03% H_2O–CO_2,[42]

and later

0.01% CH_4–0.5% CO–0.03% H_2O–CO_2

and 0.015% CH_4–0.5% CO–0.03% H_2O–CO_2,[4]

taken to high weight loss.

The initial transport pore-size distribution in this graphite was measured by Rushby[43] using the method of mercury permeametry. In this method the flow of mercury through a graphite sample for a fixed pressure drop (in the laminar flow regime) is measured as a function of mean mercury pressure. Flow only occurs through pores in which the pressure has overcome the surface tension forces. Using the capillary model the viscous flow coefficient is related to the pore radius r_0 by which flow is achieved

$$B_0(r_0) = B_0 - \frac{\pi N_T}{24q} \int_0^{r_0} r_0^4 P_T(r_0)\, dr_0 \tag{6.42}$$

The function $P_T(r_0)$ was found to obey the relationship

$$P_T(r_0) = \frac{\theta}{\phi^\theta} r_0^{\theta-1} \exp\left[-\left(\frac{r_0}{\phi}\right)^\theta \right] \tag{6.43}$$

with $\theta = 3.25$ and $\phi = 3.27$ µm. Substitution of equation 6.43 into equation 6.40 leads to

$$\left.\begin{aligned} \lambda_0 &= \frac{\pi N_T}{3q^2} \phi^2 \Gamma(\langle\theta + 2\rangle/\theta) \\ K_0 &= \frac{\pi^2 N_T}{32q} \phi^3 \Gamma(\langle\theta + 3\rangle/\theta) \\ B_0 &= \frac{\pi N_T}{24q} \phi^4 \Gamma(\langle\theta + 4\rangle/\theta) \end{aligned}\right\} \tag{6.44}$$

where $\Gamma(x)$ is the standard gamma function. The calculated ratio

$$K_0/B_0 = \frac{3\pi}{4\phi} \frac{\Gamma(\langle\theta+3\rangle/\theta)}{\Gamma(\langle\theta+4\rangle/\theta)} \tag{6.45}$$

can be compared with the measured value. Equation 6.45 for this graphite gives 6.2×10^5 m compared with the experimental value of $8.9 \times 10^5 \pm 3.7 \times 10^5$ m. The transport pore volume calculated from

$$V_T = \pi N_T q \phi^2 \Gamma(\langle\theta+2\rangle/\theta) \tag{6.46}$$

was $0.013q$ m^3/m^3 which for $q = 1$ is about 10% of the open pore volume.

The data on changes is λ, K and B with dose obtained in the series of experiments in a 0.02% CH_4–2% CO–0.03% H_2O–0.03% H_2 coolant in the Boucle Franco Britannique (BFB) in the Grenoble reactor was used to estimate the range L at $T = 673$ K and pressure 4.1 MPa. The results could be fitted using $L = 0.45$ μm, which implies that $(D/k_1)^{1/2} = 66.75$ μm (vpm)$^{1/2}$, which was then used to predict ranges in other coolant compositions. Values obtained in this way are given in Table 6.2.

Table 6.2 Diffusion length, L, and composition fraction, f, for various coolant compositions[a]

[CO] %	[CH_4] %	f	L μm
2	0.02	0.091	0.45
1	0	0	0.67
2	0.01	0.048	0.46
1	0.02	0.167	0.61
0.5	0.02	0.286	0.80
1	0.07	0.412	0.51

[a]The value of L at different temperatures and pressures is given by

$$L = \frac{66.75 \times 10^{-4}}{([\mathrm{CO}] + 10[\mathrm{CH_4}])^{1/2}} \left(\frac{41}{P}\right)\left(\frac{T}{673}\right)^{(m+1)/2}$$

where concentrations [ ] are in vpm and $m = 1.75$.

A particularly interesting prediction is that the viscous flow coefficients in the 0.07% CH_4–1%–CO–CO_2 gas composition are predicted to decrease. Although no data were available on this graphite, a result such as this had been obtained on a French nuclear graphite in an experiment in the BR-2 reactor, at least demonstrating that such a result was possible in practice.

These values of L could then be used to predict values of the gas transport parameters as a function of dose in this coolant. The comparison of the model with the data obtained in 'pure' CO_2 (BR-2 experiments) and in 0.02% CH_4–2%–CO–CO_2 (BFB experiments) are shown in Figure 6.4 and the predictions for other coolants are shown in Figure 6.5.

It is interesting that the distribution function 6.43 gives different mean values for transport pore radius from the different moments of the distribution, e.g.

$$\langle R_0^n \rangle^{1/n} = \left\langle \int_0^\infty R_0^n P_T(R_0)\,\mathrm{d}R_0 \right\rangle^{1/n} \tag{6.47}$$

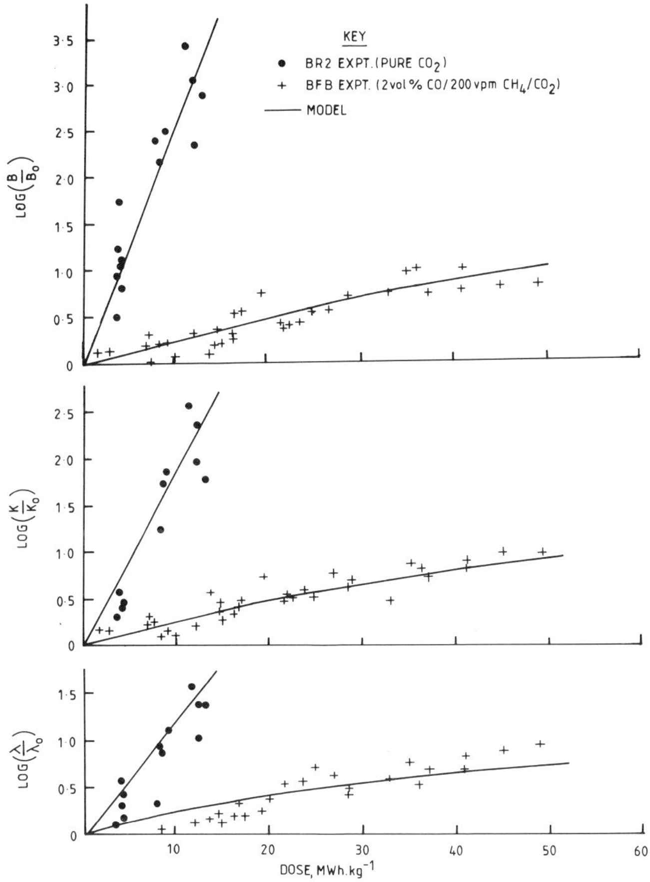

Fig. 6.4 Model curves and experimental data for moderator graphite gas transport properties

The particular values of $\langle R_0^n \rangle^{1/n}$ for this Gilso-carbon graphite are: $\langle R_0 \rangle = 2.13\ \mu m$, $\langle R_0^2 \rangle^{1/2} = 2.48\ \mu m$, $\langle R_0^3 \rangle^{1/3} = 2.88\ \mu m$ and $\langle R_0^4 \rangle^{1/4} = 3.36\ \mu m$, and it is noticeable that these values are considerably greater than the values of the range L shown in Table 6.2. This suggests that equation 6.24 could be applied to the transport pore systems in a unipore model, using for each property the appropriate value of $\langle R_0^n \rangle^{1/n}$, e.g.

$$\frac{\lambda}{\lambda_0} = \frac{1}{\langle R_0^2 \rangle} \left[\langle R_0^2 \rangle^{1/2} + \left(\frac{L}{d_0} \right) d \right]^2 \tag{6.48a}$$

in the absence of CH_4 or

$$\frac{\lambda}{\lambda_0} = \frac{1}{\langle R_0^2 \rangle} \left[R_M + (\langle R_0^2 \rangle^{1/2} - R_M) \exp\left(-\frac{fd}{2d_0} \right) \right]^2 \tag{6.48b}$$

in the presence of CH_4, with similar equations for B/B_0 and K/K_0.

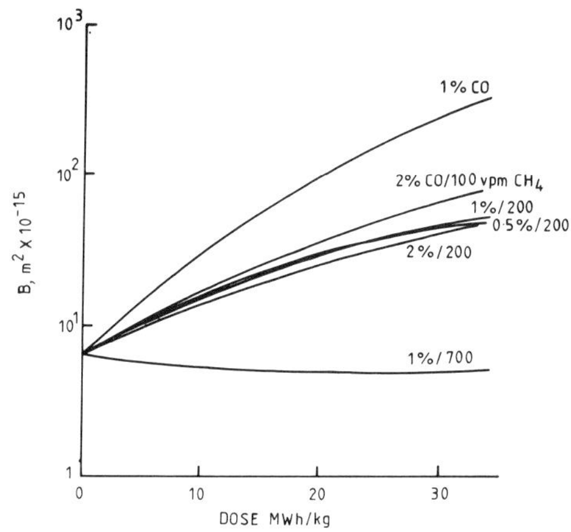

Fig. 6.5 Predicted changes of the viscous flow coefficient (B) with dose of a variety of coolant compositions

Figure 6.5 shows a comparison of these equations with calculations carried out using the full pore-size distribution for the 0.02% CH_4–2% CO–CO_2 gas composition in the BFB series of experiments and the agreement is found to be good. Johnson[44] has shown that these results are exact if $R_0 \gg L$. The pore-size distributions in an oxidized graphite sample measured using mercury permeametry have been shown to agree well with prediction in this gas composition.[45]

A simplified method of estimating changes in gas transport properties with oxidation has also been proposed.[46] It is possible to calculate three separate measures of the transport pore radius R_0 given by

$$\left.\begin{aligned} R_0 &= \left(\frac{3\pi}{4}\frac{B_0}{K_0}\right) \\ R_0 &= \left(\frac{32}{3\pi}\frac{K_0}{\lambda_0}\right) \\ R_0 &= \left(\frac{8B_0}{\lambda_0}\right)^{1/2} \end{aligned}\right\} \tag{6.49}$$

which would coincide for a unipore graphite. The average value is

$$R_0 = \frac{1}{3}\left[\frac{3\pi}{4}\frac{B_0}{K_0} + \frac{32}{3\pi}\frac{K_0}{\lambda_0} + \left(\frac{8B_0}{\lambda_0}\right)^{1/2}\right] \tag{6.50}$$

Equation 6.21 gives the pore radius as a function of dose, but for small changes in R, the efficiency factor p can be taken as constant and

$$R = R_0 \exp[\langle p(1+f) - f\rangle d/2d_0] \tag{6.51}$$

$$= R_0 \exp(\alpha d/2d_0) \text{ where } \alpha \text{ is a constant.}$$

The gas transport properties can now be written as

$$\left.\begin{aligned}\lambda &= \lambda_0 \Pi_i \exp(\alpha_i \Delta d_i / d_0)\\ K &= K_0 \Pi_i \exp[\alpha_i^{3/2} \Delta d_i / d_0]\\ B &= B_0 \Pi_i \exp[2\alpha_i \Delta d_i / d_0]\end{aligned}\right\} \quad (6.52)$$

for small dose increments Δd_i. The value of p can be parameterized in terms of L/R,[47] and the best fit to date is obtained using a value of L twice that obtained from the analysis of the BR-2 and BFB experiments, e.g.

$$L = \left\{\frac{123 \times 10^{-4}}{([\mathrm{CO}] + 10[\mathrm{CH_4}])^{1/2}}\right\}\left(\frac{41}{P}\right)\left(\frac{T}{673}\right)^{(m+1)/2} \quad (6.53)$$

This method was tested for two graphites which approximate to unipore materials (HDG and Isograph 50), which have values of R_0 of 1.18 and 1.03 μm, respectively, and which were exposed in 0.015% CH_4–0.5% CO–CO_2 to $\sim 10^4$ W h g^{-1}. The initial viscous flow coefficients were $\sim 2 \times 10^{-18}$ and 2×10^{-16} m^2. These pore sizes are such that $\alpha = 0.90$. A good fit is obtained for $d_0 = 4.14 \times 10^3$ W h g^{-1} to both, although $d_0 = 3.625 \times 10^3$ W h g^{-1} predicts changes only slightly too large. The viscous flow coefficient changes by ~100 in each case. This approximate method gives less rapid changes with dose than the more exact methods using a pore-size distribution, but gives very similar results over dose ranges of interest.[47]

A major advantage of the study of the gas transport properties is that by exposing the specimens to radiation in the presence of a small through-thickness pressure drop it is possible to ensure that the gas composition in the transport pores is the same as those in the free gas, thus making reasonable analysis possible. A similar situation applies to measuring initial oxidation rates with very low rates of energy deposition so that decomposition is small. The major problem with the study of weight-loss rate is the accurate evaluation of the appropriate open pore-size distribution. The use of gold impregnation followed by oxidation to remove the carbon[47] had shown how complex the open-pore structures of graphite really are compared with those deduced from conventional mercury porosimetry, which assumes cylindrical non-interacting pore systems. The inability of the porosimeter to model even simple pore structures has been well documented (see Dullien and Dhawan[48] and Androtsoupoulos and Manon,[49] for example). In the light of these problems and earlier work, Stephen[50] examined the use of image analysis to obtain the appropriate pore-size distribution. Samples of a number of different graphites with very different pore-size distributions were impregnated with various metals and a fluorescent compound and also evaluated using mercury porosimetry. The low melting-point alloy of bismuth known as Wood's metal was used principally together with the epoxy resin Lemix containing the fluorescent dye Brilliant Yellow R.

It was found that the resin filled the open-pore volume using only atmospheric pressure on thoroughly vacuum-baked, degassed graphite. The samples were polished to remove the surface layer where all porosity is open and filled until a representative open-pore volume was obtained. The samples were then examined using a Leitz Ortholux microscope fitted with a fluorescent illuminator giving near-ultraviolet illumination producing a green fluorescence. It is necessary to use water immersion lenses rather than oil to prevent dissolution of the resin impregnant. The images were taken directly using a small television camera and analysed with an image analyser. The pore structures were analysed in two separate ways: by measurement of the area/perimeter ratios, correcting them appropriately to take account of the two-dimensional nature of the observations (*see* Appendix), and by measurement of the area within a fixed distance of the pore edge.

In the first case, by counting and summing over a large number of pores, cumulative distributions are obtained equivalent to two-dimensional volume distributions. Best *et al.*[13] gave further details of these measurements and compared pore-size distributions obtained in different ways for several graphites. Using the image analysis data and oxidation rates taken in about ten carbon monoxide concentrations up to 3.6%, the values of p can be calculated as before and compared with the measured G value. The agreement, shown in Figure 6.6, between the calculated and predicted relationships is very good, although some of the graphites tend to lie consistently above the predicted line. Because the pore-size distributions tend to give an increase in the volume of large pores, a new value of the rate constant $K = 3 \times 10^{-13}\ \mathrm{cm^3\ s^{-1}}$ was used. Using $D = 1.09 \times 10^{-2}\ \mathrm{cm^2\ s^{-1}}$ at 2.9 MPa and 400°C gives a mean range in 1% carbon monoxide of ~2.4 μm, corresponding to a value of 1.58 μm at 4.1 MPa pressure, about twice the value obtained from the analysis of transport properties.

The second method, called image erosion, relates the efficiency p to the number of reactive species created within a distance L of the pore wall. If each pore is eroded by a distance L, the fractional change in pore area is, by definition, equal to the two-dimensional efficiency p^1 (*see* Appendix), which can be corrected with appropriate geometric factors to p. A value of the reaction constant K was assumed as before and it was then possible again to plot the

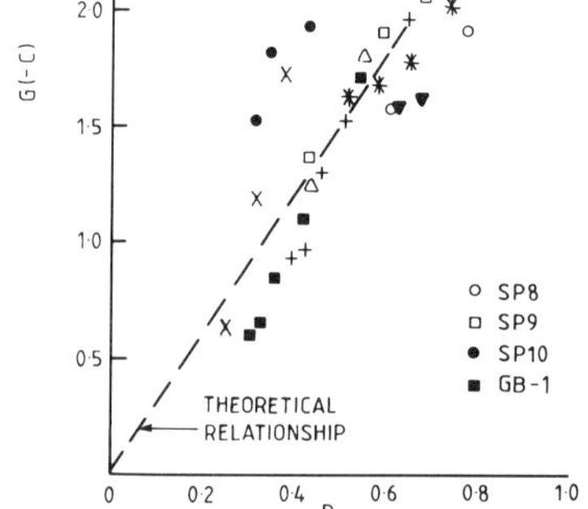

Fig. 6.6 $G(-C)$ as a function of p (after Best *et al.*[13])

measured G against the deduced p (this time for five graphites). Again, a linear relationship was found in which better behaviour was observed for at least one of the graphites tending not to fit the image analysis technique.

Given the relatively good agreement obtained between the indirect studies of pore-size changes using gaseous and liquid flows through graphite and direct measurements of pore sizes and their relation to oxidation rates, it is an obvious extension to try to predict the weight loss of graphites at high doses in addition to their other properties.

Experimental data for high doses and weight losses have been obtained in a number of gas compositions using various reactor facilities. These are as follows:

~0.1% $CO–CO_2$	BR-2 experiments
0.02% CH_4–2% $CO–CO_2$	BFB experiments
0.01% CH_4–0.5% $CO–CO_2$	DIDO–PLUTO experiments
0.015% CH_4–0.5% $CO–CO_2$	

and the results for the Gilso-carbon graphite described above are shown in Figure 6.7.

It was found in each of the high-dose experiments for this graphite that the weight loss can be represented empirically by

$$\left.\begin{aligned} x &= d_1 (\mathrm{d}x/\mathrm{d}d)_0 [\exp(d/d_1) - 1] \qquad &\text{for } d \leqslant 1.1 d_1 \\ x &= d_1 (\mathrm{d}x/\mathrm{d}d)_0 \left[3\left(\frac{d}{d_1}\right) - 1.296 \right] \qquad &\text{for } (d \geqslant 1.1 d_1) \end{aligned}\right\} \tag{6.54}$$

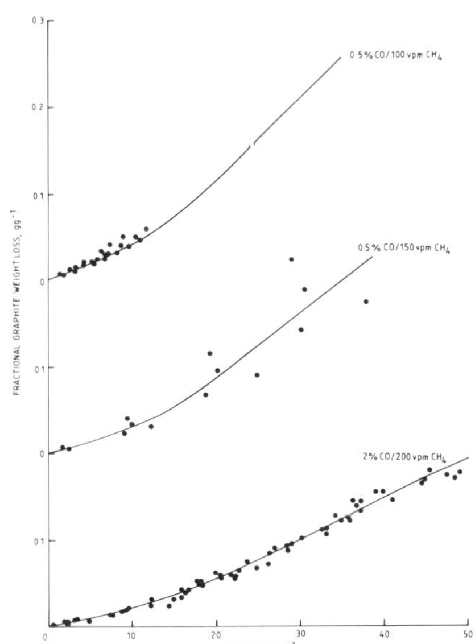

Fig. 6.7 Gilso-carbon graphite weight loss data with model calculations

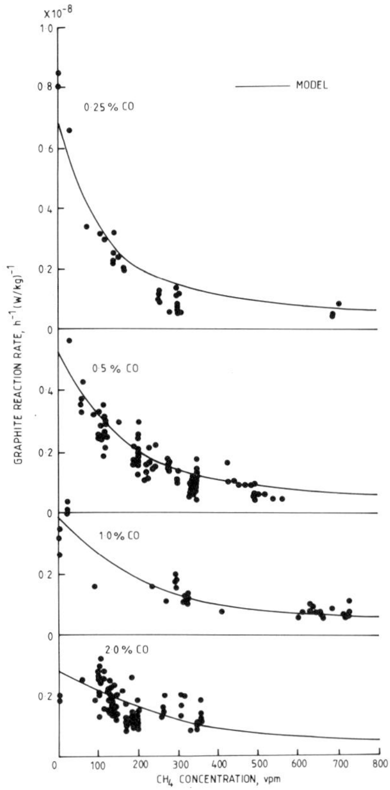

Fig. 6.8 Graphite reaction rate vs methane concentration with model calculations

where $(dx/dd)_0$ is the rate of initial weight loss (conventionally measured for doses $<2\times10^3$ W h g^{-1}) and d_1 is a dose constant with value 1.86×10^4 W h g^{-1} at 673 K and 4.1 MPa pressure. The inital oxidation rates were derived from a large number of experiments and are shown in Figure 6.8.

Given the agreement between calculated and predicted initial oxidation rates and the behaviour of the transport pores for simple oxidation models, the question posed is whether the behaviour represented by equation 6.54 can be explained.

Johnson[28] re-examined the theory for pore growth and derived the relationships for weight loss

$$x = (\rho_c/\rho_B)(V = V_0) \tag{6.55}$$

(note use of ρ_c not ρ_{He} in this work), where V is the open-pore volume after dose d, assuming that no breakthrough into closed porosity occurs. This volume can be expressed as

$$V = V_0\langle R^n\rangle/\langle R_0^n\rangle \tag{6.56}$$

where the $\langle R^n\rangle$ are the moments of the pore-size distribution, so that

$$x = (\rho_c/\rho_B)\,[\langle R^n\rangle/\langle R_0^n\rangle - 1]\,V_0 \tag{6.57}$$

Defining $\mu(n) = \langle R^n\rangle/\langle R_0^n\rangle$ it is readily shown that:

$$\left.\begin{array}{ll} \text{Diffusivity} & \lambda = \lambda_0\mu(n) \\ \text{Slip flow coefficient} & K = K_0\mu(n+1) \\ \text{Viscous flow coefficient} & B = B_0\mu(n+2) \end{array}\right\} \tag{6.58}$$

and

$$\frac{\mathrm{d}x}{\mathrm{d}d} = \left(\frac{\rho_c}{\rho_B}\right)\frac{\mathrm{d}\mu(n)}{\mathrm{d}d}$$

Johnson[28] obtained a differential equation for the moment ratios $\mu(n)$

$$\frac{\mathrm{d}\mu(n)}{\mathrm{d}d} - \frac{2nL}{\nu d_0}(1+f)\,\mu(n-1)\,\frac{\langle R_0^{n-1}\rangle}{\langle R_0^n\rangle}\,y_\nu\left[\frac{\mu(n)}{\mu(n-1)}\,\frac{\langle R_0^n\rangle}{\langle R_0^{n-1}\rangle}\right] + \frac{nf}{\nu d_0}\,\mu(n) = 0 \tag{6.59}$$

when $y_1(r) = \tanh(r/2L)$ and $y_2 = I_1(r/L)/I_0(r/L)$, which can be solved numerically for $n = 1, 2, 3, \ldots$. The behaviour for small pore sizes is as expected, e.g. the equation tends to

$$\frac{\mathrm{d}\mu(n)}{\mathrm{d}d} = \frac{n}{\nu d_0}\,\mu(n) \tag{6.60}$$

from which

$$\left.\begin{array}{l} x = (\rho_c/\rho_B)\,V_0[\exp(d/d_0) - 1] \\ \dfrac{\mathrm{d}x}{\mathrm{d}d} = (\rho_c/\rho_B)\,\dfrac{V_0}{d_0}\exp(d/d_0) \\ \lambda = \lambda_0\exp(d/d_0) \\ K = K_0\exp\left(\dfrac{\nu+1}{\nu}\,\dfrac{d}{d_0}\right) \\ B = B_0\exp\left(\dfrac{\nu+2}{\nu}\,\dfrac{d}{d_0}\right) \end{array}\right\} \tag{6.61}$$

as shown already. When the pore dimensions are large compared with the range L, equation 6.59 becomes

$$\frac{\mathrm{d}\mu(n)}{\mathrm{d}d} = \frac{2nL(1+f)}{\nu d_0}\,\mu(n-1)\,\frac{\langle R_0^{n-1}\rangle}{\langle R_0^n\rangle} - \frac{nf}{\nu d_0}\,\mu(n) \tag{6.62}$$

which gives, provided that the deposition term is small,

$$\left(\frac{\mathrm{d}x}{\mathrm{d}d}\right)_0 = (\rho_c/\rho_B)V_0\frac{2L}{d_0}(1+f)\frac{\langle R_0^{\nu-1}\rangle}{\langle R_0^{\nu}\rangle} - \frac{f}{d_0} \tag{6.63}$$

It has already been noted that there is a value of the characteristic pore dimension R for which oxidation and deposition balance and hence has stable dimensions, say R_T. This can be taken to be

$$R_T \approx 2L(1+f)/f \tag{6.64}$$

The experimental weight-loss data are corrected for the presence of deposit, which is measured using the method of differential thermal oxidation. Hence the experimental data for initial rate of weight loss should be compared with

$$\left(\frac{\mathrm{d}x}{\mathrm{d}d}\right)_0 = (\rho_c/\rho_B)\frac{V_0}{R_0(\nu)}\int_0^{R_T} R_0^{\nu} P\,\mathrm{d}R_0 \tag{6.65}$$

Johnson[28] then defined incomplete moments of the pore-size distributions of the form

$$m(n,\alpha) = \int_0^{\alpha} R^n P\,\mathrm{d}R \tag{6.66}$$

Defining

$$\mu(n,\alpha) = m(n,\alpha)/R_0(n) \tag{6.67}$$

then it can be shown that

$$\frac{\mathrm{d}\mu(n,R_T)}{\mathrm{d}d} - \frac{2nL}{\nu d_0}(1+f)\mu(n-1,R_T)\frac{\langle R_0^{n-1}\rangle}{\langle R_0^{n}\rangle}y_{\nu}$$

$$\left[\frac{\mu(n,R_T)}{\mu(n-1,R_T)} - \frac{\langle R_0^{n}\rangle}{\langle R_0^{n-1}\rangle}\right] + \frac{nf\mu(n,R_T)}{\nu d_0} = 0 \tag{6.68}$$

which can be solved numerically. The initial oxidation rate in this form is

$$\left(\frac{\mathrm{d}x}{\mathrm{d}d}\right)_0 = (\rho_c/\rho_B)V_0\frac{\mathrm{d}\mu_0}{\mathrm{d}d}(\nu,R_T) \tag{6.69}$$

If different groups of pores contribute, each describable by a distribution function, their contributions were taken to be additive, each obeying its own equation.

As we have seen already, most information of this kind can be derived from measurements of the gas transport properties λ_0, K_0 and B_0. The value of $\langle R_0^4\rangle$ for the Gilso-carbon graphite was found to be 128.14 $\mu\mathrm{m}^4$ using mercury permeametry. Taking $B_0 = 6.5\times10^{-15}\,\mathrm{m}^2$ and $\lambda_0 = 2.5\times10^{-3}$ for the same graphite, equation 6.38 gives $\langle R_0^2\rangle = 6.16\,\mu\mathrm{m}^2$. Assuming that the pore-size distribution is log-normal leads to

$$\langle R_0^n \rangle = \eta^n \exp\left(\frac{n^2\beta^2}{2}\right) \tag{6.70}$$

where η and β are parameters defining the distribution. The results obtained for $\langle R_0^2 \rangle$ and $\langle R_0^4 \rangle$ give $n = 1.832$ and $\beta = 0.552$, respectively. These values give $\langle R_0^4 \rangle = 128.14\ \mu m^4$, $\langle R_0^3 \rangle = 24.13\ \mu m^3$, $\langle R_0^2 \rangle = 6.16\ \mu m^2$ and $\langle R_0 \rangle = 2.13\ \mu m$. Equation 6.68 was solved for these values by adjusting the range L until the experimental data for the viscous flow coefficient, B/B_0, obtained in the 0.2% CH_4–2% CO–CO_2 experiments were accurately represented (actually for $L = 0.45\ \mu m$). This gave for the values of L at 673 K and 4.1 MPa pressure for varying composition

$$L = 0.45\left(\frac{2.2}{[CO] + 10[CH_4]}\right)^{1/2}\ \mu m. \tag{6.71}$$

Assuming an initial transport pore volume

$$V_{T0} = 3q^2\lambda_0 \tag{6.72}$$

gives a value of 0.015 $cm^3\ cm^{-3}$ for $q = 2^{1/2}$, and this can then be used to estimate the initial rate of weight loss from these pores and the variation with dose. The contribution to total weight loss in the 0.02% CH_4–2% CO–CO_2 coolant is shown in Figure 6.9 and to initial oxidation rates in Figure 6.10). The transport porosity alone does not account for all of the oxidation in either case. An obvious possibility to explain these differences is the presence of a coarse pore group. The presence of a group of very small pores which are highly reactive is also suggested by the rapid change in early oxidation rates in the 0.02% CH_4–2% CO–CO_2 gas composition. If this is the case, the initial oxidation rate can be represented by

$$R = R_{T0} + R_{L0} + R_{S0}\exp(d/d_0) \tag{6.73}$$

Equation 6.73 can be solved for $d = 0$ and 10×10^3 W h g^{-1} and yield $R_{30} = 4.8 \times 10^{-11}$ and $R_{L0} = 4.2 \times 10^{-10}$ $(\text{mW h g}^{-1})^{-1}$. In the absence of methane $f = 0$ and the initial oxidation rates are

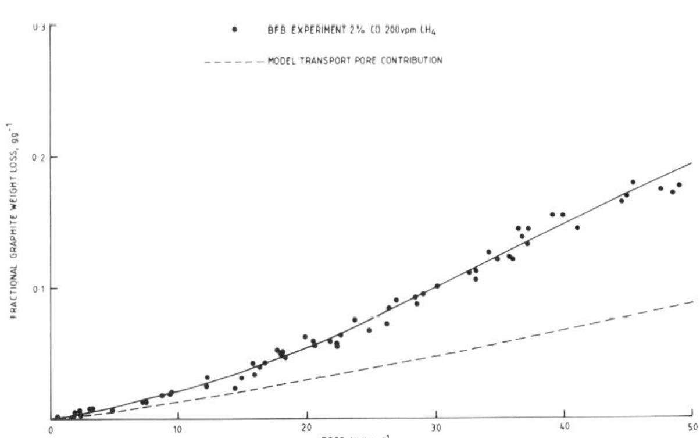

Fig. 6.9 Experimental weight loss data for 0.02% CO–2% CH_4 coolant and calculated transport pore contribution

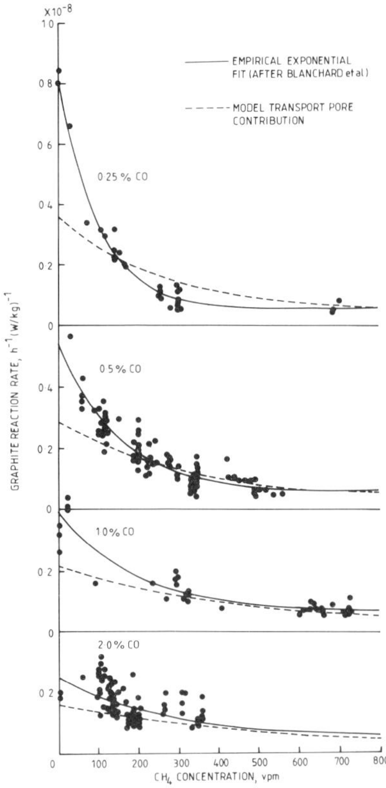

Fig. 6.10 Graphite reaction rate vs methane concentration at 4.1 MPa with empirical fit and transport pore contribution

$$\left(\frac{\mathrm{d}x}{\mathrm{d}d}\right)_0 = \left(\frac{\rho_c}{\rho_B}\right)\frac{2L}{d_0}\,V_0\,\frac{\langle R_0\rangle}{\langle R_0^2\rangle} \tag{6.74}$$

Using the initial oxidation rate curves shown in Figure 6.8, the quantity $V_0\langle R_0\rangle/\langle R_0^2\rangle$ can be calculated. The initial rates of the large pore group are obtained by subtracting the transport pore contribution and the fine pore contribution estimated from the 0.02% CH_4–2% CO–CO_2 experiments and which should be independent of composition. The value of $V_0\langle R_0\rangle/\langle R_0^2\rangle$ varies from 3×10^{-3} to 5×10^{-3} μm^{-1} with decreasing carbon monoxide concentration, with a mean value of 4.1×10^{-3} μm^{-1}. The initial rate for the large pore group in the 0.02% CH_4–2% CO–CO_2 gas composition was found to be 0.042×10^{-8} g g^{-1} (MW h g^{-1})$^{-1}$ and the values of V_0 and $\langle R_0^2\rangle/\langle R_0\rangle$ can be separated using an improvement on equation 6.74 for cylindrical pores, e.g.

$$\left(\frac{\mathrm{d}x}{\mathrm{d}d}\right)_0 = \left(\frac{\rho_c}{\rho_B}\right)V_0\left[\frac{(1+f)}{d_0}\left(2L\frac{\langle R_0\rangle}{\langle R_0^2\rangle} - L^2\frac{\langle R_0\rangle^2}{\langle R_0^2\rangle^2}\right) - \frac{f}{d_0}\right] \tag{6.75}$$

and were found to be 0.029 cm^3 cm^{-3} and 7.16 μm, respectively. Using the relationship

$$\langle R_0^2 \rangle = \langle R_0 \rangle^2 + \sigma^2 \tag{6.76}$$

where σ is the standard deviation of the pore-size distribution. The value of σ was adjusted until the initial rate curve for 0.5% carbon monoxide was fitted. The best value was found to be $\sigma = 2.0\ \mu m$, which in turn led to $\langle R_0 \rangle = 6.55\ \mu m$. Subtraction of both the transport pore contribution and the large pore group contribution from the data for the 0.02% CH_4–2% CO–CO_2 experiment leaves the contribution from the fine pore group; this is found to increase very rapidly from a negligible value to a constant rate at $\sim 20 \times 10^3$ W h g^{-1}. This result can only be fitted using slab-shaped pores which do not contain any methane. The initial value of this pore volume is found to be 1.61×10^{-4} cm^3 cm^{-3}, with an initial width of 0.0231 μm (the carbon monoxide level was assumed to be 2%).

Given the parameters for the three pore groups, it is now possible to calculate the initial rate of weight loss as a function of methane concentration for fixed carbon monoxide levels and the dose dependence of weight loss for the other coolants that have been studied. The fit in each case (Figure 6.7 and 6.8) is found to be good. However, the model presents one very obvious difficulty, namely that the total initial open-pore volume required to explain the results is 4.5×10^{-2} cm^3 cm^{-3} whereas the measured initial open-pore volume was 10×10^{-2} cm^3 cm^{-3}, i.e. about twice as large. It is not clear why half of the pore structures should be unreactive even in gas compositions with low carbon monoxide levels.

Best *et al.*[22] took a different approach to explain the same results. The studies of graphite pore structure already described were extended to samples with a significant weight loss and it was found that initially closed porosity progressively became open to the impregnant as weight loss increased. In order to model the high-dose weight loss in the 0.02% CH_4–2% CO–CO_2 gas composition, Best *et al.*[22] took the initial oxidation rate (measured for weight losses up to 1%) and then introduced a closed-pore volume of 3×10^{-2} cm^3 cm^{-3} linearly over the first 10% weight loss. The pores introduced were all assumed to be small and hence highly reactive. All pores, closed and open, were assumed to be slab-shaped. Good agreement was found up to 4.2×10^3 W h g^{-1}, but for high doses the model overestimated the weight loss. In a real porous structure it is clear that initially separate pores must eventually interact by interconnection of blind pores and the coalescence of adjacent open pores. Both of these effects will reduce the overall chemical reactivity of the graphite with increasing weight loss. For the first case, interconnection of blind pores improves the access of methane, and in the second case, coalescence of adjacent small pores yields a pore of effectively larger diameter and hence lower reactivity.

It is clearly difficult to model such effects and they must be expected to differ from one graphite to another. According to Best *et al.*,[13] given the assumptions that pores begin to coalesce when they have doubled their original width, and that coalescence results in the pore width increasing at twice the normal rate (but retaining the same pore volume by reducing the pore length), the experimental results can be modelled using the additional assumption that the gas composition is uniform throughout the specimen.

This last assumption is common to both analyses of the data (except for the fine pores in the Johnson analysis) described here and it raises a question about differences which might occur in the gas composition between blind pores and transport pores. In the experiments taken to high weight loss a pressure drop is established across a specimen such that the gas composition in the transport pores is essentially that in the external gas, but in blind pores the methane concentration would be expected to decrease along the pore and the carbon monoxide concentration to increase owing to the increasing radiolytic reaction. The depletion effects will be greater in experiments carried out at high dose rates. This may explain why some pores are non-reactive in high weight loss rate experiments and justify the Johnson approach. In reality this situation is probably too complicated to model accurately.

6.4 Discussion

A complete theory of the relationship between the pore structure of graphite and its oxidation rate in carbon dioxide due to the presence of radiation, with appropriate oxidation inhibitors, requires four separate elements:

1. a description of the creation of oxidizing species in the gas phase and their reactions in the gas phase;
2. a calculation of the flux of oxidizing and depositing species to the pore walls as a function of pore shape and size;
3. a calculation of the local gas composition in each pore and its changes with time;
4. a calculation using 1–3 of the weight loss from the material in total, and also from specific porosity such as the transport pores which control local gas concentration.

Both of the models described above simplify the creation of oxidizing species enormously and simply use a G value of about 3/100 eV. Both models consider gas-phase reactions of the oxidizing species and the estimated ranges are within a factor of about two, which is remarkably good considering the entirely different routes to their estimation. The models then differ substantially in that the Johnson model allows for a depositing species created by reaction of the oxidizing species with methane whereas the model developed by Best and co-workers represents this by an empirical function of methane concentration. It is possible using the Johnson model and knowledge of the pore-size distribution to calculate the amount of deposit to be expected, since all pores of dimensions greater than W_M or R_M show deposition exceeding oxidation and tend towards these dimensions with increasing dose. The amount of deposit present in a sample can be measured, it is believed, using the method of differential thermal oxidation. In a standardized oxidation experiment at about 400°C, the presence of deposit shows as an initial high rate of weight loss and back-extrapolation of the oxidation rates at long times to zero time permits an estimate of the total amount of deposit. Measurements of the initial deposition rate have been made in a number of gas compositions and show a fairly good correlation with the parameter f (see Figure 6.11). The changes of deposit level with dose were measured in two graphites, the French Pechiney moderator graphite and the UK Gilso-carbon graphite. The deposition increased slowly with dose and was about twice as large in the Pechiney graphite, which has about twice the open-pore

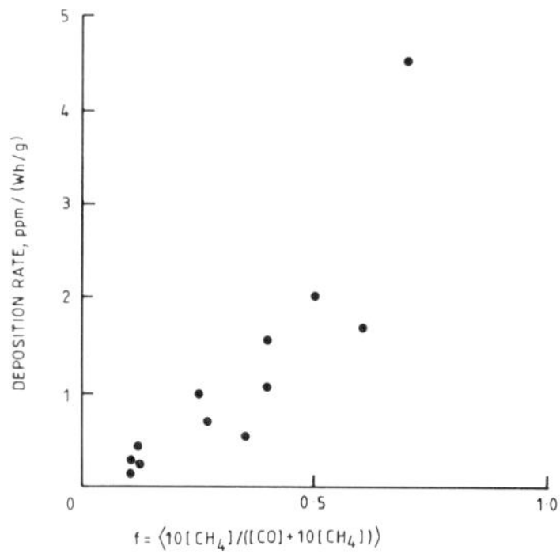

Fig. 6.11 Initial deposition rate as a function of f for Gilso-carbon graphite

volume. The gas composition was 0.02% CH_4–2% CO–CO_2. There is a large scatter on this type of observation.

Consider a pore of volume V and surface area A, then the flux of oxidizing species to the pore wall is

$$J_{OX} = \frac{V}{A}\,\rho_g E\,\frac{G_0}{100}\,p\ \mathrm{cm^{-2}\,s^{-1}} \tag{6.77}$$

where p is the 'efficiency factor' as before. If the ratio $V/A = X$, then it is readily shown that for $X > \sim 2$, $p(X)$ is virtually independent of pore shape and equal to

$$p(x) = L/X. \tag{6.78}$$

In the same way, the deposition flux is

$$J_d = \frac{V}{A}\,\rho_g E\,\frac{G_0}{100}\,(1-p)f\ \mathrm{cm^{-2}\,s^{-1}}. \tag{6.79}$$

Obviously, as noted before, pores for which

$$p(X_M) = f/(1+f) \tag{6.80}$$

do not change dimensions and correspond to W_M and R_M for the slab and cylindrical cases, respectively.

The deposition rate in a pore of volume V and surface area A referred to unit dose and unit volume is

$$\frac{dD}{dd} = A\left(\frac{V}{A}\right)\frac{m_c}{\rho_B}\,\rho_g\,\frac{G_0}{100}\,[f-p(1+f)] \tag{6.81}$$

$$= \nu\left(\frac{\rho_c}{\rho_B}\right)\frac{1}{d_0}\,[f-p(1+f)]. \tag{6.82}$$

If the fraction of an initial pore volume $V_0\ \mathrm{cm^3\,cm^{-3}}$ with X_0 in the range X_0 to $X_0 + dX_0$ is $P(X_0)\,dX_0$, the initial deposition rate is

$$\left(\frac{\mathrm{d}D}{\mathrm{d}d}\right)_0 = f\left(\frac{\rho_c}{\rho_B}\right)\frac{V_0}{d_0}\int_{X_M}^{\infty}\left(1-p(X_0)\frac{(1+f)}{f}\right)P(X_0)\,\mathrm{d}X_0$$

$$= f(\rho_c/\rho_B)\frac{V_0}{d_0}\int_{X_M}^{\infty}\left(1-\frac{p(X)}{p(X_M)}\right)P(X_0)\,\mathrm{d}X_0. \qquad (6.83)$$

The values of $X > X_M$ are such that $p(X_0)/p(X_M) \approx X_M/X_0 < 1$. For many of the pores $X_0 \gg X_M$ and it is a good approximation to take

$$\left(\frac{\mathrm{d}D}{\mathrm{d}d}\right)_0 = f\left(\frac{\rho_c}{\rho_B}\right)\frac{V_0}{d_0}\int_0^{\infty} P(X_0)\,\mathrm{d}X_0 = f\left(\frac{\rho_c}{\rho_B}\right)\frac{V_0}{d_0} \qquad (6.84)$$

proportional to f as required, and for the two coarse-grained graphites studied, also to V_0 as observed.

If the pore-size distribution for sizes above X_M is such that the oxidation flux can be neglected, then the initial deposition rate is obviously given by

$$\left(\frac{\mathrm{d}D}{\mathrm{d}d}\right)_0 = f(V_0 - V_M)\left(\frac{\rho_c}{\rho_B}\right)\frac{1}{d_0} \qquad (6.85)$$

where V_M is the pore volume below the critical dimensions W_M, R_M or X_M. For slab-shaped pores, equation 6.23 shows that the volume changes with dose as

$$V = V_0 \exp(-fd/d_0) \qquad (6.86)$$

for $W_0 \gg W_M$, the depositing situation. This gives, until the width approaches W_M, deposit

$$D = \left(\frac{\rho_c}{\rho_B}\right) V_0[1 - \exp(-fd/d_0)] \qquad (6.87)$$

in agreement, because of the value of f in the exponent, with the slow accumulation of deposit with dose.

An attempt was made to test this theory by examining the basal surfaces of highly oriented pyrolytic graphite exposed to a 0.015% CH_4–0.5% CO–CO_2 gas composition, at 4 MPa pressure and ~673 K, in the form of stacks with basal planes in contact in adjacent samples. The perfection of the pyrolytic samples was varied by using different final heat-treatment temperatures between 2200 and 3200°C.[51] All of the samples showed deposit growth on exposed surfaces of basal planes. The structures of the deposits included stalk-like material, lacy dendrites and growth pyramids of definite orientation. The samples showed net weight losses of only ~1% at 10×10^3 W h g^{-1} compared with losses of tens of percent for the polycrystalline graphites in the same experiment. Careful examination of the samples by scanning electron microscopy showed that little deposit was seen on adjoining basal surfaces separated by widths less than 20 μm, whereas deposition was copious for gap widths greater than 30 μm or on free surfaces.

The fluxes of depositing and oxidizing species within a gap between two basal surfaces were analysed using the Johnson model with the additional assumption that the depositing species could be destroyed with a reaction mean free path L_p. In a slab pore the width W_M is defined by

$$\tanh(W_M/2L) = \frac{f}{1+f}\frac{W_M}{2L} \tag{6.88}$$

for an infinite range L_p. If L_p is finite this is replaced by[52]

$$[(1+f)L_p^2 - L_x^2]\tanh(W_M/2L) - \frac{fL_p^3}{L}\tanh(W_M/2L_p) = 0. \tag{6.89}$$

In the 0.015% CH_4–0.05% CO–CO_2 gas composition $L = 0.83$ μm, $f = 0.231$ and $W_M = 8.3$ μm. If this is correct, then deposit would be visible if the gap between the pyrographite surfaces exceeds ~9 μm. In the experiments the critical gap width is apparently about twice this value. A value of L_p of about 4.4 μm would account for this difference, but this would imply that no deposition would have been observed in the normal graphite specimens in the 0.02% CH_4–2% CO–CO_2 experiments, where mercury porosimetry gives $W_M = 10.80$ μm. The larger value of the transition width W_M is probably due to methane depletion in the gap between the plates. Calculations indicate that the minimum possible value for $L_p = 9$ μm, and that this can account for the observations. A similar possibility was described by Best *et al.*,[13] but so far no attempt has been made to modify the theory presented here; an obvious conclusion would be a change in the dose dependence of deposition. However, these observations do provide *direct* evidence for the model developed above.

The net result of these studies is that simplified versions of a complex gas-phase chemistry combined with a measured pore shape can give predictions of changes in pore size which have some reality. The effects of pore coalescence can only be evaluated experimentally and so this broad understanding does not remove the need for experimental studies. In a similar way, the variations of in-pore carbon monoxide and methane concentration, which can vary with energy deposition rate, in blind and restricted access pores cannot be predicted. A number of theoretical problems remain unsolved even using the simplified chemistry described here; for instance, no proof has been given of the maintenance of arbitrary pore shapes in oxidation in pure carbon dioxide or the theory of the shape of elliptical pores as a function of dose in inhibited coolant.

The objective of these studies is, of course, the prediction of the important technical properties of reactor graphite as a function of time. The reactor graphite shows property changes due to both irradiation damage (which changes the properties of the graphite crystallites by introducing defects) and radiolytic oxidation (which changes the pore space). For most of the important properties these two effects can be treated as separate, but their magnitudes have to be evaluated experimentally. A substantial study has been mounted of the changes in Young's modulus, thermal conductivity, thermal expansion coefficient and strength of a number of graphites oxidized in different coolant compositions. The changes are, to a good approximation, independent of both the graphite and coolant composition (careful statistical analysis reveals small but significant differences) when the changes due to irradiation damage are annealed out at high temperatures. The changes can be described by:

Young's modulus	$E = E_0 \exp(-3.6x)$
Thermal conductivity	$K = K_0 \exp(-2.7x)$
Strength	$\sigma = \sigma_0 \exp(-4.0x)$

where the subscript zero denotes the unoxidized property. The thermal expansion coefficient is unchanged. Comparison with similar changes due to uniform thermal oxidation shows that radiolytic oxidation is much less damaging. These results have been discussed extensively elsewhere,[4,5] together with their use in predicting the life of reactor core bricks and are not further described here. It is difficult, however, to understand the insensitivity of the changes in properties to the coolant composition, which determines the range of pore sizes which oxidize, and to graphites where pores of different shape and size are concerned.

The current situation, to summarize, is that the radiolytic oxidation process and its consequent effects on the technically important properties of graphite are sufficiently well understood for the design and operation of carbon dioxide-cooled graphite-moderated reactors, but equally many of the scientific aspects of the process are not understood. Fortunately, these aspects can be 'lost' in the simplified theory described here.

6.5 Determination of the pore structure of graphite

The theory of quantitative stereology has been described by Underwood[53] and, with the advent of modern computing facilities, it has become possible to process the information necessary to produce useful pore-size distributions.

Assuming that the impregnation necessary to produce two-dimensional images of the pore cross-sections can be carried out, the surface is then polished away until all traces of the normally closed porosity which intersects the surface has been used. (The images of a surface immediately post-impregnation should characterize the total porosity because normally closed porosity intersects the surface. As the surface is polished away to increasing depths, eventually the impregnated porosity will tend to the normal 'open' porosity values.)

The approach taken to evaluate the properties of interest could be an integration over each observed pore, but this is generally too complex and normally the two-dimensional images are analysed in terms of a general pore model characterized by principal dimensions or orientations.[48]

Stephen[50] noted that the probabilities P for reaction of an oxidizing species with the pore wall can all be written in terms of one parameter

$$\phi_a = a\sqrt{\frac{K}{D}} \tag{6.90}$$

where a is given by the volume/surface ratio for the geometry

slab	$a = W/2$
cylinder	$a = R/2$
sphere	$a = R/3$

and for which the probabilities differ very little (a maximum of ~20%). In two-dimensional images it is the area/perimeter ratio which can be measured rather than the volume/surface ratio. In two dimensions the overall probability can be written, with the summations over all open porosity,

$$p' = \frac{\Sigma A(a)p(a)}{\Sigma A(a)} \tag{6.91}$$

where $A(a)$ is the area of open porosity with area/perimeter $= a$ and $p(a)$ is the probability already described. The error in using p' rather than p can be estimated for simple pore geometries (except for the infinite slab).

In spherical geometries, a section taken through a randomly distributed collection of spherical pores of identical radius R will give a collection of circles of radius r. In this case

$$p' = \frac{\Sigma \pi r^2 p(a)}{\Sigma \pi r^2}$$

Using $R^2 = r^2 + y^2$, where y is the distance of the section from the pore centre, and replacing the sum by an integral, gives

$$p' = \left[\frac{3}{R}\left(\frac{D}{K}\right)^{1/2}\right] \int_0^1 (1-x^2)^{1/2} \frac{I_1[(1-x^2)^{1/2}R(\frac{D}{K})^{1/2}]}{I_0[(1-x^2)^{1/2}R(\frac{D}{K})^{1/2}]}\,\mathrm{d}x \tag{6.92}$$

using $p(a)$ for a cylinder, as is appropriate. The pre-integral factor $3(D/K)^{1/2}R$ is ϕ_a^{-1} for a sphere. The correction definition of $p(a)$ for a sphere is

$$p(a) = \frac{1}{\phi_a}\left[\frac{1}{\tanh(3\phi_a)} - \frac{1}{3\phi_a}\right]. \tag{6.93}$$

Equation 6.92 can only be integrated numerically for various values of ϕ_a. The results show that:

1. the agreement with equation 6.93 is good for small values of ϕ_a;
2. as ϕ_a increases p' is less than p, such that for $\phi \approx 10$, $p' \approx 0.8p$;
3. for large ϕ_a, $p' \to (\pi/4)p$, the largest error.

The conclusion is that multiplying the value of ϕ_a measured by $\pi/4$ leads to virtually identical p and p' (maximum deviation ~3%).

Given a distribution of sphere sizes, the distribution of observed circles can be arranged into integrals over each size in the distribution.

The analysis of cylindrical pores which can be sectioned at any angle is more difficult, since the section produced on intersecting the surface is elliptical and the probability $p(a)$ has not been obtained for elliptical pores. If r is the minor axis of the elliptical section, then the major axis is $b = r/\cos\theta$ and the area/perimeter ratio is $\pi r/4E$, where E is a complete elliptic integral of the second kind (θ is the angle between the section and the perpendicular to the cylinder axis). However, the true $p(a)$ factor for cylinders can be compared with those of the two dimensional model based on slab or cylindrical geometry using

$$\phi_a = \frac{\pi r}{4E}\left(\frac{K}{D}\right)^{1/2}$$

It is shown that for a given θ the slab model gives slightly larger p' values, but that the use of a geometric factor to correct ϕ_a is still possible, although the value is a function of the aspect ratio.

For further details, see ref. 13.

6.6 Acknowledgements

I thank Dr A.J. Wickham of Nuclear Electric for permission to use material from the review of Best, Wickham and Wood.[13] My understanding of pore behaviour during radiolytic oxidation has benefited greatly from discussions with Dr P.A.V. Johnson.

6.7 References

1. DAY, R.J., WALKER, P.L. Jr and WRIGHT, C.C. 1958: *Industrial Carbon and Graphite*. London: Society of Chemical Industry, p. 348.
2. DIENES, G.J. and VINEYARD, G.H. 1957: *Radiation Effects in Solids*. New York: Interscience.
3. McKEE, D.W. 1981: in Walker, P.L. and Thrower, P.A. (eds) *Chemistry and Physics of Carbon* **16**, 1. New York: Marcel Dekker
4. KELLY, B.T. 1984: *Nuclear Energy* **23**, 256.
5. CARPENTER, E. and NORFOLK, D.J. 1984: *Nuclear Energy* **23**, 83.
6. LIND, R. and WRIGHT, J. 1965: in *International Conference on Peaceful Uses of Atomic Energy*. New York: United Nations, Vol. 9, p. 541.
7. ANDERSON, A.R. and LINACRE, J.K. 1960: in *Proceedings of Symposium on Selected Topics in Radiation Dosimetry*. Vienna, p. 609.
8. LINACRE, J.K., TAYLOR, N.K. and THOMAS, R.B. 1965: *UKAEA Report AERE-R-4806*.
9. DOMINEY, D.A. and WICKHAM, A.J. 1971: *Transactions of the Faraday Society* **67**, 2598.
10. KELLY, B.T. 1985: *Progress in Nuclear Energy* **16**, 73.
11. WICKHAM, A.J., BEST, J.V. and WOOD, C.J. 1977: *Radiation Physics and Chemistry* **10**, 107.
12. SIECK, L.W. 1978: *International Journal of Chemical Kinetics* **10**, 335.
13. BEST, J.V., STEPHEN, W.J. and WICKHAM, A.J. 1985: *Progress in Nuclear Energy* **16**, 124.

14. FEATES, F.S. and SACH, R.S. 1965: *Carbon* **3**, 261.
15. NORFOLK, D.J., SKINNER, R.F. and WILLIAMS, W.J. 1983: *Radiation Physics and Chemistry* **21**, 307.
16. PARKS, D.A. 1972: *Journal of the Chemical Society, Faraday Transactions 1* **68**, 627; 1973: **69**, 1298.
17. KELLY, B.T. 1977: *Carbon* **15**, 117.
18. FEATES, F.S. 1968: *Transactions of the Faraday Society* **64**, 3093.
19. ADAMSON, I.Y., DAWSON, I.M., FEATES, F.S. and SACH, R.S. 1966: *Carbon* **3**, 393.
20. FEATES, F.S. 1969: *Transactions of the Faraday Society* **65**, 211.
21. FEATES, F.S. and FRYER, J.K. 1971: *Carbon* **9**, 511.
22. BEST, J.V., DOMINEY, D.A. and WOOD, C.J. 1972: *Carbon* **19**, 317.
23. FEATES, F.S. and WALKER, F.A. 1971: *Carbon* **9**, 11.
24. FEATES, F.S. and KEEP, C.W. 1970: *Transactions of the Faraday Society* **66**, 3156.
25. JOHNSON, P.A.V. 1981: *Nuclear Energy* **20**, 231.
26. TAIT, J.H. 1965: *UKAEA Report AERE-R-5087.*
27. BEST, J.V. and WOOD, C.J. 1975: *Carbon* **13**, 481.
28. JOHNSON, P.A.V. 1966: *UKAEA Report ND-R-1057(S).*
29. KELLY, B.T. 1978: *Progress in Nuclear Energy* **2**, 1.
30. HARPER, A.H., personal communication.
31. HUTCHEON, J.M. and THORPE, R.P. 1966: in *2nd Conference on Industrial Carbon and Graphite.* London: Society of Chemical Industry, p. 441.
32. HUTCHEON, J.M. 1964: *Advances in Materials, Proceedings of Symposium of the North Western Branch of the Institution of Chemical Engineers, Manchester.* London: Institute of Chemical Engineering.
33. KRETCHMAN, H.K. 1957: *The Story of Gilsonite*, American Gilsonite. UT: Salt Lake City.
34. STANDING, J. and ASHTON, B. 1965: *Carbon* **3**, 157.
35. LABATON, V.Y., ASHTON, B.W., LIND, R. and TAIT, J.N. 1969: *Carbon* **7**, 59.
36. TAIT, J.N. 1966: *UKAEA Report AERE-R-5181.*
37. DOMINEY, D.A. and MORLEY, H. 1966: *Carbon* **9**, 725; *UKAEA Report AERE-R-4987.*
38. DULLIEN, F.A.L. 1979: *Porous Media—Fluid Transport and Pore Structure.* New York: Academic Press.
39. HEWITT, G.F. 1965: in Walker, P.L. (ed) *Chemistry and Physics of Carbon.* New York: Marcel Dekker, Vol. 1, p. 74.
40. KELLY, B.T., ASHTON, B.W., LIND, R. and LABATON, V.Y. 1975: in *12th Biennial Carbon Conference.* American Carbon Committee, Pennsylvania: Penn State University, p. 319.
41. BLANCHARD. A. 1980: in Dyer, A. (ed) *Gas Chemistry in Nuclear Reactors and Large Industrial Plant.* London: Heyden, p. 134.
42. BLANCHARD, A., KITCHEN, J., SCHOFIELD, P. and CAMPION, P. 1980: in Dyer, A. (ed) *Gas Chemistry in Nuclear Reactors and Large Industrial Plant.* London: Heyden, p. 129.
43. RUSHBY, M., personal communication.
44. JOHNSON, P.A.V., personal communication.
45. SCHOFIELD, P. and JOHNSON, R.A.V. 1982: in *Carbon 82, Extended Abstracts.* London: Society of Chemical Industry, p. 323.
46. KELLY, B.T., JOHNSON, P.A.V., SCHOFIELD, P., BROCKLEHURST, J.E. and BIRCH, M. 1983: *Carbon* **21**, 441.
47. LIND, R. 1974: in *Proceedings of the 4th International Conference on Carbon and Graphite.* London: Society of Chemical Industry, p. 445.
48. DULLIEN, F.A.L. and DHAWAN, K. 1974: *Journal of Colloid and Interface Science* **47**, 337.
49. ANDROTSOUPOULOS, G.P. and MANON, R. 1979: *Chemical Engineering Science* **34**, 1203.

50. STEPHEN, W.J. 1980: *Report RD/B/N4943*. Central Electricity Generating Board.
51. MOORE, A.W. 1973: in Walker, P.L. and Thrower, P.A. (eds) *Chemistry and Physics of Carbon* **11**, 69. New York: Marcel Dekker.
52. JOHNSON, P.A.V., unpublished work.
53. UNDERWOOD, E.E. 1970: *Quantitative Stereology*. New York: Addison Wesley.

7 Porosity and Strength of Carbon Materials

John W. Patrick and Alan Walker
Carbon Research Group,
Department of Chemical Engineering,
Loughborough University of Technology,
Loughborough, Leicestershire LE11 3TU, United Kingdom

7.1 Introduction

The term carbons covers a widely diverse range of materials, most of which have three features in common: they are composed predominantly of carbon, they contain pores and they exhibit brittle behaviour. Depending on the material considered, the pores may range in size from nanometres to millimetres, generally the small ones contributing significantly to the surface area and the larger ones to the volume porosity. Since the pores generally constitute the major source of defects in these materials, the strength of a carbon varies with the porosity either as a result of the variation of the amount of load-bearing material in a cross section or, often more importantly, by the larger pores acting as the critical flaws at which fracture is initiated. It is the effects of these larger pores with which this chapter is primarily concerned.

The theory of failure of brittle materials is first reviewed. For those carbons where strength is an important criteria in their use, their strength and porosity are considered. Subsequently, for those materials whose tensile strength and pore structural parameters have been quantitatively related by the authors, the experiment methods and the relationships derived are described in more detail.

7.2 Theoretical aspects of the strength of brittle materials

Brittle materials fail at stress levels far below their theoretical strengths for reasons given by the simple flaw theory of fracture mechanics. According to this, brittle materials contain inherent flaws and that, under stress, once a flaw of critical size begins to propagate, it will continue to do so with increasing speed until catastrophic failure occurs. Crack propagation at flaws depends on the fulfilment of two conditions: the local stress level at the crack tip must exceed the theoretical strength of the material and the energy available must exceed the surface energy of the two surfaces created.[1]

The concept of stress concentration was first developed by Inglis,[2] who showed that the stress at the tip of an elliptical hole was

$$\sigma = (1 + 2c/b)\sigma_0 \tag{7.1}$$

where σ and σ_0 are the local and applied stresses, respectively, and c and b are the major and minor semi-axes of the ellipse, respectively. For a narrow ellipse, i.e. a crack, this relationship, which implies that the stress concentration is shape but not size dependent, reduces to

$$\sigma = \sigma_0(2c/p)^{1/2} \tag{7.2}$$

where p is the radius of curvature at the crack tip.

After considering the energetics of the situation, Griffiths[3] deduced that crack propagation would occur when the stored strain energy exceeded the surface energy of the surfaces created. He showed that, above a critical crack size, there is more strain energy available than is required to create fresh surfaces, hence the propagation of a critically sized crack would lead inevitably and directly to failure. On this basis, the stress at failure was

$$\sigma_f = (2E\gamma/\pi c)^{1/2} \tag{7.3}$$

where E is Young's modulus, γ is the surface energy and c is one-half the length of an internal critically-sized crack.

For metallurgical coke[4] and graphite,[5] experimental evidence shows that they do not fail strictly in accord with simple flaw theory, inherent flaws growing to critical size when the material is stressed. However, simple flaw theory is a useful concept within which to consider the theoretical effect of porosity on the strength of carbons.

The variation of the strength of ceramics with porosity is stated by the empirically derived Ryshkewitch–Duckworth[6] equation

$$S = S_0 \exp(-b/p) \tag{7.4}$$

where S is the strength at fractional porosity p, S_0 is that at zero porosity and b is a constant. The form of this equation was explained by Knudsen[7] to reflect changes in the load-bearing area within the test piece. He further drew together this equation and an equation relating strength with grain size to obtain

$$S = S_0/[G \exp(-bp)]^{1/2} \tag{7.5}$$

where G is the grain diameter and appears to be identifiable with the critical flaw size in the Griffiths equation. Knudsen found that b was a pore-shape factor which decreased as the pores became more spherical. Comparing equation 7.3 with equation 7.5, the pore-free strength can be related to the term $E\gamma/\pi$ in the Griffiths equation and represents the contribution of the nature of the solid to the strength.

From this discussion, therefore, it appears that the strength of a porous carbon will be influenced by the porosity, the pore shape, the size of the critical flaws present at failure and the nature of the material. For any range of carbons, if all these factors varied and no one factor was dominant, then very careful work would be needed to explain quantitatively the variation in strength. Conversely, if one factor were dominant, or all but one factor varied little, then a strength–structure relationship could be found readily.

7.3 Nature and effects of the porosity in various carbons

Carbons whose strength is an important property for the intended use include, in order of decreasing strength, carbon fibres, carbon-fibre reinforced carbons, graphites, electrode carbons and metallurgical cokes. All are porous materials, but only for electrode carbons and metallurgical cokes have quantitative strength–pore structure relationships, based on image analysis data, been established.

Carbon fibres obtained from polyacrylonitrile consist of long sheets of graphitic layer planes aligned parallel to the fibre axis. In the transverse direction, the sheets are regarded as folded or crumpled[8] or as lying parallel to the external surface in the outer skin becoming increasingly folded in the fibre centre.[9] Both models include pores, elongated along the fibre axis, between the folded sheets of carbon layers. High strength (HS) and high modulus (HM) fibres differ in the perfection and alignment of the layers and the degree of cross-linking between them. Although a wide range of internal defects have been identified, no simple relationship has been found between fibre strength and flaw diameter/surface free energy.[10] Failure is therefore considered to arise from shear on misaligned sheets of lamellae.[11] Cross-linking in HS fibres resists shear failure, hence their higher strength.

Two-dimensional carbon-fibre reinforced carbons, carbon–carbon composites, are produced from moulded carbon-fibre cloth/resin preforms by carbonization and subsequent densification. Shrinkage of the resin on carbonization results in cracks between layers of cloth, between bundles in a layer and within bundles. The anisotropy of shape of the cracks limits attempts to measure accurately the porosity of the various voids by image analysis techniques. Nevertheless, reliable comparative values of the proportion of the voidage present in the different types of voids can be obtained,[12] as can size and shape data.[13] Interlayer and interbundle cracks are appropriately aligned to act as positions of initiation of failure in interlamellar shear or flexure, respectively.[12] Curiously, it is the flexural strength which appears to correlate with the voidage in inter-layer voids.[13]

Electrode carbons and graphites are produced by the slow carbonization to 1000 °C of pitch-bonded filler particles, graphites then being further heat treated to 2500 °C. Cathode carbons for the aluminium extraction industry usually contain at least a proportion of calcined anthracite as filler, while the fillers in anode carbons and graphites are regular-grade and premium petroleum coke, respectively. These differ in the degree of mutual alignment of the constituent graphite-like layers. Large petroleum coke particles contain cracks induced by shrinkage during calcining and these, along with devolatization pores and interparticular voids, constitute the porosity in graphites and anode carbons. No relationship between the strength and pore structure of graphites appears to have been observed but the Knudsen relationship (equation 7.5) has been shown to hold for a range of graphites, G being identified with the size of the larger filler particles.[14] This size would correspond approximately to that of the largest shrinkage cracks in the filler particles.

Strength–pore structure relationships obtained for electrode carbons and metallurgical cokes are considered below.

7.4 Strength–pore structure relationships

Although early work found that the strength of some graphites varied with grain size and porosity according to the Knudsen relationship,[14] it was not until computer-controlled image-analysis equipment (IA) became available that it was possible to collect data on the porous structure of carbons, which are often somewhat heterogeneous, in sufficient detail, and with an acceptable accuracy, realistically to seek strength–pore structure relationships. Perhaps surprisingly, the first carbon material examined by image analysis for this purpose was metallurgical coke intended for blast-furnace use.[15]

Metallurgical coke is obtained by the carbonization of granular coal in slot-type coke ovens. Coking coals soften on heating and the release of volatiles into the softened mass results in the formation of devolatilization pores. At the bottom of a blast furnace, coke is the only solid material present to support the burden above. Hence one of the most important quality requirements is for the coke to be of high strength to support this burden. Coke, however, is a brittle material, usually with porosity greater than 50 vol.%, which contains large pores, the largest often greater than 200 μm in size, several orders of magnitude larger than any other structural feature present. Thus, inevitably, it is the macroporosity which provides the largest inherent flaws in this form of carbon and which therefore constitutes the dominant factor controlling the strength properties.

Metallurgical coke differs from other forms of bulk carbon in that it is produced from a precursor, coal or a blend of coals, containing a high proportion, perhaps 80 wt%, of components which soften during the carbonization process. The non-softening component is a type of charcoal present in the coal. In contrast, industrial carbons and graphites consist of about 80 wt% of a non-softening filler material, usually petroleum coke, bound together by carbonized pitch and hence they contain interfaces between components which can vary in quality. These interfaces influence strength[16] probably by contributing to the variation of the surface energy term in equation 7.3. Hence metallurgical cokes probably contain fewer varieties of structural features which influence their strength. Therefore, although it may seem surprising to apply sophisticated new techniques to such a material, it is possible that metallurgical coke possesses a comparatively simple structure which, with hindsight, made it a particularly useful material to which to apply image analysis in the search for strength–pore structure relationships.

Similar techniques, using two different image analysis systems, have been applied to the study of electrode carbons used in the aluminium extraction industry. Such carbons require adequate strength to survive handling and to withstand the thermal stresses associated with their use. As will be seen, one of the strength–pore structure relationships developed for metallurgical coke was equally applicable to electrode carbons.

Experimental techniques

Before describing the experimental techniques in detail, it is worthwhile drawing attention to some factors which can markedly influence the accuracy of the results and which may, if not properly taken into account, mean that strength–pore structure relationships cannot possibly be obtained.

Industrial carbons are heterogeneous materials, so the first experimental requirement is that the strength test pieces and the surfaces examined by image analysis be truly representative of the material. The means of achieving representative samples depends on the heterogeneity of the material and the size of sample available.

Second, if, as in the studies described here, the image for analysis is a TV image obtained by incident-light microscopy of polished surfaces, then the surface should be polished to a scratch- and relief-free condition. Most polishing defects are regarded by image analysis systems as pores.

Finally, sufficient fields of view should be analysed so that the error associated with the measurements obtained is acceptably small. The number of fields to be used can be obtained experimentally by noting how the error on the mean value of a parameter falls as the number of fields from which measurements are taken is increased.

Materials studied

The metallurgical cokes examined consisted primarily of 25 kg samples of cokes produced in 10- and 17-ton large-scale test ovens together with a wide range of other cokes, varying in strength, produced on smaller and larger scales.[17] Where necessary, sample division had been practised to obtain a sample representative of the oven product.

Two series of electrode carbons have been studied.[18,19] Series A consisted of samples from six cathode blocks, the individual specimens studied having been heat treated to temperatures from 1100 to 2500°C. Series B contained examples of both cathode and anode carbon. The anode carbons contained calcined petroleum coke (PC) as filler and the cathode carbons were produced from pitch-bonded crushed graphite (G), graphite plus gas-calcined anthracite (GCA) or electro-calcined anthracite (ECA) filler particles.

Tensile strength determination

For all materials, the tensile strength was determined by means of the diametral compression method using a testing machine operating with a cross-head speed of 0.5 mm min^{-1}. The sizes of test pieces varied with the application as detailed below.

For metallurgical cokes, the aim was to obtain both strength and pore structural information representative of the oven charge. Hence cores 10 mm in diameter were drilled from all the available coke pieces large enough to permit this operation and sections 10 mm thick were sliced from these cores to provide the bulk sample of cylindrical test pieces. Standard sample division techniques were applied to divide these bulk samples into two sub-samples of 40–50, one for use in the tensile-strength determination and the other for image analysis. This sample preparation procedure is necessary to ensure that representative strength values are obtained and takes account of such variability as that associated with the position of the coke in the oven during production. The individual test pieces were tested in the air-dried condition.

For electrode carbons, the methods used were designed to obtain strength and pore structure data for individual specimens. For Series A carbons, 60 mm diameter cores were drilled from cathode blocks and, after heat treating each core, a 60 mm diameter by 40 mm tall cylinder was cut for tensile strength determination and an adjacent 60 mm diameter by 10 mm tall slice was cut for image analysis.

The electrode carbons in Series B consisted of various sized cores. They were further cored to obtain 15 mm diameter by 10 mm tall cylindrical test pieces, usually 15–25 in number. These were broken in diametral compression and those individual specimens with strength close to the mean value and ±1/3, 2/3 and 1 standard deviation from the mean were used for image analysis.

Sample preparation for image analysis

To prepare suitable surfaces for image analysis, specimens were embedded in epoxy resin in a 65 mm diameter mould. Using automatic polishing machines, the surface was then smoothed using various grades of silicon carbide paper and polished with alumina powder and diamond paste to achieve a scratch- and relief-free finish. If necessary, to ensure good filling of the pores with resin, the surface was smeared with resin and resmoothed before the final polishing stage.

For metallurgical cokes, two moulded blocks, each containing 25 test pieces, were prepared. One polished block, containing a 60 mm diameter carbon surface, was used for the electrode carbons in Series A. For those in Series B, the polished block contained two rectangular surfaces cut parallel to the fracture surface for each broken tensile test piece.

Image analysis procedures

Three different image analysis systems were used, a Quantimet 720 for metallurgical cokes, a Quantimet 800 for Series A carbons and a Mini-Magiscan for Series B carbons. The Quantimet systems were hard-wire based whereas the Mini-Magiscan was software based. In each case a TV image of the carbon surface, when viewed in an incident-light microscope, was used as the basis for the measurement. The TV image is captured as an array of small pixels, the brightness of each pixel being allocated to one of a number of grey levels, usually 64. Since the larger pores were of most interest, low power objective lenses were used to give pixel sizes 4.5×4.5, 1.6×1.6 and 3.3×2.9 μm, respectively, for the three systems used.

For all three systems, before measurements could be made, it was necessary to convert the grey image into a black and white image by setting a detection level such that pixels with grey levels above and below the selected level gave black and white, respectively. For the carbons examined this was achieved by manually selecting a detection level in which the white areas of bright, reflective carbon and black areas of resin-filled pores correspond accurately to those in the microscope image. The accuracy of the subsequent measurements is critically dependent on the care with which this process is carried out.

Image analysis systems are usually capable of measuring numerous parameters from each field of view. Basic parameters of interest in the present connection, obtainable by the three systems used, are listed in Table 7.1. The mean pore and wall intercepts are the mean lengths of the scan lines within the measuring field which are black or white, respectively. The values are averaged for scan lines parallel to the horizontal and vertical edges of the field. The pore height and width are the maximum dimensions of a pore in the directions parallel to the vertical and horizontal edges of the measuring field, respectively. The pore length is the maximum dimension of a pore and the pore breadth is the maximum dimension at right-angles to the direction of maximum dimension.

Table 7.1 Parameters obtainable with the Quantimet and Mini-Magiscan systems

Parameter	Symbol	Quantimet 720	Quanitimet 800	Mini-Magiscan
Total field area	TA	*	*	*
Total pore area	PA	*	*	*
Number of pores	N	*	*	*
Mean pore height	H			*
Mean pore width	W			*
Mean pore intercept	P	*	*	
Mean wall intercept	W	*	*	
Mean pore length	F_{max}	*		*
Mean pore breadth	F_{min}	*		*

These two dimensions are often referred to as the maximum and minimum Feret's diameters, respectively.

These basic measurements were used to calculate the following derived values:

Porosity	$p = PA/TA$
Mean pore size	$P = PA/[N \times (H+W)/2]$
Mean wall size	$W = (TA - PA)/[N \times (H+W)/2]$

For regularly shaped pores, the pore size and wall size should be numerically similar to the appropriate intercept size, hence the allocation of the same initial.

The subject of macroporosity measurements has been considered in Chapter 4 but it should be noted that the values obtained by image analysis for the size parameters quoted are not absolute values but are strongly dependent on the magnification used and the size of the features rejected to clean the image of instrumental noise. Thus, in general, the higher the magnification and the smaller the rejected features, the smaller are the apparent measured values of size parameters. In the studies described here, the lowest power microscope objectives available were used and features smaller than about 4 μm in linear dimension were rejected.

Strength–pore structure relationships for metallurgical cokes

For a range of test-oven and commercial cokes with volume porosities ranging from 34 to 59%, it was established[17] that there was, in general, a clear association of increasing tensile strength with increasing wall thickness and number of pores and with decreasing pore size and porosity. None of these parameters individually correlated closely with coke tensile strength but strength could be related to a combination of pore structural parameters. One example is shown in Figure 7.1. The equation for the relationship is

$$SN = (kW/P^2) - c \tag{7.6}$$

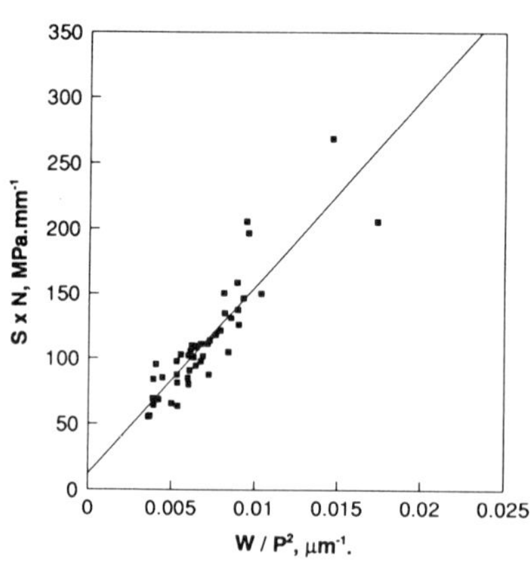

Fig. 7.1 Variation of tensile strength of metallurgical cokes with a combination of pore structural parameters according to equation 7.6

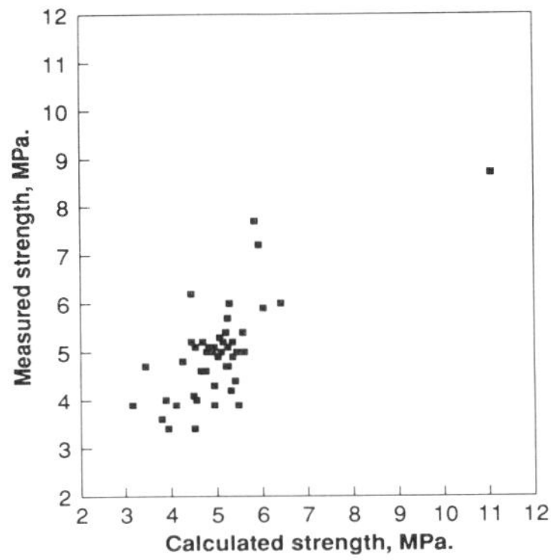

Fig. 7.2 Comparison of measured metallurgical coke strengths with those calculated using equation 7.6

where S is the tensile strength in MPa, N is the number of pores detected per mm^2 of field of view, W and P are the mean wall and pore sizes, respectively (as defined previously), and k and c are constants. This has been shown to hold for a wide range of cokes. For the series shown, the correlation coefficient is 0.88. Measured tensile strengths are compared with calculated values in Figure 7.2.

Equation 7.6 was originally thought to be purely an empirical equation but, as will be seen, the form of the relationship has application to other carbons. Since the equation did not appear to be soundly based on materials science principles, the applicability of the Knudsen equation was assessed. The maximum Feret diameter of the larger pores was identified with the critical flaw size and thus was used instead of the grain size, G, in equation 7.5. For this purpose, the Feret diameters were measured only on pores greater than $14 \times 10^3\ \mu m^2$. Applying regression analysis to the data gave the equation

$$S = 450\, F_{max}^{-1/2} \exp(-2.63p). \qquad (7.7)$$

Measured tensile strengths are plotted against calculated values in Figure 7.3.

Since Knudsen[7] recognized that the constant b in equation 7.5 was dependent on pore shape, a shape factor of $2(F_{max}/F_{min})^{1/2}$ was introduced to replace b.

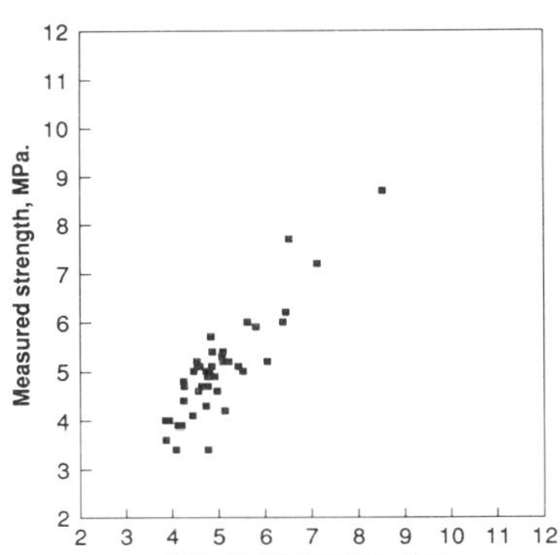

Fig. 7.3 Comparison of measured metallurgical coke strengths with those calculated using equation 7.7

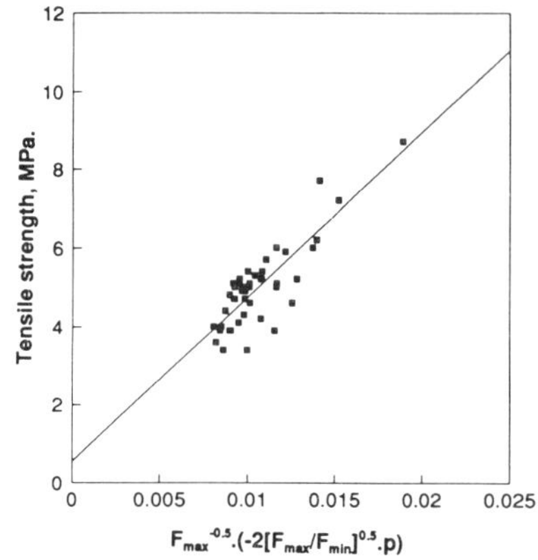

Fig. 7.4 Variation of tensile strength of metallurgical cokes with a combination of pore structural parameters according to equation 7.8

This factor is based on the stress concentration at the tip of an elliptical slot at right-angles to the applied stress. Equation 7.5 then becomes

$$S = KF_{\max}^{-1/2} \exp[-2(F_{\max}/F_{\min})^{1/2}p] \tag{7.8}$$

p being the fractional porosity.

For the cokes examined, Figure 7.4 shows a plot of coke tensile strength against $F_{\max}^{-1/2} \exp[-2(F_{\max}/F_{\min})^{1/2}p]$. The correlation coefficient is 0.83 and the value of K calculated by regression analysis is 470. Measured and calculated strengths are compared in Figure 7.5.

Strength–pore structure relationships for electrode carbons

For the electrode carbons in Series A, with porosities ranging from 16 to 50%, again no individual parameter could be used to explain the strength changes.[18] The data were therefore treated according to equation 7.6 with the result shown

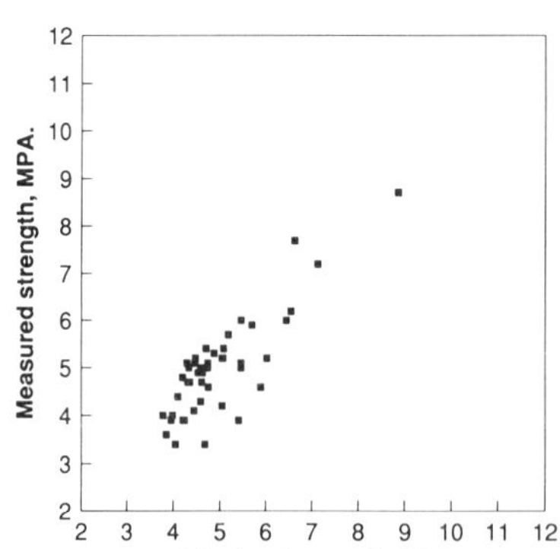

Fig. 7.5 Comparison of measured metallurgical coke strengths with those calculated using equation 7.8

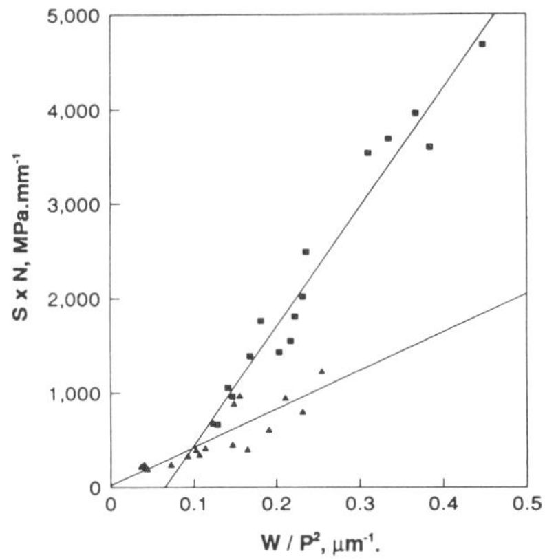

Fig. 7.6 Variation of tensile strength of Series A electrode carbons with a combination of pore structural parameters according to equation 7.6

in Figure 7.6. As is evident from the two lines drawn, it is believed that the data are best considered in two groups. One group consists of carbons containing ECA as filler and the other GCA and/or graphite. For the ECA-containing materials, the data are represented by the regression equation

$$SN = 4048(W/P^2) - 23 \tag{7.9}$$

and the regression equation for the graphite-containing carbons is

$$SN = 12587(W/P^2) - 823. \tag{7.10}$$

The correlation coefficients for the two equations are 0.99 and 0.85, respectively. The accuracy with which the equations calculate the tensile strengths of the carbons is illustrated in Figure 7.7.

The different values of the coefficient in equations 7.9 and 7.10 imply differences in the contribution of the nature of the solid material to the strength. Examination of the surfaces of fractured tensile test pieces in a scanning electron

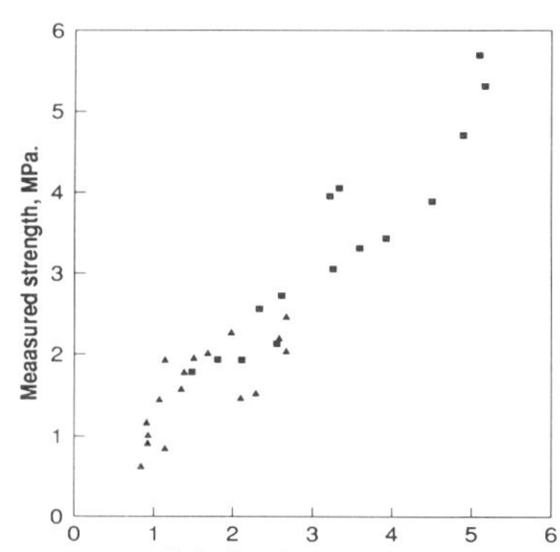

Fig. 7.7 Comparison of measured strengths of Series A electrode carbons with those calculated using equation 7.6

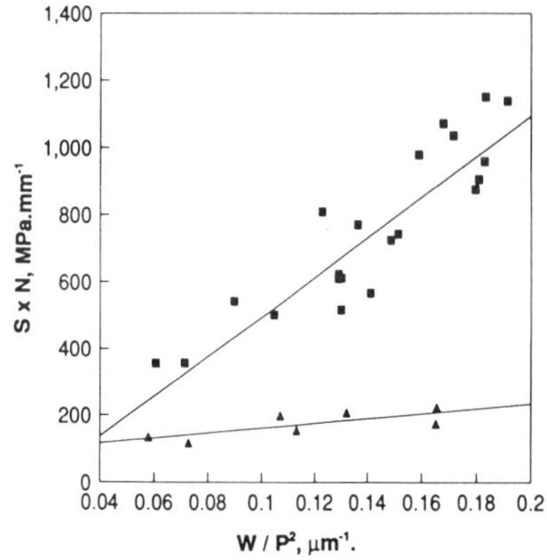

Fig. 7.8 Variation of tensile strength of Series B electrode carbons with a combination of pore structural parameters according to equation 7.6

microscope showed that there was a difference in the bonding at the binder/filler interfaces in the two series of carbons.[18] Bonding in the ECA-containing materials was poor so that, in the fracture surface, evidence of interfacial failure was seen. In the other series of carbons, interfacial bonding was good and the fracture crack passed through the filler particles rather than around the binder/filler interface.

For the electrode carbons in Series B with volume porosities ranging from 10 to 22 vol.%, the behaviour according to equation 7.6 was similar to the data obtained before and was again best regarded as belonging to two groups (Figure 7.8) according to the filler used to produce the carbon.[19] For these carbons, the mean maximum Feret diameters of all the pores detected varied from 32 to 52 μm and the mean minimum Feret diameters from 26 to 43 μm. These pore structural data could not be related to the tensile strengths by the modified Knudsen equation, perhaps because of the use of measurements taken from all the detected pores. However, for each carbon studied, the data could be fitted to the original Knudsen relationship when the grain size was replaced by the maximum Feret diameter. The relationships obtained by linear regression, for both equations 7.6 and 7.7, for this series of carbons are summarized in Table 7.2. Measured strength values and those calculated according to the four relationships obtained using equation 7.7 are compared in Figure 7.9.

Table 7.2 Strength–pore structure relationships for Series B electrode carbons

Electrode No.	Strength × pores/mm² (SN)	Correlation coefficient	Strength	Correlation coefficient
1	$55.4 + 4804\, W/P^2$	0.84	$139 F_{max}^{1/2} \exp(-7.83p)$	0.82
2	$143 + 1082\, W/P^2$	0.50	$43 F_{max}^{1/2} \exp(-4.73p)$	0.74
3	$-293 + 7530\, W/P^2$	0.75	$419 F_{max}^{1/2} \exp(-15.2p)$	0.79
4	$-210 + 6221\, W/P^2$	0.87	$351 F_{max}^{1/2} \exp(-16.2p)$	0.84
1+2+4	$-19.2 + 5515\, W/P^2$	0.96		

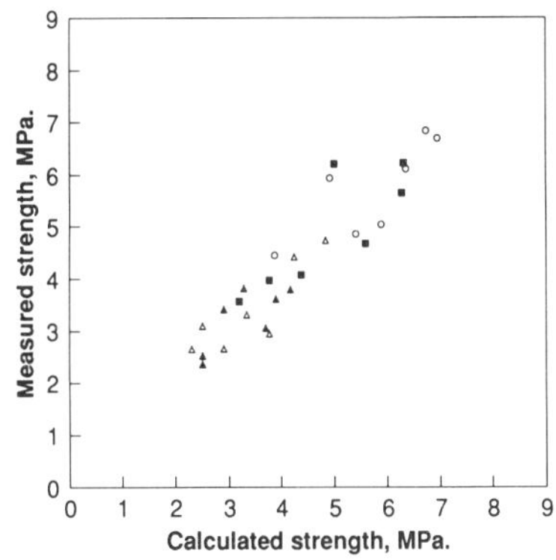

Fig. 7.9 Comparison of measured strengths of Series B electrode carbons with those calculated using equation 7.7

7.5 Concluding remarks

Many of the applications of carbons, especially those produced in quantity, require that the carbon should possess adequate strength. Since most carbons are porous materials, it is inevitable that the pore structure plays a role, in some cases a dominant role, in controlling their strength. Porosity always influences the strength of brittle materials, including carbons, in reducing the amount of material bearing the applied stress. In carbons, pores may not always act as inherent critical flaws but, if not, they act as positions of initiation of microfissures from which the critical flaws develop. Other factors, which influence the Young's modulus and the surface energy, affect the strength of a carbon. Since, apart from the studies described here, very few studies successfully relating the strength and pore structure of carbons have been published, it appears that, in many cases, the pore structure does not play so dominant a role that relationships can be obtained from consideration of only the strength and pore structural parameters.

Of the relationships described in this paper, the equation

$$SN = (kW/P^2) - c$$

was derived from consideration of the trends between the tensile strength of metallurgical cokes and the individual pore structural parameters. The relationship has proved very useful in studies of coal blend formulation for carbonization.[17] It was originally considered to be a purely empirical relationship since it could not be reconciled with the Griffiths view of brittle fracture. However, the relationship now appears to hold for other types of carbon and for pore structural data obtained using different image analysis systems. The system used, which can influence the magnitude of the parameters for the same carbon, has an influence on the constant k but it also appears that k varies with the nature of the bonding between components in the carbons examined. Taking a general view, it seems probable that the constant in equation 7.6 is a materials constant dependent on the nature of the pore-free carbon matrix. As such, its magnitude can be expected to depend on many factors, the quality of the bonding between components being important when considering poly-granular carbons.

The Knudsen relationship (equation 7.5) is more firmly based on modern fracture mechanics views, the form of the relationship now having been deduced theoretically.[20] Replacing the grain size in the original equation by the maximum dimension of the pores allows the application of the relationship to be widened to metallurgical cokes and to electrode carbons. For the electrode carbons in Series B however, unlike equation 7.6, the Knudsen equation does not separate the carbons according to the mode of failure as deduced from SEM study of tensile fracture surfaces. Identifying b in the Knudsen equation with a stress concentration factor is in general accord with Knudsen's views on the effect of pore shape. However, for the metallurgical cokes to which the modified equation was applied, the stress concentration factor varied only from 2.62 to 2.76. Hence further confirmation with materials giving a wider range of values is still required.

7.6 References

1. DAVIDGE, R.W. 1979: *Mechanical Behaviour of Ceramics.* Cambridge: Cambridge University Press.
2. INGLIS, C.E. 1913: *Transactions of the Institute of Naval Architects* **55**, 219.
3. GRIFFITHS, A.A. 1920: *Transactions of the Royal Society, London, Series A.* **221**, 163.
4. PATRICK, J.W. and WALKER, A. 1987: *Journal of Material Science* **22**, 3589.
5. SMITH, M.C. 1972: in Greenstreet, W.I. and Battle, C.C., Jr (eds) *Proceedings of the Conference on Continuum Aspects of Graphite Design, Gatlinburg, 1970.* USAEC Conference Report 70115. Washington: USAEC, p. 475.
6. DUCKWORTH, W. 1953: *Journal of the American Ceramic Society* **36**, 68.
7. KNUDSEN, F.P. 1959: *Journal of the American Ceramic Society* **42**, 376.
8. KLEMAN, M. and SADOC, J.F. 1979: *Journal of Physics Letters* **40L**, 569.
9. BENNETT, S.C. and JOHNSON, D.J. 1979: *Carbon* **17**, 25.
10. REYNOLDS, W.N. 1973: *Physics and Chemistry of Carbon* **2**, 1.
11. REYNOLDS, W.N. and SHARP, J.V. 1974: *Carbon* **12**, 103.
12. GAO, F., PATRICK, J.W. and WALKER, A. 1993: *Carbon* **31**, 103.
13. GAO, F. 1993: *PhD Thesis.* Loughborough University of Technology.
14. KIBBS, R.J. 1967: *Journal of Nuclear Materials* **24**, 174.
15. PATRICK, J.W., SIMS, M.J. and STACEY, A.E. 1976: in *Abstracts, Carbon 76, 2nd German Carbon Conference.* Baden-Baden: Deutschen Keramischen Gesellschaft, p. 211.
16. MARTIN, N., OGDEN, G.N., PATRICK, J.W. and WALKER, A. 1992: *Carbon* **30**, 487.
17. PATRICK, J.W. and STACEY, A.E. 1983: *ISS Transactions AIME* **3**, 1.
18. PATRICK, J.W., SORLIE, M. and WALKER, A. 1989: *Carbon* **27**, 469.
19. PATRICK, J.W. and WALKER, A. unpublished data.
20. BUCH, J.D. 1983: in *Extended Abstracts, 16th Biennial Conference on Carbon.* San Diego: American Carbon Society, p. 400.

8 Application of Active Carbons for Gas Separation and Respiratory Protection

Alan Bailey
Royal Military College of Science
Cranfield University
Shrivenham
Swindon
Wilshire SN6 8LA
United Kingdom

8.1 Introduction

Charcoals (carbonized wood) have been known and used from early times. Records show that the Egyptians used the material for medical purposes and for purifying water and foods. The first use of charcoal in a 'gas mask' for air purification, however, was not until the Great War of 1914–18.[1] Since that time, the range of precursor materials for the manufacture of gas-mask charcoals or, to use a better term, 'activated carbons', has been extended to include a wide range of materials of vegetable origin, coal, lignite and peat, newspaper, textile waste and some recycled polymers.

Activated carbons owe their useful adsorbent properties to their extensively developed porous structure. Commercial carbons may well have a porosity equivalent to a surface area in the range 800–1500 $m^2 g^{-1}$. This surface area is contained predominantly within micropores having effective diameters smaller than 2 nm and it is now conventional[2] to classify the complex network of pores within a carbon granule into micropores (diameters <2 nm), mesopores (diameters 2–50 nm) and macropores (diameters >50 nm). The structure of an activated carbon may be visualized as stacks of poorly developed aromatic sheets ('crystallites') distributed and cross-linked in a random manner, separated by disorganized carbonaceous matter and inorganic matter (ash) dependent on the precursor material. This structure can be seen in the electron micrograph of a cellulose charcoal in Figure 8.1.

8.2 Sources and preparation

Activated carbons can be formed by the pyrolysis of any carbonaceous material which does not undergo melting or have a plastic phase during decomposition at temperatures up to 850 °C. A very extensive but narrow incipient pore structure is present after pyrolysis,[3] but this needs to be developed by activation with carbon dioxide or steam in the temperature range 800–900 °C to produce a useful

Fig. 8.1 Micrograph of a cellulose char showing the poorly developed stacks of aromatic sheets (magnification × 2 000 000)

active carbon. In the activation process the number of aromatic sheets in the 'crystallites' is reduced and a slit-like micropore structure is developed. As the extent of activation or 'burn-off' is increased, the mean pore size increases and the slot-like nature of the pores becomes less evident.

The formation of a porous structure in the final carbon is prevented if the material passes through a plastic or molten phase during carbonization so the addition of swelling inhibitors (e.g. 1–2 wt% borax) to coal feed stocks prior to pyrolysis is frequently practised. This enables a wider range of coals to be used for active carbon production.

For respirator and air filter applications it is an advantage for the active carbon to be as dense as possible so that the maximum adsorption capacity can be contained in the minimum volume. Dense carbonaceous materials such as coconut shells, peach stones, coal and sometimes wood and lignite are therefore chosen as active carbon precursors.

Alternatively, active carbons can be prepared from cellulosic materials by a 'chemical' method. Char formation on pyrolysis of such materials is illustrated in figure 8.2.[4]

In the early stages of pyrolysis, at temperatures up to 350°C, two competing reaction routes are possible. One route is by the depolymerization of cellulose to form low molecular weight products such as laevoglucosan, which may undergo further breakdown to form methanol, acetone, ethanoic acid and similar products. Also on further heating, the laevoglucosan will form tars of considerable complexity. The other mechanism of decomposition is the dehydration of cellulose with the formation of a carbon–carbon double bond. With further reaction a conjugated system develops and the char structure is consolidated by aromatization through a Diels–Alder-type condensation reaction. Since relatively little molecular movement takes place during these char-forming reactions, the molecular structure of the starting material exerts considerable influence on the resulting pore structure of the char.

The addition of Lewis acids to the raw material enhances the dehydration reaction and hence the yield of char. Additionally, suppression of tar formation prevents blocking of the incipient pore structure of the char, making the activation process easier. Some dehydration reaction promoters, such as zinc chloride and phosphoric acid, are able to swell the cellulose material before it is

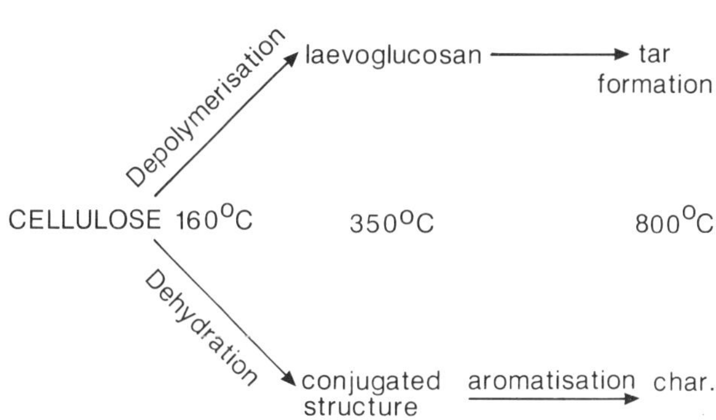

Fig. 8.2 Competing processes in the pyrolysis of cellulose

carbonized, resulting in a char with enhanced microporous structure after carbonization which may not require further activation with steam or carbon dioxide. Since cellulosic material of all kinds (sawdust, newsprint, peat, etc.) is plasticized by the addition of these Lewis acids, the precursor mix may be extruded under high pressure to form cylinders of diameter 2–3 mm which yield grains of dense porous active carbon on carbonization. These have a morphology ideal for use as respirator carbons. This process is used extensively in Eastern Europe and in China for the production of charcoals from peat.

In recent years a novel form of active carbon, active charcoal cloth, has found a number of important applications in air purification. The material is manufactured by the pretreatment of viscose rayon with solutions containing zinc chloride followed by carbonization and further activation with carbon dioxide.[5]

8.3 Charcoal parameters affecting filter performance

To be able to consider the characteristics of an activated carbon which affect its filter performance and hence to control the development of those characteristics during its preparation, it is useful to have some understanding of how respirator and other air purification filters perform. A useful but simple treatment was given by Hinshelwood and co-workers[6] in 1946.

Active carbon filters

Gases and vapours enter the activated carbon in the filter, driven by a diffusion process of some kind, and are then adsorbed. Filters are normally assessed for effectiveness by a test in which air containing a known concentration (C_0) of a test vapour is passed through the filter until a given (low) vapour concentration (C_p) (usually much lower than the toxic concentration of the vapour) is detected in the effluent air. The time to this penetration is the 'penetration time' (t_p) and is a measure of the filter performance. Hinshelwood and co-workers[6] showed that

$$t_p = \frac{1}{KC_0}\left\{\ln\left[\exp\left(\frac{KWd}{L}\right) - 1\right] - \ln\left(\frac{C_0}{C_p} - 1\right)\right\}$$

where W is the capacity of the active carbon for the vapour, K is the rate of adsorption and L is the linear velocity of the air stream. The equation can be much simplified to

$$t_p = \frac{W}{C_0 L}(d - d_c)$$

where d is the depth of the active filter bed and d_c is the minimum bed depth to reduce the influent vapour concentration, C_0, to C_p, and is called the critical bed depth. This may be written as

$$d_c = \frac{L}{WK} \ln\left(\frac{C_0}{C_p}\right)$$

Filter performance is thus determined by the adsorptive capacity of the carbon and the kinetics of adsorption. Both of these factors are determined by the extent of the carbon's porous system and pore diameter. The other important filter parameter is its air flow resistance. This is controlled by the charcoal grain size and its packing efficiency; these parameters also affect the kinetics of adsorption.

Physisorption capacity of active carbons

Activated carbons of high capacity can only be obtained by reacting the carbonized material with an oxidizing gas such as steam or carbon dioxide. Initially (below 10 wt% burn-off), disorganized carbon is removed from the char, thus exposing blocked pores. Further activation involves the progressive removal of the aromatic sheet system producing wider pores. The steady increase in pore diameter (as exemplified by heat of immersion measurements made in the author's laboratory) with increasing 'burn-off' is illustrated in Figure 8.3 for a cellulose carbon activated with steam at 850°C. The increasing accessibility of the structure, first to the smaller benzene molecules, then with activation to the larger silicone oil molecules, can clearly be seen as an increase in appropriate heat of immersion.

By 30 wt% burn-off, the smaller benzene molecule has access to almost the entire pore system and surface area of the carbon but greater burn-off (~50 wt%) is required to procure full access for the much larger silicone oil molecule. It is interesting to note the evidence for slot-like pores since the largest dimension of the benzene molecule and that for silicone oil are similar but benzene, a planar molecule, can gain access to the carbon structure at lower burn-offs than the spherical silicone oil.

The development of porosity using the chemical activation method can also be seen in Figure 8.3. A far smaller burn-off with steam is required to produce a given level of accessability to the pore system for any particular molecule when the cellulose precursor has been treated with zinc chloride.

Both steam and carbon dioxide develop only microporosity in chars in the early stages of activation but by about 70 wt% burn-off steam has produced a wide pore-size distribution whereas carbon dioxide has developed mainly microporosity over the whole range of activation burn-offs.[7] The carbons produced after activation with carbon dioxide had a lower pore volume than an equivalent steam-activated carbon, but approximately the same surface area. Activation with oxygen can be carried out but it is difficult to control the exothermic reaction and to obtain reproducible results. Pore volumes up to

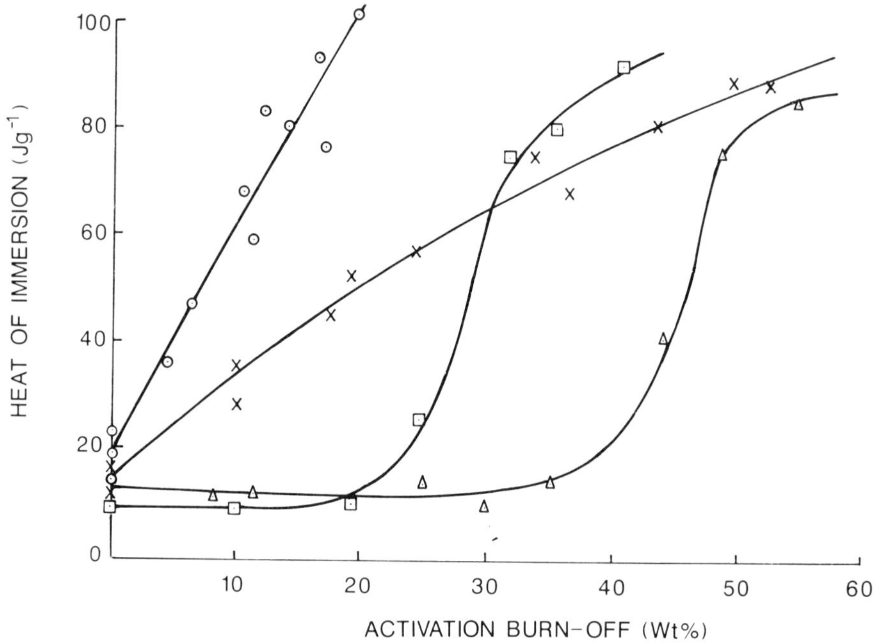

Fig. 8.3 Heats of immersion of cellulose charcoals at various extents of burn-off (all charcoals were steam activated). (○) Benzene, $ZnCl_2$ impregnated; (×) benzene, unimpregnated; (□) silicone fluid, $ZnCl_2$ impregnated; (△) silicone fluid, unimpregnated

0.8 $cm^3\ g^{-1}$ and 'surface areas' of 1800 $m^2\ g^{-1}$ can result from extensive activation, but a typical medium-activity commercial respirator filter carbon made from coconut shell has the capacities shown in Table 8.1[8] and can be packed into a filter bed at a dry bulk density of 0.486 $g\ cm^{-3}$. The filter has a saturation capacity for physically adsorbed vapours of approximately 0.24 cm^3 of vapour (as liquid) per cm^3 of filter bed. The capacity of the carbon for any other vapour at any concentration can be calculated from a knowledge of the adsorption isotherm for one vapour, by means of the Dubinin–Radushkevich equation.[9]

In general, as the boiling point of the substance falls, the adsorption capacity of the carbon and the protection time of the filter falls (Figure 8.4). The normal boiling point of a substance is a good measure of its potential for physisorption and a rule of thumb is that at ambient temperatures (~20°C) charcoal filters should not be used for protection against the vapours of substances boiling below 60°C unless the charcoal has been specially modified to give it a particular affinity for the substance under consideration.

Table 8.1 Adsorption capacities at saturated vapour pressure

Vapour	Molar volume $cm^3\ mol^{-1}$	Capacity $cm^3\ g^{-1}$
Benzene	89	0.456
Chlorobenzene	102	0.471
Carbon tetrachloride	96	0.452
1-Bromobutane	105	0.508

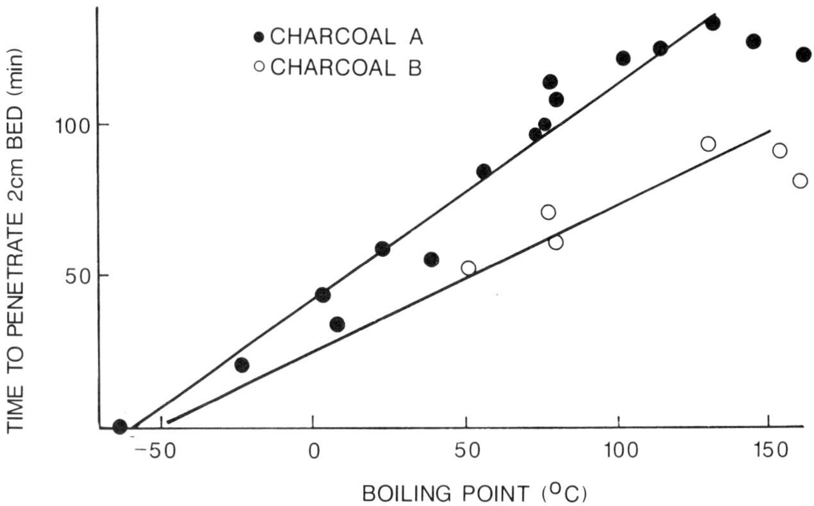

Fig. 8.4 Time to penetrate a 2 cm bed of charcoal versus liquid boiling point

Chemisorption capacity of respirator charcoals

To cope with vapours and gases from materials with boiling points significantly lower than 60 °C, where the physisorption forces are weak, active carbons are modified by the incorporation of reactive impregnants into their structure. The role of the carbon modified in this way ranges from that of a simple large area support for a reactant or catalyst to being an essential participant in chemisorption reactions to immobilize the gas or vapour being removed from the air. Often the precise mechanisms are not understood and the impregnation process is a 'black art'. To retain the non-specific physisorption characteristics of respirator charcoals, the nature and extent of impregnation must necessarily be limited. An account of commercially important impregnated respirator charcoals is presented below.

Copper-impregnated carbons

Copper-impregnated carbons for the removal of hydrogen cyanide, hydrogen sulphide and arsine have been produced commercially since World War II.[6,10] Suitable coals are finely ground and 2–5 wt% of copper oxide, sulphate or carbonate is intimately mixed with the coal powder. The mixed powder is then pressed into briquettes about 5 cm in diameter and about 2.5 cm thick using pressures of 1.5–2 cm^{-2}. The briquettes are then carbonized, steam activated, crushed, sieved and graded for use in air filters. Incorporation of copper into the raw material prior to carbonization ensures that the physical adsorption capacity of the carbon is not affected as it is when impregnants are added to the active carbon after its manufacture. X-ray diffraction of the carbons shows that the copper is initially present as small (~0.1 μm) metal crystallites thought to be distributed over the surface of the mesopores.

With time and exposure to air, the copper is converted into oxide which is still active against hydrogen cyanide and hydrogen sulphide. On long-term storage, however, any sulphur present in the carbon, whether inorganic or organic, will react with the impregnant, ultimately to form copper sulphate. This is able to convert hydrogen cyanide into cyanogen, which is not retained by carbon, with the result that the charcoal is no longer suitable for respiratory protection. The charcoal activity can be restored by heating the material to eliminate the sulphur as sulphur dioxide gas and to produce fresh copper crystallites. Coppered charcoals are capable of removing between 1 and 2 wt% of hydrogen cyanide and sulphide whilst still retaining their non-specific adsorption affinity for vapours removed from the air by the physical adsorption process.

Whetlerized carbons

The history of Whetlerized carbons is well documented.[10] When impregnants to the carbon are added before carbonization, the retorts and other equipment become contaminated, restricting their subsequent use for the preparation of charcoals for other purposes, especially for use in the food industry. Over the years great efforts have been made to impregnate charcoals after the activation stage has been completed, usually by spraying them with aqueous solutions of metal salts and organic reactants followed by drying in a rotary kiln.

The early copper-impregnated charcoals for protection against hydrogen cyanide and hydrogen sulphide produced this way at the end of the Great War were called 'Whetlerites' after their developers, J.C. Whetzel and E.W. Fuller. The original product, Whetlerite A was prepared by precipitating hydrated copper oxide on the parent charcoal by the action of hot caustic soda on a slurry of copper sulphate solution and charcoal followed by drying in a rotary dryer. Later, Whetlerite AS was prepared in a similar way but with 0.1–0.5 wt% silver nitrate added to the impregnating solution. The silver supported on the carbon was found to oxidize trivalent phosphorus and arsenic compounds to the corresponding pentavalent states.[f]

The reaction products have increased physisorption affinity on the charcoal. The range of toxic gases removed by the treated carbons can be further extended by the addition of Cr^{6+} to the charcoal from ammoniacal solution. The product, Whetlerite ASC, is currently extensively used in gas masks in many nations. Although Cu^0 and Cu^+ chemisorb hydrogen cyanide, Cu^{2+} ions will oxidize some of the cyanide to the weakly physisorbed cyanogen $(CN)_2$. In the presence of the Cr^{6+} ion, the cyanogen is hydrolysed through cyanoformamide to oxamide, a solid product which is firmly retained by the charcoal. However, with exposure to air and moisture, the ASC charcoal ages[11] as the Cr^{6+} is reduced on the carbon surface to Cr^{3+} which is inactive in the hydrolysis of cyanogen. Thus, when aged ASC charcoal filters are challenged with hydrogen cyanide, cyanogen gas breaks through the filter first, as shown in Figure 8.5.

Because of the toxic nature of chromium, its replacement with vanadium, molybdenum and zinc has been studied[6,10] and the use of zinc impregnations for charcoals used for the removal of hydrogen cyanide from tobacco smoke has found extensive application in the production of cigarette filter tips.[12–15]

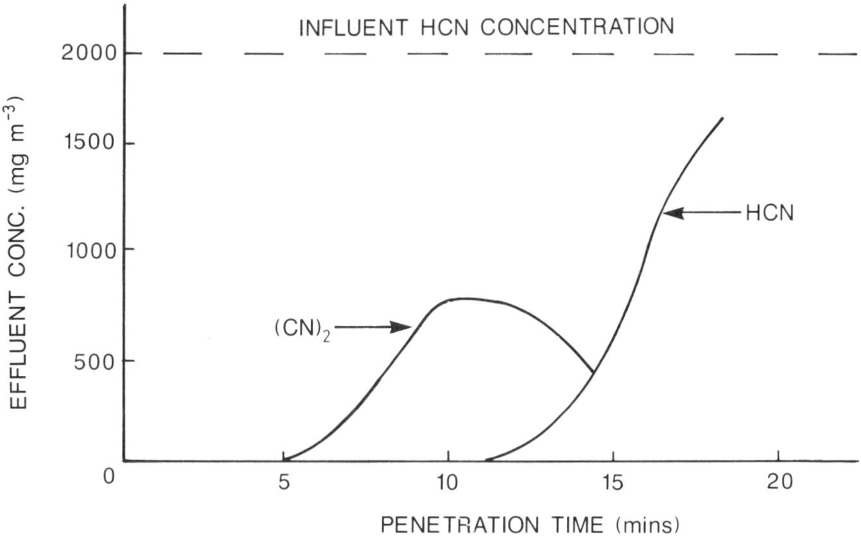

Fig. 8.5 Penetration of HCN and $(CN)_2$ through filters containing aged ASC charcoal

The impregnation of ASC and other copper-containing respirator charcoals with a variety of amines has been used to enhance their ability to remove cyanogen chloride from air. Increasing the loading of amine reduces the capacity of the charcoal to other physically adsorbed molecules, but the reactivity towards cyanogen chloride increases in proportion to the amount of amine present. For many years, pyridine was the amine of choice but triethylenediamine has been found to be more effective and is now used in commercial and military respirator filters.[16] The charcoal is not simply a carrier for amine, but Baker and Poziomek[17] showed that a charcoal impregnated with 4-vinylpyridine was 25% more effective than a pyridine-impregnated charcoal in adsorbing cyanogen chloride, even through 4-vinylpyridine is less basic than pyridine. The orientation of the adsorbed molecules must be more favourable in the former case, giving the cyanogen chloride greater accessibility to the basic nitrogen atoms. The capacity of this carbon for adsorbing water was also reduced to two thirds of that of the pyridine carbon, indicating that the degree of oxidation of the charcoal surface prior to impregnation may well influence the adsorption of 4-vinylpyridine.

Other impregnations for the removal of hydrogen cyanide

Smith and Alder[18] have investigated the impregnation of active charcoal cloth with a variety of transition metal salts to remove cyanide by complexing. They concluded that the salts were present as a thin film of solution in water adsorbed by the carbon and that reaction took place in the film itself. The role of the anion was important in that the salt solution should be as basic as possible to favour the dissociation of HCN to CN^- and that the salt should be as water soluble as possible. These considerations led to the choice of cobalt and nickel acetates as the complexing reagents of choice.

Precious metal catalysts on activated carbon are finding increasing use. Van der Smissen[19] has patented a zinc–palladium catalyst for hydrogen cyanide and hydrogen sulphide removal from air and platinum–palladium catalysts on charcoal are very effective in the oxidation of carbon monoxide to carbon dioxide for lightweight escape hoods for use in fire situations.

Respirators for the removal of radioactive iodine and methyl iodide arising from coolant release and ventilation in nuclear installations widely use charcoals impregnated with triethylenediamine (TEDA) and KI_3, either separately or together.[16,17,20] Whereas the interaction with TEDA is reversible and the respirator service life is adversely affected by temperature, the service life of filters containing KI_3 is enhanced by operation at high temperatures.

Adsorption of water by charcoals

In realistic use, respirator filters are exposed to air containing water vapour in addition to the toxic material to be removed and hence there will be competition between the two components for adsorption from air on the charcoal surface. Water, having a low polarizability, will have only a weak interaction with a non-polar charcoal surface. At similar relative vapour pressures, larger and often polar toxic molecules will be much more strongly adsorbed. However, the water is likely to be present at much higher concentrations than the toxic material and, since the respirator may have been in use and exposed to moist air for some considerable time prior to exposure to the toxic vapour, the micro- and mesopores of the charcoal may already be filled with water. The kinetics of adsorption of the toxic vapour are therefore likely to be determined by the kinetics of water displacement and this is likely to be a considerably slower process than the adsorption of the toxic vapour by the charcoal in spite of the fact the adsorption is considerably impeded (for low-affinity materials) by the presence of moisture (Figure 8.6). For this reason, national respirator standards require performance testing to be carried out at high humidities with filters pre-equilibrated with water.[21] Further, the adsorption of water by activated charcoals is significantly affected by storage under moist conditions. Hall and Sing[1] showed water isotherms obtained on a fresh sample of activated charcoal and the same charcoal after storage for 18 weeks at 45°C under a relative humidity (RH) of 80%. It can be seen that the aged sample has a much higher capacity for water at humidities up to 70% and that more water is retained on desorption at low humidities. This higher affinity for water of the aged charcoal reflects in shorter protection times for filters filled with the aged charcoal.

This behaviour is generally explained[22–24] in terms of the generation of water adsorption centres which have a greater adsorption affinity for water than the normal non-polar charcoal surface. This arises from the chemisorption on the charcoal surface of water and/or oxygen following exposure to moist air.[25] The change in numbers (a_0) of these adsorption centres for the storage of a fresh nutshell charcoal stored at 80% RH and 20°C is shown in Figure 8.7. The value of a_0 was obtained using the Dubinin–Zhaverina equation. No changes in charcoal pore structure can be detected by nitrogen adsorption isotherm measurements after this reaction.

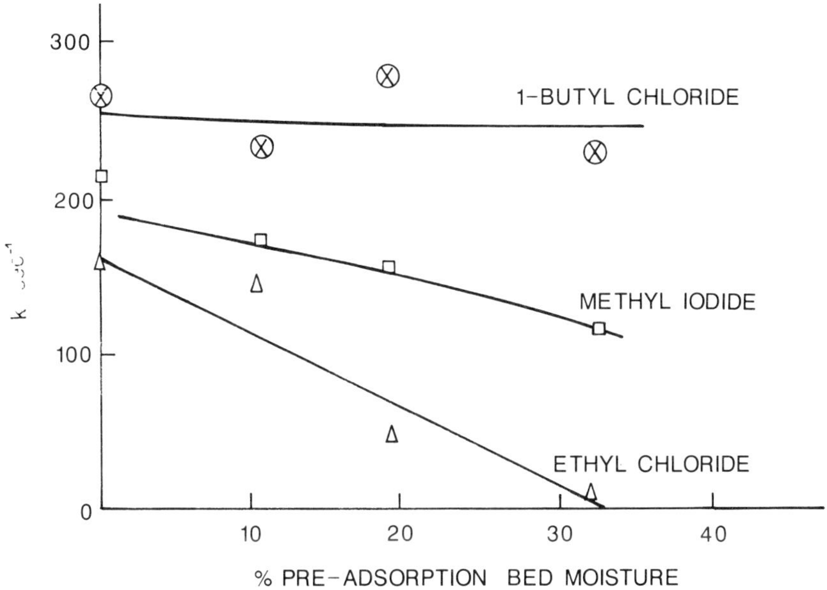

Fig. 8.6 Kinetics of adsorption as a function of charcoal moisture content for three substances of differing adsorption affinity

These findings have led to several attempts to control the generation of water adsorption centres to produce active charcoals for air filters which are less susceptible to exposure to moisture. It has been shown by selectively blocking sections of the charcoal porous system[26] that the adsorption centres for water are probably situated at the edges of the aromatic sheets in the charcoal structure.

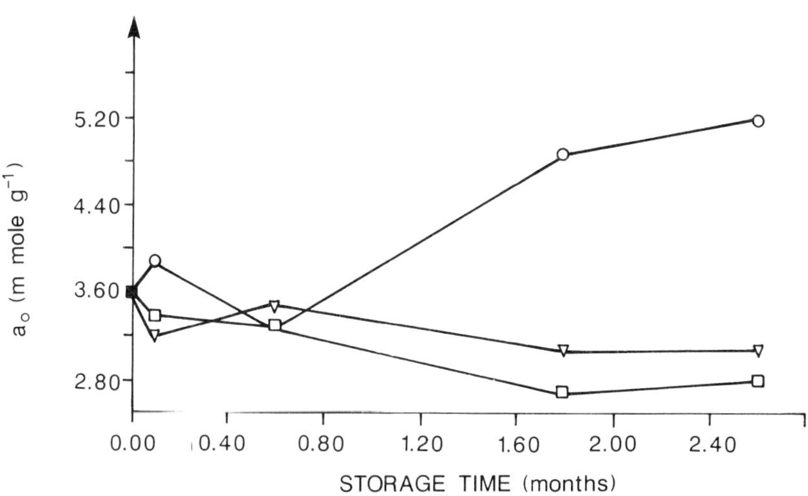

Fig. 8.7 Increase in water adsorption sites (a_0) with storage time under various storage conditions: (○) moist air; (▽) dry nitrogen; (□) dry air.

Once these sites are filled it seems likely that further water molecules add to the adsorbed water and the resulting clusters grow to fill the micropore spaces between the aromatic layers. The adsorption sites can be eliminated by heating the charcoal to 400°C in an inert or reducing atmosphere, but they are again slowly regenerated on further exposure to air and moisture. Holmes[27] attempted a more permanent solution to the problem by chemically modifying the charcoal through the introduction of specific functional groups onto the charcoal surface. Directly fluorinated carbons proved to be unstable but modification with the selective fluorinating agents hexafluoropropane and 1,1-difluoroethene yielded stable hydrophobic carbons which adsorbed vapours from humid air much more efficiently than similar untreated carbons. Some ageing did take place in the long term, however. A more permanent change to the carbon surface was effected by treatment with carbonyl chloride and chlorine, followed by solvent extraction to remove any adsorbed reagents and reaction products. These carbons had an enhanced hydrophobic character which, more importantly, resisted ageing on exposure to moist air.

Charcoals treated with oxidizing agents such as nitric acid, nitrogen dioxide, ozone and aqueous solutions of halogens showed enhanced adsorption of water due to the increased number of polar adsorption centres.

8.4 Displacement of adsorbed water

Although the surface chemistry of charcoals is important in determining the quantity of water adsorbed from air at a particular relative humidity, the kinetics of displacement of the water by the incoming larger toxic molecules to be filtered from the air are determined by the pore structure of the charcoal. Lawrie and co-workers[26] showed that filters loaded with charcoals having a high proportion of mesoporosity were considerably less affected by moisture than filters filled with charcoals having less well developed mesoporosity. The presence of mesopores in the carbon gave rise to filters having penetration times 25% longer than similar filters loaded with charcoals of equal surface area but having less well developed mesopores.

The progressive displacement of water by the incoming vapour molecules has been described by Hall and Holmes[28] and Alder and co-workers.[29] Often, in practical usage, the displaced water cannot be removed by the air stream as quickly as it is being generated since the air becomes saturated with water vapour. The water therefore begins to accumulate in the transport (meso)pores of the charcoal and as liquid in the interparticulate spaces in the filter bed.[26] If the mesopore system has sufficient capacity then the displaced water will have less effect on the filter protection time than if it is accommodated in the interparticulate spaces where it effectively blocks the entrance of vapour molecules into the charcoal grains.

Often, charcoals manufactured from peat and lignite have a high iron content and are found to have a particularly well-developed meso- and macropore system[30] Rodríguez-Reinoso *et al.*[31] showed that iron present in the charcoal precursor material can enhance the development of mesoporosity in the bonization and activation process.

8.5 Charcoal grain size and packing

Charcoal grain size influences filter performance in two ways. As the grain size is increased, the resistance of the air filter decreases and the specific surface area of the charcoal grains in the filter decreases. The relationship is described by Darcy's law, which indicates that the resistance to flow is directly proportional to the specific surface area of the granules in a packed bed. The rate of vapour adsorption for a filter, however, is proportional to the specific surface area of the charcoal granules in the filter, hence the improvement in adsorption performance gained by using finer grains has to be balanced by the filter designer with the increased resistance to airflow through the filter.

The influence of the two parameters is shown in Figures 8.8 and 8.9 and Table 8.2. Note that specific surface area is a function of the reciprocal of the grain radius. The efficiency of utilization of the adsorption capacity of the charcoal falls as the critical bed depth increases.

Table 8.2 Influence of grain size on adsorption rate and critical bed depth

Grain size (BSS)	Mean diameter mm	Critical bed depth, d_c mm	Adsorption rate constant s^{-1}
14–25	0.80	5	630
10–14	1.45	5.75	530
7–14	2.15	13.25	322
$\frac{3}{16}$–7	3.60	19.50	289

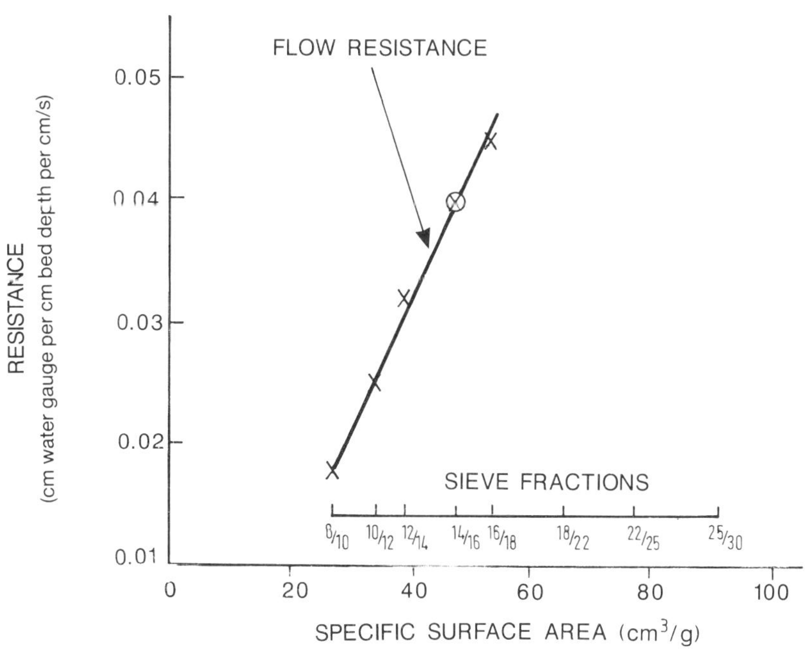

Fig. 8.8 Dependence of flow resistance on granule specific surface area

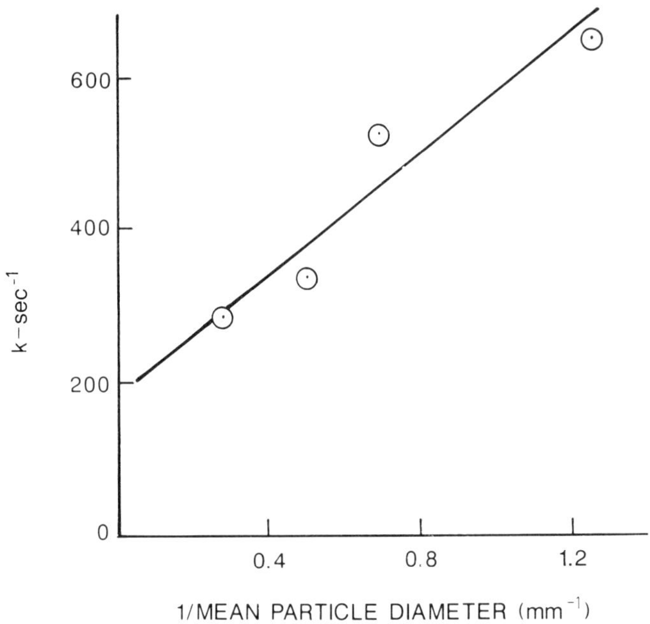

Fig. 8.9 Rate of bromobutane vapour adsorption with charcoal grain size

To obtain the maximum packing density of charcoal in a filter bed and to ensure maximum reproducibility in performance, a number of vibration techniques to settle filter beds are used. A novel packing technique developed by the British Ministry of Defence is the so-called 'snowstorm' method,[32] where a uniform distribution of granules falling vertically is collected in a filter body until it is full. In this way, beds with packing densities equivalent to spherical close packing can be achieved.

8.6 Limitations on the use of the respirator and other charcoal air filters

There is a tendency by the user to regard the adsorption of gases and vapours on air filter charcoals as 'irreversible'. This is not so. Respirator manufacturers usually recommend that filters be changed after use. In another application, increased consciousness of the harmful effects of solvents has led to extensive use of portable fume-hoods fitted with charcoal filters. All charcoal filters, besides having a finite adsorption capacity, behave as a 'chromatographic column' with a long retention time. However, even if only a tiny fraction of the bed capacity is filled with adsorbed vapour, the continued passage of air through the filter after its exposure to vapour will eventually lead to elution of the adsorbed material. The characteristic retention time for the filter will be dependent on the adsorption affinity the material has for the carbon (approximated by its boiling point). The displacement of the adsorbed material will also

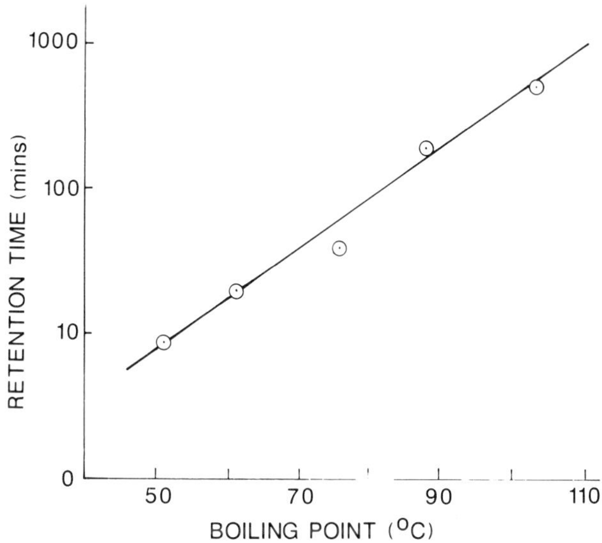

Fig. 8.10 Influence of boiling point on retention time

be hastened if the air relative humidity is high. These effects are illustrated in Figures 8.10 and 8.11 where the filters were doped with adsorbed material to 0.1% of their capacity and air drawn through the filters.

The displacement effect is more marked if a low boiling point material is displaced from a filter by a material of higher boiling point.[33] This phenomenon can be turned to advantage in the separation of mixtures of vapours of differing boiling point. As the mixed vapours enter the filter, the 'heavier' vapour will be adsorbed on the front of the filter bed and the 'lighter' vapour further into the filter. As time progresses, the lighter vapour will breakthrough the filter and the effluent will have the influent concentration of light vapour. With further exposure of the filter to the vapour mixture, the adsorption front of the 'heavier' vapour advances through the filter, displacing the already adsorbed 'light' vapour so that the effluent concentration of light vapour now rises considerably above that in the influent air. Eventually breakthrough of the heavy vapour takes place and a mixed effluent ensues. This process is used commercially for the recovery and separation of chlorofluorocarbon refrigerants.

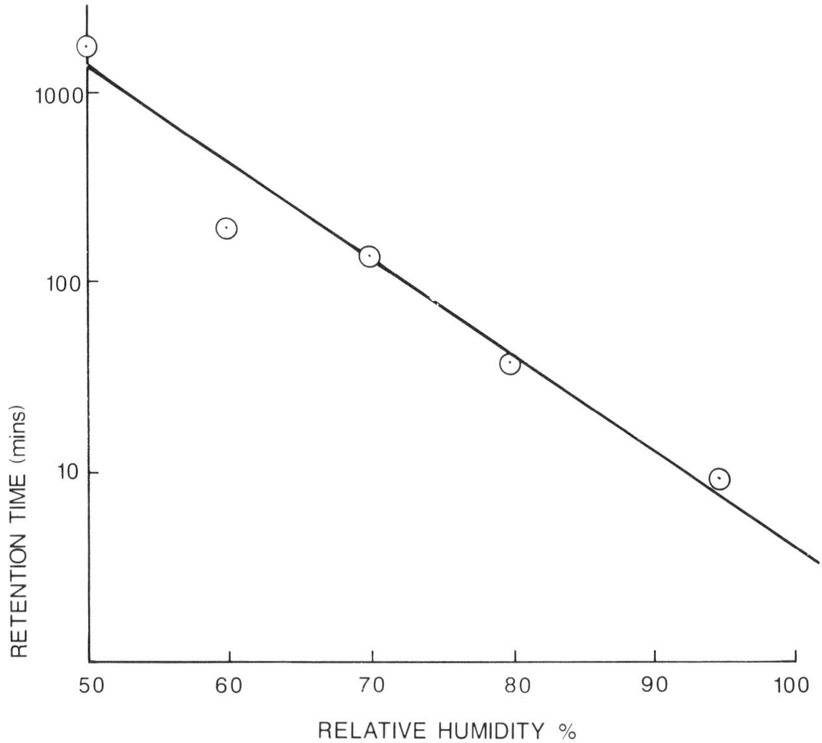

Fig. 8.11 Influence of relative humidity on retention time

8.7 References

1. HALL, C.R. and SING, K.S.W. 1988: *Chemistry in Britain* **24**, 670.
2. IUPAC 1972: Manual of Symbols and Terminology. Appendix 2, Part 1, Colloid and Surface Chemistry. *Pure and Applied Chemistry* **31**, 578.
3. MAGGS, F.A.P. 1953: *Research Correspondence* **6**, 13.
4. BACON, R. and TANG, M.M. 1964: *Carbon* **2**, 211.
5. BAILEY, A. and MAGGS, F.A.P. 1972: British Patent No. 1301101.
6. DANBY, C.J., DAVOUD, J.G., EVERETT, D.H., HINSHELWOOD, C.N. and LODGE, R.M. 1946: *Journal of the Chemical Society*, 918.
7. TOUTKOW, K., SIEMIENIEWSKA, T., CZECKOWSKI, F. and GANGOWSKA, A. 1977: *Fuel* **56**, 121.
8. BAILEY, A. and SMITH, M.E. 1980: paper presented at the International Conference on Filtration and Respiratory Protection, Copenhagen.
9. GREGG, S.J. and SING, K.S.W. 1982: *Adsorption, Surface Area and Porosity*. New York: Academic Press.
10. USA, PB158505, 1964: *Military Problems with Aerosols and Non-persistent Gases*. 1964: Summary Technical Report of Div. 10, National Defence Research Committee.
11. ALVES, B.R. and CLARK, A.J. 1986: *Carbon* **24**, 287.
12. KEITH, C.H., NORMAN, V. and BATES, W.W. Jr, 1966: US Patent No. 3251365.

13. KEITH, C.H., NORMAN, V. and BATES, W.W. Jr, 1969: US Patent No. 3460543.
14. KEITH, C.H., NORMAN, V. and BATES, W.W. Jr. 1967: US Patent No. 3355317.
15. KIEFER, J.E. 1971: US Patent No. 3618619.
16. WOOD, G.O. 1985: *American Industrial Hygiene Association Journal* **46**, 251.
17. BAKER, J.A. and POSIOMEK, E.J. 1974: *Carbon* **12**, 45.
18. SMITH, S.J. and ALDER, J.F. 1988: *Carbon* **16**, 701.
19. Van der SMISSEN, C.E. 1985: German Patent No. DE3405100.
20. BILLINGS, B.H.M., DOCHERTY, J.B. and BERVAN, M.J. 1984: *Carbon* **22**, 83.
21. British Standard 2091. 1969: *Specification for Respirators Against Harmful Dusts and Gases.*
22. EVANS, M.G. and BILLINGS, B.H.M. 1984: *Journal of Chemical Physics* **81**, 779.
23. ADAMS, L.B., HALL, C.R., HOLMES, R.J. and NEWTON, R.A. 1988: *Carbon* **26**, 451.
24. HALL, C.R. ADAMS, L.B. and HOLMES, R.J. 1988: in McEnaney, B. and Mays, I.J. (eds) *Carbon '88, Proceedings of the 5th International Conference on Carbon, Newcastle, September 1988.* Bristol: IOP Publishing, p. 137.
25. EVERETT, D.H. and WARD, R.J. 1986: *Journal of the Chemical Society, Faraday Transactions I* **82**, 2915.
26. BAILEY, A., LAWRIE, G.A. and WILLIAMS, M.R. 1990: in *Proceedings of International Conference on Carbon.* Paris: Groupe François d'étude des Carsones; Lawrie, G.A. 1989: *PhD Thesis.* Cranfield Institute of Technology (RMCS).
27. HOLMES, R.J. 1991: *PhD Thesis.* Brunel University.
28. HALL, C.R. and HOLMES, R.J. 1989: *Adsorption Science and Technology* **6**, 83.
29. ANSTICE, P.J.C., HALLIDAY, N. and ALDER, J.F. 1991: *Adsorption Science and Technology* **8**, 107.
30. BANSAL, R.C., DONNET, J. and STOECKLI, F. 1988: *Active Carbon.* New York and Basle: Marcel Dekker.
31. RODRÍGUEZ-REINOSO, F., LINARES-SOLANO, A., MOLINA-SABIO, M. and PEREZ-LLEDO, I. 1985: in *17th Bienniel Conference on Carbon.* Lexington, Kentucky: American Carbon Society, p. 237.
32. HARRIS, E. 1948: British Patent No. 606867.
33. YOON, Y.H., NELSON, J.H., LARA, J., KAMEL, C. and FERGEAU, D. 1991: *American Industrial Hygiene Association Journal* **52**, 65.

9 Activated Carbons—Production and Applications

Frank Derbyshire, Marit Jagtoyen and Michael Thwaites
Center for Applied Energy Research
University of Kentucky
3572 Iron Works Pike
Lexington
Kentucky 40511-8433
USA

9.1 Introduction

The term 'activated carbon' defines a group of materials with highly developed internal surface area and porosity and hence a large capacity for adsorbing chemicals from gases or liquids. Activated carbons are extremely versatile adsorbents of major industrial significance and are used in a wide range of applications which are concerned principally with the removal of species by adsorption from the liquid or gas phase, in order to effect purification or the recovery of chemicals. They also find use as catalysts or catalyst supports. The strong market position held by activated carbon adsorbents relates to their unique properties and low cost compared with that of possible competitive adsorbents.

Activated carbons can be produced from virtually any carbonaceous solid precursor, both naturally occurring and synthetic. Process economics normally dictate the selection of readily available, low-cost feedstocks. Common examples of commercial feedstocks are materials of botanical origin (e.g. wood, coconut shell and nut kernels) and degraded or coalified plant matter (e.g. peat, lignite and all ranks of coal). The price of the end-product is dependent on the degree of processing required to attain the desired adsorptive properties. The more commonly used activated carbons are produced in neo-commodity volumes by relatively straightforward processing methods, and in powdered, granular or preformed shapes (pellet, extrudate).

The adsorptive properties of activated carbons derive from their extensive internal pore structure, which presents a high surface area for the adsorption of molecular species. Activated carbons can essentially be considered as solid sponges. The processes used to create high porosity from a low-porosity starting material are known as activation. Surface areas as high as 3000 $m^2 g^{-1}$ have been produced, although a commercial activated carbon will have a surface area typically in the range 1000–1500 $m^2 m^{-1}$. These values are much higher than those measured for other porous solids, and begin to explain the attraction and diverse uses of these materials. The microporosity of porous carbons derives from two structural considerations: the small crystallite size of the pseudo-graphitic building blocks and the lack of crystallite alignment.[1] Poor alignment is associated with the presence of voids or microporosity. If the precursor is thermoplastic, the development of a fluid phase during carbonization facilitates crystallite alignment and a loss of porosity. Hence an essential requirement is that carbonization takes place in the solid phase. The precursor must be intrinsically thermosetting or be converted from thermoplastic to thermosetting behaviour by some process technique (the suppression of fluidity development in bituminous coals by oxidation is a particular example).

Differences in the pore-size distribution affect the adsorption capacity for molecules of different sizes and shapes, and this is one of the criteria by which carbons are selected for a particular application. Micropores (below 2 nm diameter) contribute most to the high surface areas of activated carbons and provide high adsorptive capacities for molecules of small dimensions such as gases and common solvents. Mesopores (2–50 nm diameter) are important for the adsorption of larger molecules, such as colour bodies, and probably provide much of the adsorption surface for carbons impregnated with chemicals. It is considered that macropores (above 50 nm diameter) are normally unimportant for adsorption and that their main role is to serve as feeder pores for the transport of adsorbate molecules. However, in some applications they can exert a more significant influence on adsorptive behaviour.[2]

Although the selection of processing conditions can allow some degree of latitude in controlling the structure and properties of the final product, the properties of activated carbons are determined to a large extent by the nature of the precursor, and also by the particular processing technology. The differences in the adsorptive characteristics of carbons produced from different starting materials are such that, in many applications, one carbon type will prove to be the most appropriate.

Little pertinent information exists in the open literature on the fundamental relationships between the structures of the precursor and carbon product, or the process chemistry. Although the manufacture of activated carbons is a long-established industry, it has not experienced anything like the same level of growth as, for example, did the coking industry. Most of the world production is in the hands of a relatively small number of manufacturers. Each of these tends to use different raw materials and/or different process methods, the result of which is that the properties of the activated carbons offered are very different.[3] This means that there are only a few market segments where there is serious competition. Consequently, there has been no great incentive for major investments in research and development. This position is beginning to change as applications become more diverse, concern over environmental issues continues to grow and demand for higher performance adsorbents continues to increase.

This chapter is concerned with the production and applications of activated carbons and provides a perspective on historical production and use, markets, processing methods, industrial equipment, and applications. Useful references are the texts by Hassler[4] and Bansal *et al.*[5] and the reviews by Houghton and Wildman,[6] Wilson[3] and McCabe.[7] The authors would also like to acknowledge the use of documents, written by J. Wildman of Sutcliffe Speakman Carbons, that have not been disseminated in an accessible published form. Throughout this chapter these publications have been widely used as sources of information.

9.2 Historical production and use

The history of the technology of activated carbon synthesis and the industrial needs that provided the driving force are particularly relevant to considering modern practice, as many of the processes now in use, and possibly many future developments, have their origins in earlier discoveries. For convenience, the salient points of this history are chronicled in tabular form.

Date	Significance
3750 BC	Earliest known use by the Egyptians and Sumerians. Wood chars (charcoal) used for the reduction of copper, zinc and tin ores in the manufacture of bronze. Charcoal also used as domestic smokeless fuel.
1550 BC	The first recorded application of charcoal for medicinal purposes was cited in Egyptian papyri.[8] The principal use appears to have been the application of charcoal to adsorb odorous vapours from putrefying wounds and from within the intestinal tract.

460 BC	Hippocrates and Pliny record the use of charcoal to treat a wide range of complaints including epilepsy, chlorosis and anthrax.
450 BC	Recent studies of the wrecks of Phoenician trading ships suggest that drinking water was stored in charred wooden barrels. This practice was certainly still in use in the 18th Century for extending the use of potable water on long sea voyages. Hindu documents of the same period (450 BC) also refer to the use of sand and charcoal filters for the purification of drinking water.
157 AD	Claudius Galen wrote 500 medical treatises, many of them referring to the use of carbons of both vegetable and animal origin for the treatment of a wide range of diseases.
297 AD	Roman Emperor Diocletian ordered the total destruction of all scientific books within the Roman Empire, thereby setting back the progress of science within Europe by a least 1000 years.
18th Century	
1773	The specific adsorptive powers of carbons (i.e. charcoal) were recognized by Scheele, who measured the volumes of various gases that could be adsorbed by carbons derived from different sources.[9]
1777	It was reported that heat effects are associated with the adsorption of gases by charcoal.[10] This is significant in that it led later to the 'condensation theories of adsorption'.
1785	Lowitz reviewed the known abilities of charcoals to adsorb odours emanating from medical conditions and published accounts of their ability to adsorb vapours from a range of organic chemicals.[11] In addition, he studied the effectiveness of charcoal in decolorizing various aqueous solutions and, in particular, its commercial application to the production of tartaric acid. This appears to be the first systematic account of the adsorptive powers of charcoal in the liquid phase. At this time, too, the developing sugar refining industry was looking for an effective means of decolorizing raw sugar syrups. However, the wood charcoals available at this time were not particularly effective in this role,[11] presumably because their porosity had not been developed beyond the extent produced by carbonization.
1793	Kehl discussed the use of chars for the control of odours from gangrenous ulcers and discovered that carbon prepared from animal tissues could be used for removing colours from solution.[12]
1794	An English sugar refinery successfully used wood charcoal for the decolorization of sugar syrups but kept the method of preparing the carbon a secret.[13]
19th Century	
1805	Gruillon introduced the first large-scale sugar refining facility in France using ground and washed wood charcoal for decolorizing the syrups.[14]
1805–08	Delessert successfully demonstrated the use of charcoal for decolorizing sugar-beet liquor.[15] He was directly responsible for the growth of the sugar beet industry in France. By 1808 all sugar refineries in Europe used charcoal as a decolorizer.[15]
1811	Figuier discovered the greatly enhanced decolorizing capability of bone char compared with wood char.[16] The sugar refining industry was quick to substitute bone char for wood char. Methods of regenerating the bone char by heating were discovered, and shortly afterwards a granulated bone char was developed which could be much more readily regenerated.
1815	By this date most of the sugar refining industry had switched to the use of granulated bone char as a decolorant.
1817	Joseph de Cavaillon patented a method of regenerating used bone chars.[17] The method was not entirely successful.

1822	Bussy[18] demonstrated that the decolorizing properties of carbons were inherent to the source material and also depended on the thermal processing and the particle size of the finished product. He demonstrated that carbonization at too high a temperature, or for too long, reduced the adsorptive properties and that porosity was important, although he had no means to measure this factor. He also heated blood with potash to produce a carbon with 20–50 times the decolorizing power of bone char. This is the first recorded example of producing an activated carbon by a combination of thermal and chemical processes.[18]
1841	Schatten systematized the use of a hydrochloric acid wash prior to heating in the regneration of bone chars.[19] This effectively removed the mineral salts adsorbed on to the carbon. He also introduced in Germany the first continuous process vertical kiln for the manufacture and regeneration of bone chars.[19]
1854	Stenhouse described the successful application of carbon filters for removing vapours and gases in the ventilation of London sewers.[20]
1862	Lipscombe prepared a carbon to purify potable water.[21]
1865	Hunter discovered the excellent gas adsorbent properties of carbons derived from coconut shells.[22]
1868	Winser and Swindells heated paper mill waste with phosphates.[23] Many of their disclosures are relevant to processes now in industrial use. They were not developed on a commercial scale because of engineering difficulties.
1881	Kayser first used the term 'adsorption' to describe the uptake of gases by carbons.[24]
20th Century	
1901	Von Ostrejko set the basis for the commercial development of activated carbons through processes involving (a) the incorporation of metallic chlorides with carbonaceous material before carbonization and (b) the mild oxidation of charred materials with carbon dioxide or steam at raised temperatures.[25]
1911	Marketing of first industrially produced activated carbon, 'Eponit' (trade name), by the Fanto Works, Austria. They adopted von Ostrejko's gasification approach in the manufacture of Eponit from wood. It was marketed as a decolorizer for the sugar-refining industry.[13] Until this time the principal user of active carbons was the sugar-refining industry and manufacturers produced active carbon by their own secret or patented process. All the principal types of reactor with the exception of the fluidized bed were by then in use.
1913	Wunsch heated a mixture of Eponit and zinc chloride and found that the decolorizing capacity of the reactivated material was greatly increased.[26] Wunsch went on to prepare other active carbons with superior properties to those of Eponit by substituting wood and other carbon precursors for Eponit in his process.[26] The zinc chloride process was used for many years, sometimes in combination with steam or carbon dioxide activation. It has now been largely superseded by the use of phosphoric acid as the chemical activant.
1914–18	The introduction of poisonous gases into the battlefields of the First World War gave great impetus to the development of large-scale production methods for adsorbent carbons suitable for use in military respirators. Granulated carbons, with adequate adsorptive powers and providing low resistance to air flow through the respirator canister, were developed by activating wood chips with zinc chloride. These were the first manufactured carbons with reliably controlled adsorptive and physical properties. A group studying under Chaney in the USA examined a wide range of precursors intended to produce adsorbents for gas-mask canisters. They determined that coconut shell yielded the best combination of required characteristics in the resulting carbon.[27]
Post-1918	The wartime developments to mass-produce activated carbons with closely controlled characteristics lead to a post-war expansion in the commercial production and application of carbons. In Europe considerable progress was made in the manufacture of active carbon from new raw materials. Coconut and

	almond shells, activated with zinc chloride, yielded active carbons with high mechanical strengths and adsorptive capacities for gases and vapours.
1935–40	In Czechoslovakia two varieties of pelletized carbons were produced from sawdust by zinc chloride activation, for the recovery of volatile solvents and for the removal of benzene from town gas. Mecklenburg, working on the dynamics of adsorption, and Kubelka, who interpreted sorption phenomena on active carbon by the mechanism of capillary condensation, developed a method for calculating the distribution of pore diameters in active carbon. They published a theoretical interpretation for the service time of a filter charged with active carbon. This theory and its mathematical development are now obsolete, but at the time represented a significant contribution in the field.

The advances made in the manufacture of granulated carbons during the First World War found a number of post-war applications in the purification of gases or in the extraction of vapours from gas streams. The process was applied to the recovery of liquid petroleum from natural gas, the extraction of benzene from manufactured coal-gas and the recovery of volatile solvent vapours from various industrial processes.

Since these early days, industry has developed a much wider range of active carbons for both liquid- and vapour-phase applications. Over the last two decades there has also been a developing interest in other carbon forms (e.g. activated carbon fibres and microbeads) and applications continue to be developed in response to growing demands for environmental protection and for emerging technologies, such as the catalytic reduction of NO_x, natural gas storage and solar cooling.

9.3 Markets for activated carbon

The world annual production of activated carbon in all forms is estimated to be in the region of 300 000–400 000 tonnes. About 55% of this total is in powder form, about 35% in granular form and the remainder is produced as pellets or extrudates. According to one reference,[5] about 80% of the total production (powder, granular and formed carbon) is used in liquid-phase applications and the remaining 20% (granular and formed carbons only) in gas-phase applications. The highest consumption is in Japan and the USA; per capita annual consumption is 0.5 kg in Japan, 0.4 kg in the USA, 0.2 kg in Europe and 0.03 kg in the rest of the world.

Consumption is increasing at a reported rate of about 7% per annum.[28] Growth areas are in air purification, solvent recovery and groundwater treatment and new applications are emerging, particularly those related to environmental protection, that will tend to sustain the level of demand. A list of some important areas of activated carbon use is given in Table 9.1.

Granular or formed carbons are more desirable for continuous or cyclic processing in fixed-bed operations, such as solvent-recovery plants or in water treatment. The pressure drop over the bed can be controlled by adjustment of the particle-size distribution. For the same reason, granular or formed carbons are used in respirators, where a balance is required to ensure adequate contact time for the removal of hazardous substances and to allow the wearer to breathe

Table 9.1 Applications of activated carbon

Liquid phase:	
Decolorization	Industrial water treatment
Potable water treatment	Metals recovery (Gold)
Groundwater treatment	Chemical and pharmaceutical purification
Gas phase:	
Solvent recovery	Military: respiratory and suiting
Dechlorination	Gas purification
General air conditioning	Cigarette filters
Air purification	Catalysis

freely. Granular carbons can also be removed from fixed-bed adsorbers for regeneration. In some instances, manufacturers now supply adsorber modules. This allows either on-site regeneration of the carbon or the replacement of the module. Powdered carbons find more intermittent use and often on a once-through basis. Their advantage over large particles is that there is less diffusional resitance to adsorption and, hence, much higher adsorption rates are attained. For the same reasons, there is a developing interest in the use of activated carbon fibres for certain applications.

The approximate average selling price for granular or formed carbons is between US$1500–4000 per tonne. For powder carbons, it is between US$500–1200 per tonne. The actual price will vary with the level of activity (surface areas commonly fall in the range 800–1500 $m^2 g^{-1}$), the strength and attrition resistance and whether impregnants have been added to enhance adsorption selectivity.

9.4 Manufacture

Outline of process methods

Activated carbons in the form of granules, extrudates, powders, fibres or beads can be produced from suitable precursors by either thermal or chemical routes. The thermal and most common route involves a two-step process of carbonization followed by activation via partial gasification. The sequence is intended to preserve and further develop the blueprint for pore structure which is inherent to the raw material (and which differs with raw material type). For the most part, the reactions are thermally driven. The second commercial route involves the reaction of a precursor, usually cellulosic, with chemical reagents such as zinc chloride or phosphoric acid in a single heat-treatment stage. In the absence of any formally agreed terminology, these two process routes will be referred to as thermal and chemical activation, respectively. The general features of both processes are described below. The precise details of the methods used by different manufacturers are not publicly available, for proprietary reasons.

Table 9.2 Commercial carbon precursors (source: Bansal *et al.*[5])

Precursor	Use (%)
Wood	35
Coal	28
Lignite	14
Coconut shell	10
Peat	10
Other	3

Thermal processing

Precursor selection

The principal commercial carbon feedstocks and the proportions in which they are used are summarized in Table 9.2. The selection of the precursor essentially determines the range of adsorptive and physical properties that can be attained in the activated carbon products. Important considerations to be made in selecting a source include cost, availability, consistency of quality, and, particularly for coals, peat and lignite, the mineral matter and sulphur contents.

Process

A generalized schematic diagram of the thermal processing route is shown in Figure 9.1. All of the possible steps that may be involved in manufacture of

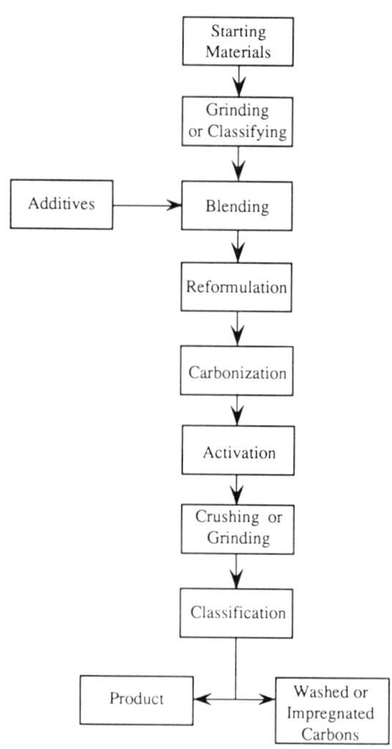

Fig. 9.1 Thermal processing route

activated carbons are shown, although only some of them will apply to any particular process.

The preparation of the precursor depends to a large extent on its nature and the desired form of the end-product. Coconut shell, nutshells, anthracite coal and certain woods can be carbonized in lump or granular form to produce hard carbons. Precursors which tend to form soft carbons have to be reconstituted at some stage, often using a binder, in order to produce strong granular or formed carbons. Alternatively, they may be used directly for the production of powdered activated carbons.

Bituminous coals are an important source of granular or formed activated carbons. In their natural condition, many of these coals possess thermoplastic properties and, as already noted, an essential prerequisite is that the carbon precursor is thermosetting. Thermoplasticity in native bituminous coals is destroyed by promoting cross-linking reactions, and this can be achieved by air oxidation prior to or during carbonization (using an air bleed), or by reaction with chemical additives such as boric acid. It may also be desirable to induce oxidative cross-linking when using a thermoplastic binder, such as pitch, even if the precursor itself is thermosetting. Blending different precursors has also been used to control thermoplasticity.[3,6,7,29]

Before carbonization, bituminous coals are first ground to a particle size less than about 5000–10000 nm and then reconstituted by briquetting under pressure, in some cases using a binder. This method of preparation eliminates the macroscopic heterogeneity in the coal and produces a more homogeneous and mechanically strong carbon precursor.

Carbonization is conducted at 600–650 °C in an essentially inert atmosphere (except, as noted, where an air bleed is introduced to effect cross-linking during the early stages of heat-up). Between 400 and 600 °C most organic solids undergo reactions leading to the loss of hydrogen and the formation of free radicals which condense to form a rigid cross-linked solid char.

The carbonization process causes some increase in porosity, although this is generally insufficient for practical use, and serves to modify the pore structure inherent to the precursor, as opposed to creating it. This is why the structure and properties of porous carbons are so strongly dependent on starting material.[30] Similarly, the activation process helps to develop further this structure by the selective gasification of carbon at 850–950 °C in steam or CO_2 or mixtures of these (in industrial furnaces, the gas mixture can be complex).

The removal of carbon atoms both increases the average size of the micropores already accessible to the gas and opens up closed porosity. The principal changes in pore structure with progressive gasification concern the micropores (<2 nm diameter), while changes in the mesopore (2–50 nm diameter) and macroporosity (>50 nm diameter) are small.

The production of the desired final product size ranges may be achieved by size reduction and grading before or after carbonization, after activation or a combination of these. The precursor size is not so critical for carbonization, as this process tends to proceed uniformly through the material. Activation is a heterogeneous, solid–gas reaction that can be diffusion limited, resulting in a higher degree of activation at the particle periphery than at the core. The activation of large pieces may result in a wide gradation of activity between the centre and outside.[7] The product size ranges formed by subsequent granulation will then possess a wide distribution of activities, whose average properties may appear acceptable to a given application, but whose performance in service may be inadequate. Consequently, in almost all processes, the feed to the activation kiln is as close to the final desired size as possible. If the precursor is carbonized in a larger size, it is usually granulated before activation in order to ensure that activation proceeds as uniformly as practicable.

Following activation, there is a final screening and de-dusting operation. There may also be an additional size reduction step in this series of operations. However, even if there is not, attrition in the activation furnace and during subsequent handling will inevitably produce a proportion of powdered activated carbon. Bearing in mind the difference in sale price between granular and powdered activated carbon, there is a reasonable incentive to consider the recycle and reconstitution of the fines rather than their direct sale.

The screened carbons may then be further processed to remove mineral constituents by leaching or to add chemical species by impregnation. The reduction in mineral matter content can increase the adsorption capacity through exposing more adsorption surface, can remove unwanted contaminants that may solubilize in certain applications and can elevate the ignition temperature of the carbon by removing metals, such as alkali metals, that can catalyse combustion.

For completeness, brief mention is made of using alkali catalysts to increase the rate of activation. The high reactivity of coal and lignite chars in air has been related to the concentration of calcium and magnesium, originally present as mineral matter.[31] Sodium and potassium are also known to catalyse carbon oxidation[1]; prior removal of soluble alkali metals can elevate the ignition temperature of activated carbons.[32] The catalysed gasification of coal chars using added catalysts has been the subject of earlier reviews.[33,34] It seems likely that potassium salts, such as K_2CO_3, are used industrially to catalyse the steam activation of carbons[7,35] although, for catalyst recovery and economy, a subsequent leaching stage must be added to the process.

Chemical activation

The second commercial route to producing porous carbons involves the reaction of a precursor with chemical reagents at temperatures in the range 500–900°C. The reagents most commonly used in industry are $ZnCl_2$ and phosphoric acid. The precursors are usually cellulosic materials, wood being the most common.[3,7,30] An example of a typical process flow sheet for the chemical activation process is shown in Figure 9.2.

The zinc chloride process was the most common method of chemical activation up to 1970.[36] Zinc chloride is added as an aqueous solution to the precursor and mixed and kneaded at a low temperature, around 130°C. The low-temperature reactions are reported to exert a large influence over the properties of the final product. The precursor and chemical activant are mixed and kneaded. During this stage there may be extensive digestion of the starting material. Optionally, the mixture may then be extruded before carbonization at temperatures between 600 and 850°C. The process economics depend heavily upon the efficiency of zinc chloride recovery. McCabe[7] reported a recovery efficiency of only 80–85%. Because of problems of corrosion, limited reagent recovery efficiency and environmental disadvantages, the use of $ZnCl_2$ is decreasing in favour of phosphoric acid.

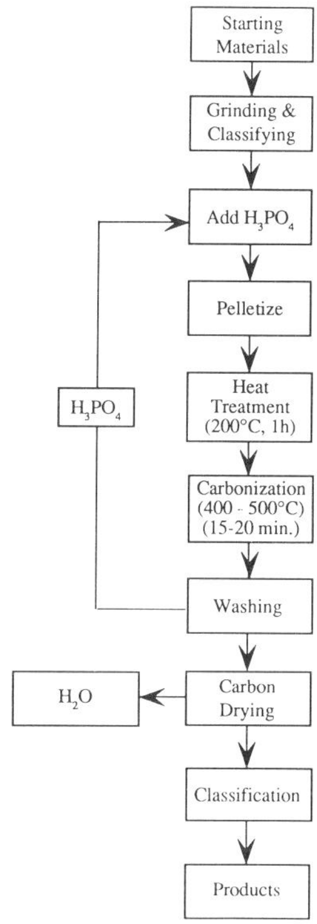

Fig. 9.2 Typical process flow sheet for the chemical activation process

In the phosphoric acid process, the temperature of activation is comparatively low, around 400–500 °C, and volatile losses are low. Phosphoric acid can be recovered at a high concentration by means of multi-stage extraction, and it is reported that there is no necessity for concentration after repeated use.[36] Norit (UK) use phosphoric acid to produce activated carbons from wood at a plant in Scotland. The plant has an efficient recovery system which is one of the most important unit operations in the process.[37]

With either chemical, the belief is that the relevant reactions primarily involve attack on the cellulosic structures in the precursor.[7,30] Water is eliminated and cross-linking and increased aromatization are promoted during carbonization. The effect of cross-linking is to bond otherwise volatile products into the precursor structure before reaching pyrolysis temperatures, with a consequential increase in the carbon yield and reduction in tar formation. The promotion of cross-linking is somehow related to the development of an extensive pore structure without the need for subsequent activation in steam or CO_2. Cross-linking is believed to inhibit shrinkage during heat treatment and help to create the micropore system.[30] Another aspect of this process is that while most of the chemical additive is removed from the resulting carbon by leaching for recovery and recycle before the product is dried, some of the chemical is incorporated into the carbon, which will, to some extent, modify its adsorptive properties.[37]

The pore-size distribution of the carbons is reported to be dependent on the ratio of chemical activant to precursor. A high ratio leads to wider microporosity and more developed meso- and macroporosity.[38,39] The optimum ratio of reagent to precursor with respect to adsorptive properties and yield is dependent on the precursor composition; for wood-fractions cellulose and lignin this ratio was found to be 1:1 and over 4:1, respectively.[40] Other work has shown that the development of porosity in carbons from wood and biomass is dependent on the ratio of cellulose to lignin.[40,41]

Normally, chemically activated carbons from wood are produced with high pore volume and low bulk density (~0.27 g cm^{-3}); these carbons are in powdered form and find application in liquid-phase adsorption, e.g. water purification. Attractive features of wood, and other biomass precursors, are low ash contents (0.3–1.1 wt%) in comparison with coals and lignite (2–15 wt%), and very low sulphur contents. Research on other precursors has shown that high-quality activated carbons can be produced from olive stone, coconut shell and coal, using phosphoric acid, and these have bulk densities around 0.48 g cm^{-3}.[35,42,43]

The bulk density of carbons produced from wood can be increased to~0.48 g cm^{-3} by compression or extrusion of the material during manufacture, so that there is the potential to tailor the properties of wood-based activated carbons to suit particular applications. Wood fractions have been used as effective binding agents for the production of formed carbons, as also has lignin.[42] Ammonium lignosulphate has been used as a binder to produce hard shaped carbons from soft carbonaceous raw materials such as wood and peat.[44]

Although commercial processes have been limited essentially to the use of zinc chloride and phosphoric acid, and wood precursors, there is much broader potential for this synthesis route. Brief mention will be made here of some of the published findings.

Much of the literature is concerned with $ZnCl_2$ activation. Studies of the activation of almond shells, nut shells, agricultural waste and different woods have shown that the main factors controlling porosity development are the lignin to cellulose ratio in the precursor, reagent to precursor ratio, heat treatment temperature and heating rate.[45–48] There have been relatively few studies on the synthesis of activated carbons from wood using other chemical reagents. Gonzalez *et al.*[49] studied the activation of olive wood with sulphuric acid and found that surface areas of 470 $m^2 g^{-1}$ were developed at 600–800 °C, the carbons being dominantly microporous (<2 nm pore diameter). Shasizadeh *et al.*[50] showed that useful grades of active carbon could be obtained from cellulosic wastes using acid catalysts such as $ZnCl_2$, phosphoric and boric acid.

The application of chemical activation to coals has been still more limited. A patent awarded to Carborundum describes a process for the production of activated carbon from sub-bituminous coal using acid reagents (5 wt% of concentrated H_2SO_4, H_3PO_4 or HCl).[51] The impregnated feed is heat treated to 450 °C in inert atmosphere followed by steam activation at 800–900 °C.

Alkali reagents can also be applied to chemical activation. Very high surface area activated carbons have been produced from precursors such as petroleum coke and coal using KOH and NaOH.[52] High surface area carbon fibres can be prepared from 'charcoal cloth' (viscose rayon cloth carbonized to 800 °C) by chemical activation with alkali hydroxides at 500 °C.[53] A surface area of about 2800 $m^2 g^{-1}$ is obtained at a weight ratio of KOH to carbon of 6. Another process involves the preparation of activated carbons from lignin using NaOH. A surface area of 1600 $m^2 g^{-1}$ was obtained at a weight ratio of alkali to lignin of 2:1 and heat treatment to 500 °C.[54]

Durie and Schafer[55] found that activated carbons with surface areas in excess of 1000 $m^2 g^{-1}$ could be produced by the pyrolysis of the potassium salt of a brown coal. The obstacle in the development of activated carbons from brown coal is the concurrent achievement of structural hardness and surface area. Brown coal is a naturally soft material and produces a weak char on carbonization. This process was improved to produce carbons with high surface area and excellent physical hardness.[38] The carbons were prepared by kneading brown coal with KOH solution followed by extrusion, drying and activation. At a weight ratio of KOH to dry coal of 0.4, carbons were obtained with an iodine number of 1265 and a Takeda hardness of 91%.

Wright[56] proposes a method for producing a highly mesoporous carbon (pore diameters in the range 2–50 nm) by impregnating a cellulosic fibrous material with a liquid medium containing phosphate and an alkali metal, then carbonizing and activating the impregnated fibrous material by heating in a suitable atmosphere. The resulting carbons have at least 10% of the pore volume in mesopores and in one instance as much as 80% of the total pore volume in mesopores.

9.5 Commercial production equipment

Early activated carbons were produced by batch processes in which the raw material, usually wood, was carbonized in a closed retort with a limited air supply. The products will have possessed low surface areas, as the process primarily involved carbonization, and the extent of activation must have been small. Since 1914, production methods have essentially involved sequential carbonization and activation in steam or CO_2 using continuous processing methods.

The various types of continuous reactor can be distinguished by those in which carbonization and activation are conducted in a single stage, and those in which they are conducted in sequential stages using different reaction regimes. The former configuration is more or less obsolete, and all modern processes recognize the advantages of separating these steps. However, there are certain advantages offered by the single-stage reactor, and they are also considered in this section.

An example of one of the first continuous reactors is illustrated in Figure 9.3.[57] Pulverized carbonaceous material is fed into the thermally insulated inclined chamber A into which an air blast B is introduced to effect partial combustion and carbonization at about 800°C. The hot carbon particles pass into chamber C and contact a flow of steam introduced at D. The steam partially gasifies (activates) the carbon particles and the gaseous products of oxidation exit the chamber at E. Because of the low gas velocity in the chamber A, the carbon particles gradually fall to the bottom of the reactor chamber where they are extracted by the feed-screw F.

A second example of a single-stage reactor, the vertical furnace, is illustrated in Figure 9.4. The feed material in granular or shaped form is fed at the top of the furnace and moves downward under gravity through the central tube A. During its passage downwards it is heated, and if the feed material is raw it is carbonized by the heat recovered in a brick checkerwork regenerator from the combustion of the (synthesis) gases produced by gasification. Volatile products

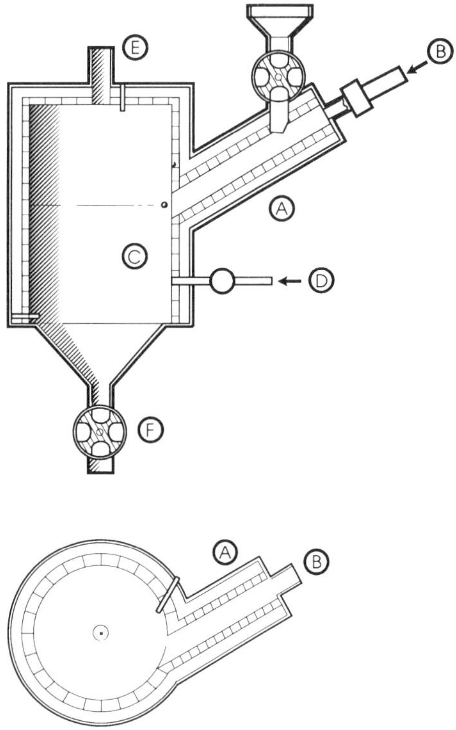

Fig. 9.3 Early continuous reactor

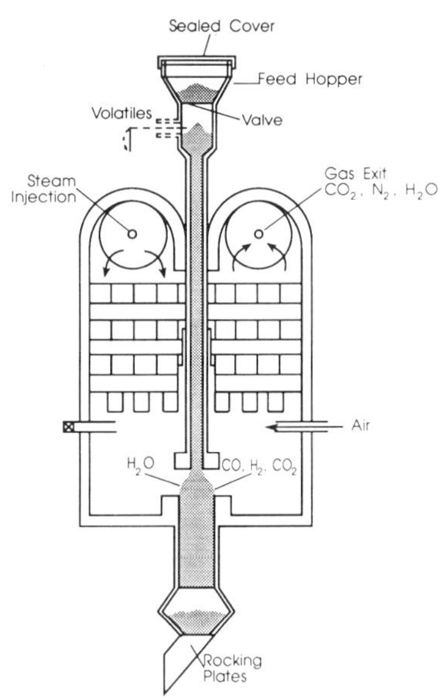

Fig. 9.4 Vertical single-stage reactor

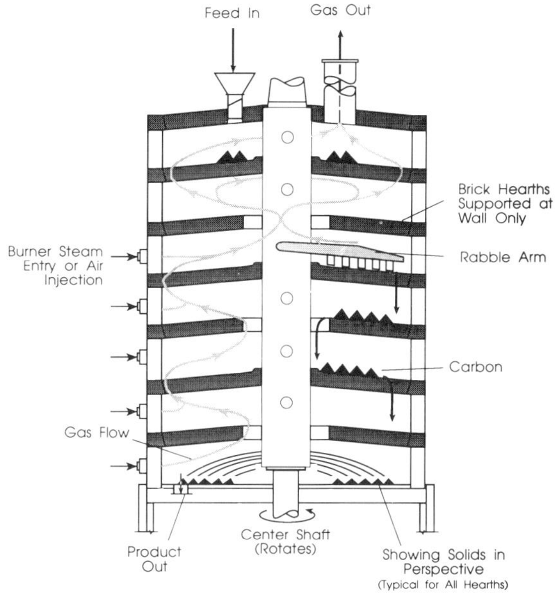

Fig. 9.5 Vertical multiple hearth furnace

pass out through the top of the reactor. The temperature increases as the activation zone is approached. Gasification takes place in the hearth area where the hot feed is contacted with steam. The steam and combustion air flows are cycled to help maintain constant activation temperature. The extent of activation is determined by the residence time, which is controlled by the rate of product withdrawal. To avoid the development of excessive pressure drop across the tube, the feed size is rather large. This can lead to uneven activation from the exterior to the interior of the particles.

The vertical multiple hearth furnace (Figure 9.5, consists of separate circular hearths each of which may be maintained under different conditions of temperature and steam flow. The carbon, in granular form, gradually progresses through the system by the combing action of a series of rabble arms that are connected to a central shaft. Apertures allow the passage of carbon to lower hearths. The steam and combustion gases pass over the carbon, whose surface is constantly disturbed by the movement of the rabble arms, allowing uniform activation. However, owing to the relatively poor gas–solid contacting, steam and fuel consumption tend to be high. Because of the difficulty of removing volatile and condensible hydrocarbons, this type of reactor is usually used for activation of pre-carbonized material or for the regeneration of spent carbons.

Rotary kilns are the most versatile and widely used furnaces. They can be used for carbonization, activation or regeneration, are readily scaled up and are used in a wide range of sizes; commercial furnaces may be 30 m in length and 2–3 m diameter. In this type of reactor, illustrated in Figure 9.6, the precursor is fed in granular or extrudate form. The feed must be of some minimum size to avoid its entrainment in the effluent gases. The kiln is usually inclined at a slight angle to ensure that the carbon gradually progresses from the feed to the discharge end. During their passage through the kiln, the carbon particles are continually raised by baffles fixed to the inner walls of the kiln, creating a falling curtain which

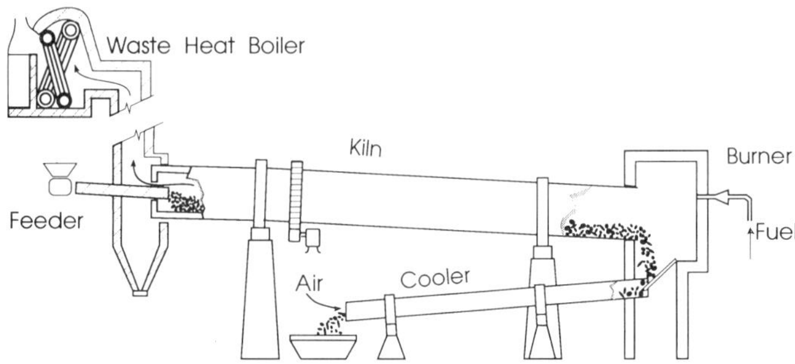

Fig. 9.6 Rotary kiln

promotes good gas–solid contact. The movement of particles is gentle, which minimizes attrition. Fuel gas, air and steam are injected along axial tubes to different parts of the kiln to control reaction conditions along its length. The kilns tend to be long to achieve the required residence time. The reactor is more or less energy self-efficient owing to the combustion of volatile materials and/or product gases. Once conditions are in balance, the reactor requires little externally supplied heat.

Fluidized-bed reactors can also be used for carbonization, activation and regeneration. This type of reactor was used in a process developed to produce activated carbon from anthracite (Figure 9.7).[58] The advantages are that the rates of heat and mass transfer are very high. However, the reactor suffers from the disadvantages that the residence-time distribution of the carbon particles is very broad, producing a broad distribution of activities, and attrition levels are high, which means that the yield of granular product is reduced at the expense of producing lower value powdered carbon.

9.6 Applications

Some examples are provided in this section of the application of activated carbons for liquid-phase and gas-phase adsorption. This is not intended to be a

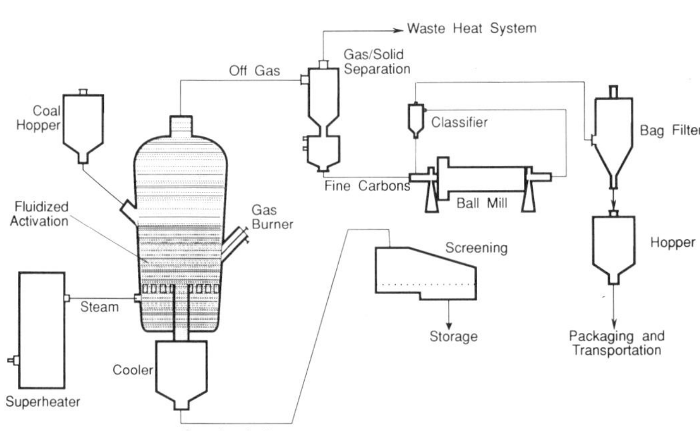

Fig. 9.7 Fluidized-bed reactor

comprehensive treatment, but rather to highlight important areas and provide reference sources for further reading. In addition, a number of applications are considered in other chapters of this book.

Liquid-phase applications

Three of the largest uses of activated carbons for liquid-phase adsorption are discussed below. In addition, activated carbons are used in various other applications that include the purification of chemicals, pharmaceuticals and foods (one example in food processing is the decaffeination of coffee), medicinal uses such as the treatment of ailments of the digestive system by the adsorption of bacterial toxins and in dialysis apparatus to purify the blood of patients suffering from kidney failure or poisoning.

Water treatment

One of the largest uses of activated carbons is in water treatment. Powdered activated carbons have been used for more than 50 years to treat public water supplies for the removal of compounds that adversely affect taste and odour, arising from the presence of natural organic compounds. The addition of powdered carbon to the flocculation stage of a treatment process also helps to resolve seasonal problems due to algal growth as well as pollution from accidental spillages.

In recent year there has been an increase in the level of synthetic organics in public water supplies. These compounds are often produced through chlorination. With the deterioration of the quality of potable water supplies, there has been a growing increase in the use of chlorine for water treatment. A side effect is the formation of by-products such as halogenated organics that are deleterious to water quality (taste and odour). For example, the reaction of chlorine with humic acids in groundwaters can produce chlorophenols, and halomethanes are the most common product of chlorination. As a consequence, the use of granular activated carbons (GAC) for water treatment has increased. In Europe, GAC has been used in potable water treatment plants for a long time. The use of GAC has been much more limited in the USA. A detailed text on the use of GAC for water treatment has been written by Clark and Lykins.[59]

Although granular carbons are more expensive than powdered carbons, they may be more cost effective if the usage rate is high, since they can be regenerated and re-used.[60] In fact, services are now beginning to emerge to increase the convenience of dealing with spent carbons and their regeneration. Essentially, the carbon is supplied by the manufacturer in a module that serves as an adsorber and as a shipping container to facilitate transportation to a regeneration facility.[28] A recent paper considers the use of activated carbon fibres in fixed beds for water treatment.[61] The main advantage lies in that the small diameter of activated carbon fibres essentially removes the mass transfer limitations of GAC, allowing rapid adsorption kinetics. Similar high adsorption rates can be achieved with powdered carbon, but this presents problems in handling and regeneration.

Hundreds of synthetic organic chemicals, such as pesticides and other pollutants, have been identified in drinking water supplies. GAC treatment is a feasible technique for the removal of low-solubility constituents such as trihalomethanes (THM), detergents, pesticides, herbicides, trace metals and polyaromatic hydrocarbons, and is effective for improving taste and odour. It is, however, ineffective for removing high-solubility contaminants and inorganic salts (e.g. nitrates). For the adsorption of polar compounds such as phenol, research has shown that the carbon surface chemistry is more relevant than the total available surface area.[62]

Activated carbons are now used on a much larger scale than ever before. In some European countries, especially France, the sand in large filters has been replaced with activated carbon to control the level of organics in potable water supplies. A new technique used in France is to combine ozone treatment with sand and activated carbon double filtration. The ozone treatment is applied before the second filtration with activated carbon.

Important properties of a granular activated carbon for water treatment are:

- adsorptive capacity and selectivity;
- ability to withstand thermal reactivation;
- resistance to attrition losses during transport and handling.

GAC has been used in both industrial and municipal waste water treatment. There are examples of many successful applications of GAC in advanced waste treatment plants, but also poor results.[59] Generally, the problems have been of a mechanical nature rather than relating to the adsorptive properties of the carbon, and many of the problems occurring in waste water treatment (e.g. excessive slime growths and H_2S release) do not occur in water purification.

Activated carbon is also used in the purification of industrial process water, cooling water and water to run steam turbines. Demineralized water is used in these applications to prevent damage to equipment from scale formation. The ion exchangers used to demineralize process water are extremely sensitive to organic compounds and are protected by pretreating the water with activated carbon.[63]

A consideration of the dimensions of some pollutants illustrates that activated carbons can feasibly remove many of the impurities occurring in water; small organic molecules with low solubility have sizes in the range 0.6–0.8 nm and can be adsorbed in micropores; colour bodies and humic acids have dimensions (1.5–3.0 nm) that will favour their adsorption in mesopores.

Bacteria have diameters of 500–2000 nm and can be accommodated in the macropores and also in the interstices between carbon particles. The bacterial colonization of activated carbons can readily occur in liquid-phase systems and there are many recorded observations of the enhanced adsorption capacities of activated carbons used for biodegradable pollutants in waste waters. Filter life can extend far beyond that which can be attained by purely physical adsorption and, in such cases, is due to colonization of the carbon matrix by bacterial populations.[2]

In other applications such as the purification of water, bacterial colonization is neither acceptable nor easily avoidable, since the adsorption of organics provides an enriched substrate for bacterial nutrition. Bacterial growth in beds can be prevented by steam injection at temperatures higher than 80°C for prolonged periods. Domestic water filters are also subject to this phenomenon. Protection is afforded by the addition of a chemical agent which is capable of preventing bacterial growth within the matrix. The principal additive in current use is silver, which has pronounced bactericidal properties at high concentrations.[64]

Decolorizing

A second major use of activated carbons is in the removal of colour from sugars and foods. Prior to the First World War, the main incentive for the development of adsorbent carbons was provided by the sugar industry for the removal of colour from raw sugar. Activated carbons are now employed in the production of most sugar and sweetener products to purify either the raw sugar or the sweetener through the removal of high and low molecular weight colour bodies.[65] The colour originates partly from the raw material and is partly formed during the refining process. The carbon is also used to remove colour precursors, such as hydroxymethylfurfural. Of lesser importance is the removal of high molecular weight foaming agents (protein materials). Another similar and important application is in the treatment of edible oils to remove components that adversely affect colour, taste and odour.

In sugar processing, a typical procedure is to treat the sugar liquor with powdered activated carbon. The carbon is subsequently separated by filtration. After purification the carbon has some residual adsorption capacity as the adsorption–desorption equilibrium does not permit maximum loading. The carbon is therefore often used again in a two-stage countercurrent system, where the raw liquor is first treated with once-used carbon and then, in a second step, with virgin carbon.[65]

Active carbons for gold recovery

A relatively recent application of activated carbons is in the recovery of gold from its ores. The subject has been discussed at some length by Bansal et al.[5] Because of its commercial importance, most of the information about the technology is proprietary. Basically, the process involves the treatment of finely ground ore with a very dilute solution of sodium cyanide and oxygen. The gold is oxidized to Au(I), which is obtained in solution as $Au(CNO)^{2-}$ by complexation with cyanide ions. Other metallic impurities present in the ore are also oxidized and form cyanide complexes, and are leached together with the gold. The gold is recovered from the liquid by adsorption on activated carbons which can be added directly to the cyanide pulp: the carbon-in-pulp or CIP process.[66,67] Adsorption can be most efficiently achieved when the carbon moves countercurrent to the cyanide solution.

The gold-loaded carbon is removed by screening. The carbon is treated to recover the gold and is then recycled after undergoing some regenerative steps. Gold is recovered by elution using either solutions of metal salts, such as K_2CO_3–KOH, or organic solvents, such as a mixture of acetone, methanol and ethanol.[5]

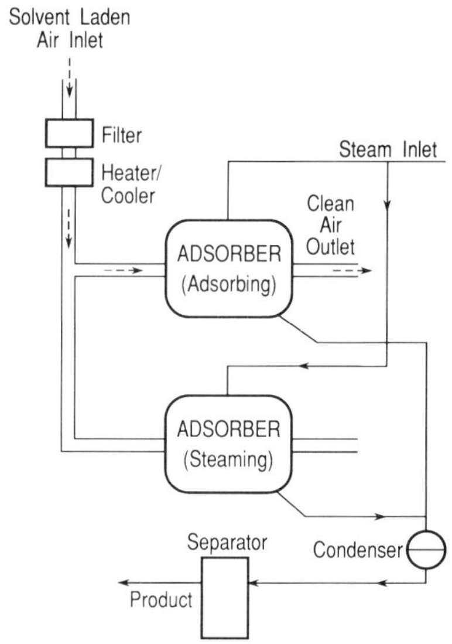

Fig. 9.8 Activated carbons recovering solvents by adsorption and desorption

Gas-phase applications

Solvent recovery

An important industrial use of activated carbons is in solvent recovery. The volatility and solvent properties of organic compounds make them suitable for a great many industrial processes including dry cleaning, the manufacture of paints, adhesives, polymers and explosives, and in printing and extracting oil from seeds. The highly volatile nature of many of these solvents also creates unacceptable problems if they are vented to the atmosphere; they can create health, fire and explosion hazards and also adversely affect the environment. By collecting the vapour and recovering the solvent for recycling, these problems can be resolved and considerable economies can be made.

Activated carbons provide an effective means for the recovery of solvents by adsorption and subsequent desorption, usually in a cyclic process, as depicted in Figure 9.8. Solvent-laden air is directed through a bed of activated carbon until the solvent breakthrough point is almost attained. The effluent air stream is then directed to a second adsorber unit while the first is regenerated. The solvent is recouped by desorption in the regeneration stage by passing low-pressure steam through the bed in a direction counter to the air flow. The steam–solvent mixture is condensed and separated by distillation or decantation and the bed dried and cooled. The solvent is then recirculated back into the process.

There are three basic types of solvent recovery plant design: the flat bed and the annular bed adsorbers (Figures 9.9 and 9.10), which operate on a cyclic basis, and the continuous rotary adsorber (Figure 9.11), formerly used in the printing industry, in which at any one moment in time a segment of it is either adsorbing, desorbing or being dried and cooled. The first two types of adsorber use granular or extruded carbons. Activated carbon fibres could be particularly

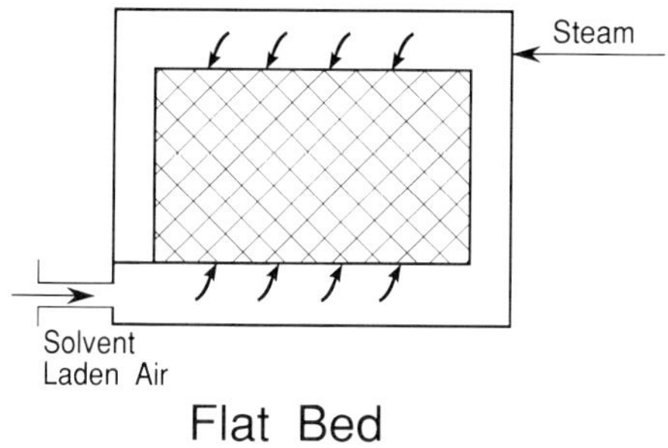

Fig. 9.9 Flat-bed adsorber

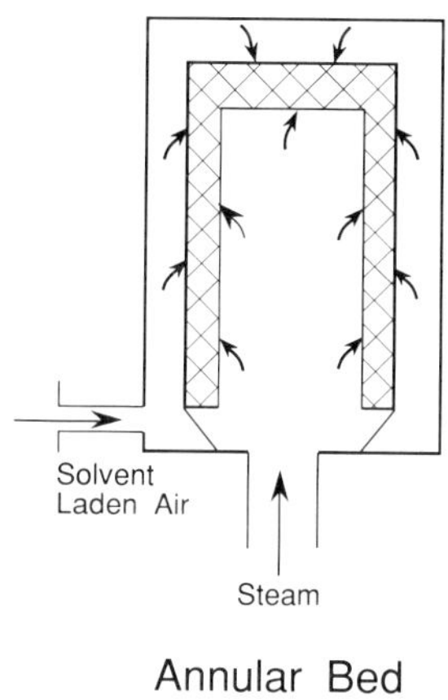

Fig. 9.10 Annular-bed adsorber

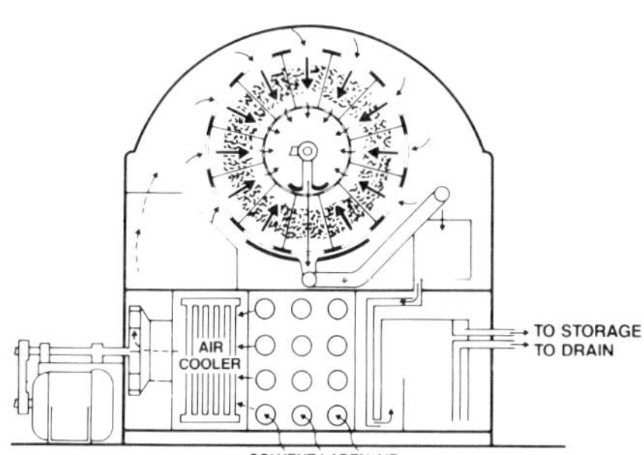

Fig. 9.11 Continuous rotary adsorber

suitable for rotary adsorbers as the minimal diffusion limitation that they present to gas and vapour adsorption will permit accelerated adsorption and desorption kinetics, and hence short cycle times.

Representative solvents reclaimed by solvent recovery include acetone, pentane, methylene chloride, methyl ethyl ketone, tetrahydrofuran, white spirit, benzene, toluene, xylene and petroleum spirit. Solvent interactions with the carbon surface can make the recovery of certain solvents, such as ketones and chlorinated hydrocarbons, difficult. Ketones can be oxidized to varying degrees, releasing heat and increasing the rate of oxidation. If the accumulation of heat cannot be suppressed and sufficient oxygen is available, the situation can lead to ignition of the carbon bed. The susceptibility of the carbon bed to auto-ignition can be reduced by removing soluble sodium and potassium salts from the carbon.

Chlorinated hydrocarbons can be hydrolysed to form highly corrosive hydrogen chloride, leading to rapid degradation of the materials of construction used in the adsorber. These reactions are further assisted by the temperature and moisture changes experienced during adsorption and regeneration of the carbon bed. The promotional effect of the carbon surface is such that even apparently stable solvents such as carbon tetrachloride can be readily hydrolysed.

Protective filters

Military The consequences of gas warfare were first realized in the trenches in World War I when the Germans released chlorine gas on the Allied Forces. The allies quickly developed a means of protection from inhaling the 'choking' gas in the form of gas masks containing granular activated carbon.

The continued threat of chemical warfare has led to the development of much more sophisticated gas masks. Because of the wide range of offensive gases that can be used, the activated carbon is required to remove gases by both physical adsorption (e.g. nerve gases) and chemical adsorption (e.g. hydrogen cyanide, cyanogen chloride, phosphine and arsine). To accomplish this the activated carbon is impregnated with a complex mixture of metal compounds that include copper, chromium, silver and sometimes organic species. These sophisticated carbons are also utilized in filters for underground shelters and inlet filters for armoured vehicles and other forms of military transport.

A second type of threat is in the form of percutaneous nerve gases. Protection against these gases can be afforded by suiting that contains activated carbon in a form that allows rapid and effective adsorption (fine granules or fibres).

Industrial The technology used by the military has been adopted by industry to provide protection to workers and the general public against hazardous vapours and gases that may be encountered in certain processes. As with the military, this involves protective respiratory devices, filters and clothing.

To adsorb contaminants from effluent air, an activated carbon filter may be combined in series with a ventilation unit. With toxic effluents such as sulphur dioxide and hydrogen sulphide, it is necessary to impregnate the activated carbon for maximum adsorption efficiency.

Nuclear Nuclear-grade activated carbon beds are applied to the ventilation of nuclear power reactors to adsorb accidental leakages of radioactive iodine (in elemental form and as methyl iodide), krypton and xenon from the reactor core or heat exchanger systems. Because methyl iodide is not as readily adsorbed as iodine under the high humidity conditions encountered in filter service, it is necessary to impregnate the carbon with potassium iodide, potassium triiodide or triethylenediamine.

Nuclear-grade activated carbons are produced from either coconut shell or coal-based precursors and are highly microporous. Long contact times delay the passage of the radioactive krypton and xenon through the filter until their radioactive decay is complete and the off-stream gas is environmentally safe. These noble gases are not readily adsorbed physically or chemically. Regeneration of the carbon is initiated only after radioactive decay of the products is complete.

General air conditioning

Granular activated carbon filters, placed in series with air conditioning equipment, are commonly used for the removal of disagreeable noxious trace contaminants from populated environments that require the recycling of air as a means of ventilation. Such environments include airports, hospitals, submarines, office blocks and theatres. The recirculation and purification of the air have the added economic advantages of reducing heating and refrigeration costs. A further example is the use of activated carbons in household kitchen cooker hoods, which are used to adsorb cooking odours.

Gas purification

Utilization of landfill gas A new application of activated carbons relates to the recovery and purification of methane from landfill gas. Methane can be released in substantial amounts (55 vol.%) as landfill gas during the anaerobic decay of buried organic municipal waste. This makes its processing into high BTU gas a viable potential source of energy. The remainder of the landfill gas consists of carbon dioxide (45 vol.%) and an abundance of undesirable trace contaminants.

This potential source of energy has been recognised by Air Products and Chemicals in the USA, who have developed an economic means to producing methane from landfill gas at recoveries above 90%.[68] The process is illustrated in Figure 9.12. It essentially involves first passing pressurized landfill gas through a bed of activated carbon to remove the trace impurities. The emergent flow of pure carbon dioxide and methane is then passed through a second bed of activated carbon maintained at high pressure which selectively adsorbs the

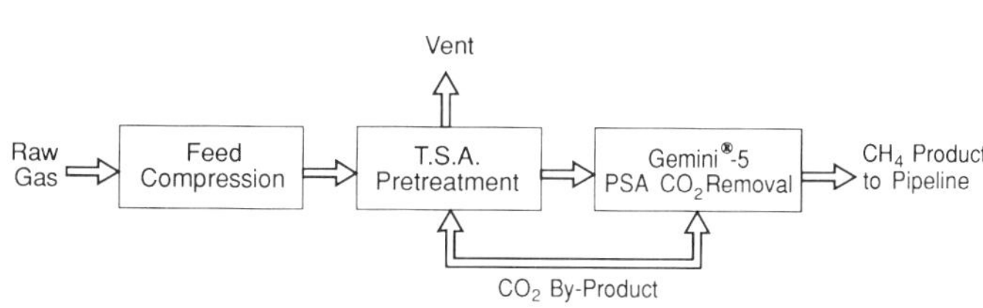

Fig. 9.12 Process for producing methane from landfill gas

carbon dioxide, allowing the production of high-purity methane. The second bed is regenerated by reducing the pressure. The released carbon dioxide may then be channelled back as a hot gas to regenerate the first bed by stripping desorbed impurities which are carried away with the hot gas and incinerated.

Mercury vapour adsorption Mercury exists in trace quantities in fossil fuels (i.e. coal, oil and natural gas), the combustion of which in power- and heat-generating plants can release toxic mercury vapour. The use of granular activated carbon impregnated with sulphur has been shown to be effective in the removal of mercury from contaminated gases,[69] with the formation of mercury-(II) sulphide on the carbon surface. A further development lies in applying this as a pretreatment stage in the processing of natural gas to remove mercury vapour.

Evaporative loss control device (ELCD)

To conform with increasingly stringent government air pollution control standards, it has become necessary to use activated carbons as adsorbents in motor vehicles to prevent the emission of volatile petroleum constituents from stationary vehicles. The carbon is contained in a canister (ELCD) located between the fuel tank and carburettor. The vapours are subsequently desorbed by fresh air entering the intake manifold. A secondary benefit is the conservation of fuel by this process.

Catalysis

The high internal surface area and surface activity of activated carbons make them candidates for use as catalysts and as catalyst supports, in addition to their more conventional use as adsorbents. The use of activated carbon as a catalyst support is discussed in Chapter 10, and will not be considered in detail here. Examples are in the synthesis of chlorofluorocarbons, terephthalic acid, vinyl chloride and sulphuryl chloride. Carbon-supported catalysts also show some advantages for hydrodesulphurization. A useful review of this topic was provided by Abotsi and Scaroni.[70]

Brief mention will be made of a specific use of carbon as an adsorbent for the removal of SO_2, and also as a catalyst for the reduction (and hence removal) of NO_x by ammonia from flue gases. This NO_x reduction process has the economic advantage of operating at lower temperatures (i.e. at about 100°C) than alternative conventional selective catalytic reduction (SCR) techniques using vanadium–titanium catalysts (which operate typically between 350 and 420°C).

9.7 Acknowledgement

The authors express their thanks to Marybeth McAlister for her assistance with the preparation of the manuscript and in proofreading and ensuring the accuracy of references for this chapter.

9.8 References

1. WALKER, P.L. 1986: *Carbon* **24**, 379.
2. WILDMAN, J. and DERBYSHIRE, F.J. 1991: *Fuel* **60**, 655.
3. WILSON, J. 1981: **60**, 823.
4. HASSLER, J.W. 1974: *Purification with Activated Carbon.* New York: Chemical Publishing.
5. BANSAL, R.C., DONNET, D.B. and STOEKLI, F. 1988: *Active Carbon.* New York: Marcel Dekker.
6. HOUGHTON, F.R. and WILDMAN, J. 1971: *Chemical and Process Engineering* **52**(5), 61.
7. McCABE, W.L. 1946: in *Summary Technical Report of Division 10, National Defense Research Committee.* Washington, DC, Chapter 3.
8. JOACHIM, H. 1890: *Papyrus Ebers* **8**, Berlin.
9. McBAIN, J.W. 1932: *The Sorption of Gases by Solids.* London: Rutledge, p. 1.
10. SCHEELE, C.W. 1894: in *Ostwalds, Klassiker der exakten Wissenschaften*, (ed) Ostwald, W., Akademische Verlagsgesellschaft seest 8 perfig, **58**, Leipzig.
11. LOWITZ, 1786: *Crell's Chemische Annalen* **1**, 211.
12. KEHL, D.M. 1793: *Observations et Journal sur la Physique, de Chemie et d'Histoire Naturelle et des Arts.* Paris: Tome XLII, p. 250.
13. DEITZ, V.R. (ed.) 1944: *Bibliography of Solid Adsorbents.* Washingdon, DC: Lancaster Press.
14. GRUILLON, M. 1805: in *Dictionnaire Technologique.* 1832: Vol. XX, p. 107.
15. DELESSERT, B. (1805–08), 1944: in Deitz, V.R. (ed.) *Bibliography of Solid Adsorbents.* Washington, DC: Lancaster Press, p. xvii.
16. FIGUIER, M. (1811), 1944: in Deitz, V.R. (ed.) *Bibliography of Solid Adsorbents.* Washington, DC: Lancaster Press, p. xvii.
17. DE CAVAILLON, J. (1817), 1944: in Deitz, V.R. (ed.) *Bibliography of Solid Adsorbents.* Washington, DC: Lancaster Press, p. xxi.
18. BUSSY, A. 1822: *Journal de Pharmacie et des Sciences Accessoires* **8**, 257.
19. SCHATTEN, (1841), 1874: in Stammer, C. *Traite Complet Theoretique et Pratique de la Fabrication du Sucre.* Paris: E. Lacroix, p. 248.
20. STENHOUSE, J. (1854), 1861: *Chemical News* **3**, p. 78.
21. LIPSCOMBE, F. 1862: British Patent, 2887.
22. HUNTER, J. 1865: *Journal of the Chemical Society* **8**, 285.
23. WINSER, F. and SWINDELLS, J. 1868. British Patent, 835.
24. KAYSER, H. 1881: *Annalen der Physik und Chemie* **12** (3), 526, **14**, 450.
25. VON OSTREJKO, R. 1901: German Patent 136792.
26. SMISEK, M. 1970: *Active Carbon: Manufacture, Properties and Applications.* Amsterdam: Elsevier.
27. CROWELL, B. 1919: *Transactions Electrochemical Society* **36**, 91.
28. IRVING-MONSHAW, S. 1990: *Chemical Engineering.* Feb., 43
29. DERBYSHIRE, F.J., DAVIS, A. and LIN, R. 1989: *Energy and Fuels* **3**, 431.
30. JÜNTGEN, H. 1977: *Carbon* **15**, 273.
31. JENKINS, R.G., NANDI, S.P. and WALKER, P.L. 1973: *Fuel* **52**, 288.
32. WILDMAN, J. 1988: in *Proceedings of the International Conference on Carbon. Newcastle upon Tyne, UK.* p. 185.
33. McKEE, D.W. 1981: *Chemistry and Physics of Carbon* **16**, 1. New York: Marcel Dekker.
34. JÜNTGEN, H. 1983: *Fuel* **62**, 234.
35. LAINE, J., CALAFAT, A. and LABADY, M. 1989: *Carbon* **27**, 191.
36. SHINOMIYA, T., TARUI, Y. and ENOMOTO, S. 1972: United States Patent 3835064.
37. *Introduction to Activated Carbon*: sales brochure, Amersfoort: Norit.

38. GUY, P.J., VERHEYEN, T.V., FELBER, M.D., HENG, S. and PERRY, G.J. 1990: in *Proceedings of the 4th Australian Coal Science Conference.* Brisbane, p. 380.
39. JAGTOYEN, M., THWAITES, M., STENCEL, J., McENANEY, B. and DERBYSHIRE, F.J. 1992: *Carbon* **30**, 1089.
41. EUSUF, M., HASAN, M., KHAN, R. and HAQUE, M.S. 1977: *Bangladesh Journal of Scientific and Industrial Research* **12**(3–4), 180.
41. CHAMBERS, H.H. 1941: *Industrial Chemistry* **17**, 299.
42. MACDOWALL, J.D. 1990: European Patent 0423967 A2.
43. DERBYSHIRE, F.J., JAGTOYEN, M., McENANEY, B., RIMMER, S.M., STENCEL, J.M. and THWAITES, M. 1991: in *Proceedings of 1991 International Conference on Coal Science.* Butterworth-Heinemann, IEA Coal Research, p. 480.
44. KOVACH, J.L. 1975: United States Patent 3864277.
45. CATURLA, F., MOLINA-SABIO, F. and RODRÍGUEZ-REINOSO, F. 1991: *Carbon* **29**, 999.
46. TORREGROSA, R. and MARTIN-MARTINEZ, J.M. 1991: *Fuel* **70**, 1173.
47. TEZ, Z. 1989: *Wasser* **72**, 283.
48. USMANI, T.H., AHMAD, T.W. and AHMED, S.Z. 1989: *Pakistan Journal of Scientific and Industrial Research* **32**, 282.
49. GONZALES, J., CALAHORRO, C.V., GUIJOSA, A.N. and SERRANO, V.G. 1986: *Adsorption Science and Technology* **3**, 41.
50. SHASIZADEH, F., McINTYRE, C., LUNDSTROM, H. and LUNG-FU, Y. 1973: *Proceedings of the Montana Academy of Sciences* **33**, 65.
51. MURTY, H.N. 1977: United States Patent 4032476.
52. WENNERBERG, A.N. and GRADY, T.M. 1978: United States Patent 4082694.
53. AUDLEY, G.J. 1989: European Patent 0312395 A2.
54. YAMAGUCHI, T. 1990: European Patent 0328040.
55. DURIE, R.A. and SCHAFER, H.N.S. 1979: *Fuel* **58**, 472.
56. WRIGHT, J.E. 1989: European Patent 0311364.
57. SANDERS, M.T. 1933: US Patent 1913340.
58. NESS, T. 1981: in Granger, L. and Gibson, J. (eds) *Coal Utilisation: Technology, Economics and Policy.* Leeds: King's English Book Printers, p. 156.
59. CLARK, R.M. and LYKINS, B.W. Jr 1989: *Granular Activated Carbon, Design, Operation and Cost.* MI: Lewis, Chelsea.
60. HUTCHINS, R.A. 1980: *Chemical Engineering.* Feb. 101.
61. SUZUKI, M. 1991: *Water Science Technology* **23**, 1649.
62. MAHAJAN, O.P., MORENO-CASTILLA, C. and WALKER, J.R. 1980: *Separation and Science Technology* **15**, 1733.
63. *The Many Faces of Activated Carbon*: sales brochure, Amersfoort: Norit.
64. COLLINS, D.A., TAYLOR, L.R. and TAYLOR, R. 1967: *The Development of Impregnated Charcoals for Trapping Methyl Iodide at High Humidity.* Warrington: UKAEA TRG Report 1300(W), UKAEA.
65. ASBECK, T.M.W., GOUWEROK, M. and POLMAN, E. 1981: *Starch/Starke* **33**(11), 378.
66. ZANDRA, J.B. 1950: *Process for recovery of gold from activated carbon by leaching and electrolysis.* R.I. No. 11672, Washington, DC: US Bureau of Mines.
67. ZANDRA, J.B., ANGEL, A.K. and HEINEN, H.J. 1952: *Process for recovery of gold and silver from activated carbon by leaching and electrolysis.* R.I. No. 4843, Washington, DC: US Bureau of Mines.
68. GOLDEN, T. 1990: *Energeia.* Lexington, KY: Newsletter of the University of Kentucky, Center for Applied Energy Research, Vol. 1, No. 3.
69. SINHA, R.K. and WALKER, P.L. 1972: *Carbon* **10**, 754.
70. ABOTSI, G.M.K. and SCARONI, A.W. 1989: *Fuel Processing Technology* **22**, 107.

10 Carbon as a Catalyst Support

Francisco Rodríguez-Reinoso
Departamento de Química Inorgánica
Universidad de Alicante
Alicante
Spain

10.1 Introduction

Catalytic processes are of tremendous importance in modern industrialized society. Altogether, they account for as much as 90% of the chemical manufacturing processes in use throughout the world. The role of the catalyst in these processes consists in increasing the rate and controlling the selectivity of the chemical reactions involved. Many of these catalysts consist of metals or metal compounds supported on an appropriate support, the basic role of which is to maintain the catalytically active phase in a highly dispersed state. However, it is well documented[1–9] that the role of the support is not merely that of a carrier; it may actually contribute catalytic activity and it may react to some extent with other catalyst ingredients during the manufacturing process. Further, the interaction between the active phase and the support surface can affect the catalytic activity. A well-known example of the role of these metal–support interactions is that of sulphide catalysts applied in hydrodesulphurization processes; they are most often supported on γ-Al_2O_3 because of its ability to disperse effectively large amounts of metal sulphide. The use of alumina, however, has a considerable drawback in that the alumina surface causes unwanted metal–support interactions which lower the hydrodesulphurization activity of the catalyst because a fraction of the precursor oxide phase remains as such after sulphiding due to the interaction between the amorphous alumina and the transition metal oxide.[10–12] It was later found[13–15] that the use of a less reactive support such as SiO_2 or even carbon, being a much more inert support material, could be more suitable as support for sulphide catalysts. Many studies were carried out aimed at reducing the active phase–support interaction by modification of the alumina carrier, which allowed some regulation of the interaction.[16–18]

The same example of supported sulphide catalysts can also be used to show the effect of a support on catalyst deactivation. The decline in catalytic activity can be attributed to several factors, including coke and metal deposition and chemisorption of molecules from the feed on the catalytically active surface. It has been shown that alumina-supported sulphide catalysts, because of the strength of the surface acid sites, have a much higher coking propensity than comparable carbon-supported catalysts, with lower support acidity and weak active phase–support interactions.[19–21].

The above are just two typical examples of how important the selection of the support in heterogeneous catalysis is. The last 20 years have seen a tremendous increase in the interest in carbon as catalyst support and this chapter aims to show the advantages and drawbacks of the different forms of carbon for such application. Special attention will be given to the effect of carbon porosity and chemical surface composition on the dispersion of the active phase, the catalytic activity and maintenance and selectivity of the catalyst in relation to other supported catalysts. The chapter does not include a discussion of carbon as a catalyst, nor does it include the poisoning of catalysts by carbon except for a few examples of the effect of carbon supports in catalyst deactivation. These topics are well covered in the literature.[22–27]

Because of the relatively large number of papers and patents on carbon-supported catalysts that have appeared in the last 20 years and the wide range of carbons and reactions in which they have been used at the research or industrial level, this chapter will present – after an introduction to catalyst supports and specifically to carbon supports – a section devoted to carbon-supported metals and another section devoted to carbon-supported metal compounds, both centred on the catalysts which have been more extensively studied.

10.2 Catalyst supports

General

The most important characteristic of any industrial catalyst is the chemical composition, but other factors such as surface area, stability and mechanical properties are also very important. For this reason, many of the chemical industries use supported catalysts. There are a number of advantages in spreading the catalyst on a support: the active phase can be highly dispersed throughout the pore system of the support; in the case of metal catalysts a large active surface is produced per unit weight of metal used. On the other hand, a supported catalyst provides for flow of gases through the reactor and then diffusion of reactants through the pores to the active phase; the support can also improve dissipation of reaction heat, retard the sintering of the active phase and increase the poison resistance.

The concept of a support has changed from the early one in which it was considered an inert substance that provided a means of spreading out an expensive catalyst ingredient such as platinum for its most effective use, or a means of improving the mechanical strength of a weak catalyst, to a more complex concept in which it is realized that the support may actually contribute catalytic activity and it may react with other catalyst ingredients during manufacture.[3] A good example of a catalyst support acting as more than a platform for the active catalytic component is that provided by dual-functional catalysts, where the so-called support has a catalytic function that makes it as important as the metal(s) it carries (e.g. catalytic reforming of petroleum naphthas by platinum supported on an acidic support).

The selection of a support is based on a series of desirable characteristics[2,3]:

1. *Inertness.* The support should not catalyse unwanted reactions but it cannot be assumed that because the support is inert it will remain so after the addition of the active phase. Examples of the support actually contributing catalytic activity can be found in the literature.[2,3]
2. *Stability under reaction and regeneration conditions.* A knowledge of the possible reaction of the support with its environment during use or regeneration is essential. Thus, if a high-temperature gas-phase oxidation is going to take place the support must not be consumed; the chemical stability should also be tested for regeneration with steam, oxidation, etc. On the other hand, the possibility of compound formation between the support and the active phase is also undesirable in many cases.
3. *Mechanical properties.* Attrition and crushing resistance, hardness, compressive strength, pelletability, etc., are very important in catalyst supports.
4. *Physical form.* The catalytic process will suggest the most suitable physical form since granular or pelleted supports are suitable for packed reactors involving vapour-phase reactions and powder supports may be suitable for liquid-phase batch processes; the grain size is a compromise between low particle size with a possible improved catalyst performance and a coarse powder that can be more easily separated by filtration.

5. *Surface area.* High surface area is usually, but not always, desirable. Low surface area supports are usually catalytically inert whereas supports with large surface areas may become involved in the catalytic process. The former are either non-porous or may have a meso-macroporous structure and they are useful in cases in which slow diffusion through a well-developed pore structure is undesirable. High surface area supports may have an influence on the catalytic reaction by either direct participation or by the effect on diffusion rates of reactants and products.
6. *Porosity.* When reaction rates become comparable with rates of diffusion of reactants or products within the pore system, catalytic activity and selectivity are modified by changes in the pore structure of the support. On the other hand, the porous structure is essential for the preparation of the catalyst (in fact, the pore-size distribution of the support may be severely changed as a consequence of the spreading of the active phase on its surface), since it will condition the distribution of the active phase. However, the chemical nature of the surface groups of the support is also very important during both the preparation and the use of the catalyst.

Of a wide range of possible useful materials that can be used as catalyst supports, in practice only three combine the above characteristics in an optimum way and account for most industrial uses: alumina, silica and activated carbon. The combination of useful properties of aluminas makes them generally the first choice for catalyst supports although under some specific experimental conditions silica is to be preferred. These two are the most common supports used in industrial catalysts.

Carbon supports

Today carbon occupies only a relatively small portion of the catalyst support market (although the proportion of carbon-supported catalysts is rapidly growing in the brochures of catalyst manufacturers), the basic reason being that its potential has not yet been fully exploited. Thus, although the use of carbon presents some specific problems, there are characteristics of carbon that are very valuable and not attainable with any other support. It is true that carbon supports can only be used for catalytic reactions under conditions where the carbon is non-reactive; for instance, it has recently been shown[28] that above 700 K carbon-supported iron catalysts exhibit large weight losses in hydrogen due to methane formation, indicating gasification of carbon with hydrogen. This means that carbon-supported iron catalysts with high dispersion can be used in hydrogenation reactions only at temperatures below 700 K, when carbon is stable. Inertness to oxygen is restricted to temperature ranges below 500 K. Another possible inconvenience of carbon supports, the fact that regeneration is difficult, may be an advantage since the active ingredient (specially if it is expensive) can be recovered by burning away the support.

The carbon surface has a unique character, offering unparalleled flexibility for tailoring catalyst properties to specific needs. This flexibility arises from a porous structure which determines its adsorption capacity, a chemical structure which influences its interaction with polar and non-polar molecules and a number of active sites (in the form of edges, dislocations, etc.) which determine the chemical reactions of its surface with heteroatoms. Some of the clear advantages of carbon as support may be summarized as follows:

- Carbon is resistant to acidic or basic media, which is not true of other supports; this means that if the catalytic reaction is to be conducted in a corrosive medium harmful to other supports, the only solution is to use a carbon support.
- The carbon structure is stable at very high temperatures. Thus, whereas the cavities in zeolites collapse if heated much above 1000 K, some carbon sieves are stable up to 1700 K. The high-temperature stability of carbon minimizes sintering of the active phase on the surface of the support.
- The pore structure of a carbon support may be tailored to obtain the pore-size distribution needed for each practical application if the precursor and the preparation methods are chosen correctly. Further, the support may be prepared with different physical forms from powder to pellets, making carbon-supported catalysts useful for both fixed and fluidized beds.
- Carbons can be produced with ion-exchange properties to increase the precursor adsorption and to increase the dispersion of the active phase.
- Carbon can be made more or less hydrophilic depending on the precursor and preparation conditions; in any case, carbon exhibits much less hydrophilic character than zeolites or other oxide supports.
- From a practical point of view, an advantage of carbon is that expensive catalytic metals are readily recoverable from spent catalysts by burning away the carbon support.
- Finally, carbons can be prepared at low cost (e.g. activated carbon) as compared with traditional oxide supports such as alumina, silica or zeolites.

On the other hand, it should also be mentioned that carbon can be an active catalyst by itself and it is used as such for some industrial reactions such as the formation of phosgene, sulphuryl chloride and chlorinated alkenes and alkanes, and in many oxidation reactions such as those of H_2S to sulphur, SO_2 to H_2SO_4 and NO to NO_2.[28] However, this catalytic activity of carbon does not prevent its use as a catalyst support, as will be shown below.

One final aspect about carbon, especially activated carbon, as a catalyst support is the presence of inorganic impurities arising from the mineral matter in the precursor; they can have an important and undesirable catalytic effect or act as catalyst poison. The solution is either to use a pure precursor or to remove the impurities by special de-ashing procedures even though this would mean an increase in the cost of the support. There is then a clear advantage in using lignocellulosic materials such as olive stones as precursors for activated carbons to be used as catalyst supports.[29–31]

Although the preparation and characterization of porous carbons are dealt with appropriately in other chapters of this book, a few practical aspects of the two more common carbon supports, activated carbons and carbon blacks, will be introduced here.

Activated carbon is prepared by carbonization and activation of a large number of raw materials of biological origin such as coconut shells, wood, peat and coal; occasionally the preparation is carried out by incorporation of an activating agent ($ZnCl_2$, H_3PO_4, etc.) with the raw material and carbonizing and activating in one operation.[32–38]

The total porous structure of an activated carbon is formed by a wide range of pore sizes, although for practical reasons they are classified in three groups, following the IUPAC definitions,[39] according to their width:

micropores: less than 2 nm;

mesopores: between 2 and 50 nm;

macropores: more than 50 nm.

The macropores of activated carbon act as transport pores enabling the molecules to reach smaller pores situated in the interior of the carbon particle. Thus, the macropores are not important from the point of view of surface area, which is very low, but they affect the rate of admission of the adsorptive by the carbon. The mesopores, which branch from the macropores, serve as passages to the micropores; although the surface area of mesopores is relatively low in most activated carbons, some may be prepared with a well-developed mesoporosity. The micropores constitute the largest part of the internal surface and, consequently, most of the adsorption takes place within them. At least 90–95% of the total surface area of a conventional activated carbon corresponds to the micropores.

Activated carbons, having important contributions from the three groups of pores, are considered general adsorbents and, consequently, they are useful as supports for catalysts. The surface area, pore-size distribution and chemical nature of the surface are conditioned by the raw materials and the experimental conditions used in the manufacturing process.[32–38] Thus, the BET surface area can be as high as around 3000 $m^2\ g^{-1}$ for some special activated carbons, the more common figure being around 1000 $m^2\ g^{-1}$, the largest proportion of pores being micropores.

In addition to the physical structure, activated carbons have a chemical structure which is also very important since the adsorption capacity is determined by the physical or porous structure but is also strongly influenced by the chemical nature of the surface. Thus, whereas in graphite the adsorption capacity is determined mainly by dispersion forces, the random ordering of the graphitic crystallites in activated carbons results in the creation of active sites (unpaired electrons and incompletely saturated valences) which influence the adsorption behaviour. Activated carbons are also invariably associated with appreciable amounts of heteroatoms such as oxygen and hydrogen (in addition, they may be associated with atoms of chlorine, sulphur, etc.) preferentially bonded at the active sites. Carbon–oxygen surface groups are by far the most important in influencing the chemical surface behaviour of an activated carbon.[38] Consequently, the adsorption behaviour of an activated carbon cannot be interpreted on the basis of surface area and porosity alone. Carbons having similar surface area but prepared by different methods or given different activation treatments show markedly different adsorption characteristics. It is easy then to understand how important this aspect of activated carbons is when preparing catalysts by impregnation with, for instance, a metal compound precursor.

Although the features of activated carbon make it an attractive catalyst support, there are cases in which the predominantly microporous structure is an important drawback in practical applications; the micropores are not accessible to large molecules present in some industrial feedstocks. In most activated carbons, enlarging the meso- and macropore structure will be at the expense of their mechanical properties (resulting in materials with low crushing strength), their bulk density and their pore volume and surface area. However, in the last few years there has been a considerable amount of work directed toward the preparation of activated carbons with large surface areas and pore volumes outside the narrow micropores and success has been attained by some research groups,[40–43] this being an attractive solution for their use as catalyst supports.

There is another problem with the use of activated carbons as supports for catalysts. Most of them contain significant amounts of materials other than carbon, levels of ash in the range 3–10 wt% being very common. On the other hand, activated carbons are prepared for uses other than catalyst supports so that no specific requirements are necessary. One solution is to carry out an exhaustive leaching in hydrochloric and hydrofluoric acids to reduce the inorganic impurities (e.g. ref. 44). Another solution is to prepare activated carbons from very pure precursors, although this is not always commercially feasible. The group of Rodríguez-Reinoso[45–52] has prepared activated carbons from lignocellulosic materials such as olive stones, almond shells and fruit stones, the final ash level being lower than 1 wt%. Since, on the other hand, these materials lead to activated carbons in which the porous texture may be tailored by suitable selection of the preparation conditions,[40] they are very appropriate for use as catalyst supports. Some examples will be given in the following sections. At this stage it is sufficient to show in Figure 10.1 the very different adsorption isotherm shapes (reflecting the different pore-size distributions) and BET surface areas that can be obtained in activated carbons prepared from the same lignocellulosic precursor by changing the preparation conditions.

The generic term carbon black refers to a group of industrial products consisting of furnace blacks, channel blacks, thermal blacks and lamp blacks.

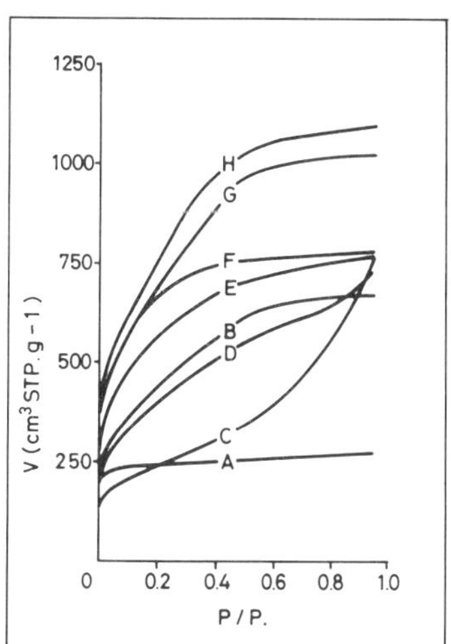

Fig. 10.1 Different adsorption isotherm (butane, 293 K) shapes for carbons prepared from a common precursor

They are materials composed essentially of elemental carbon in the form of near-spherical particles of colloidal sizes, coalesced mainly into particle aggregates, obtained by partial combustion or thermal decomposition of hydrocarbons. Carbon blacks available from the two main manufacturing processes (furnace and channel) have average diameters from about 8 to 500 nm and a porosity that varies with the specific preparative method employed; surface areas range from a few to over 1000 $m^2 g^{-1}$. The preparation techniques, physical and chemical properties and potential applications of these carbons have been described in detail (e.g. ref. 196).

Carbon blacks have been used in many studies as supports for a wide range of catalysts with very satisfactory results on the laboratory scale[53–58]; however, the small particle size is an important drawback for industrial applications and, consequently, the last few years have seen the development of so-called carbon black composites based on a patent by Schmitt *et al.* in 1976.[59] The composites are made with carbon black particles embedded in a carbon binder such as polyfurfuryl alcohol; after extrusion and pelletizing the mixture is subjected to carbonization to convert the binder into carbon. The selection of carbon black spheres is based on the porosity and pore-size distribution desired in the carbon composite. When small pores of a narrow pore-size distribution are desired, carbon black spheres of small particle size are selected. When large pores are desired, spheres of large particle size are selected. When a wide range of pore sizes are desired, mixtures of carbon black spheres of varying particle size are selected. In this way a carbon composite with good mechanical strength in conjunction with high purity and a large surface area in meso- and macropores can be prepared, these being properties needed in most industrial processes using supported catalysts. This type of carbon black composite has actually been used in many studies published in the last few years (e.g. ref. 59).

10.3 Carbon-supported catalysts

The preparation of supported metal catalysts is one of the most important aspects of today's petroleum and chemical processing technology. These metals are dispersed on high surface area supports to maximize the surface area of the metal component, since the rate of a chemical reaction is, in general, proportional to the number of surface atoms available.

Supported metal catalysts have been prepared and are currently being used in many different processes. However, the preparation of these catalysts is still considered more of an art than a science and reproducibility of different preparations remains a problem. The last few years, however, have seen a considerable amount of work directed toward the understanding of the preparation process and the effect of variables such as support material, metal compound and deposition technique.

Carbon supports are used primarily for noble metal catalysts, although the last few years have seen a growing interest in supporting metal compounds such as sulphides and halides.

Of the different forms of carbon, activated carbon constitutes the largest contributor to the carbon-supported catalyst market. Carbon black (both ungraphitized and graphitized) and graphite are used as supports in a much smaller proportion. Although in some cases graphite is used to support metals, the largest catalytic application is the use of graphite intercalation compounds; many of these compounds have been used in laboratory reactions such as hydrogenation, isomerization and Fischer–Tropsch synthesis.[22] In the following, attention will be focused on the use of carbon blacks and activated carbons as supports for catalysts.

Carbon-supported metals

The metals most commonly supported on carbon are those of Groups 8–11 of the Periodic System (Fe, Co, Ni and Cu Groups), although the largest volume corresponds to platinum, palladium, rhodium, ruthenium and iridium, as shown in the brochures of the more important catalyst manufacturers. In order to show the behaviour of two very different situations, the examples presented here concern platinum and iron supported on different types of carbon.

Case 1. Platinum catalysts

The general effect that the different variables used in the preparation of supported catalysts – nature of the support, metal precursor, impregnation method, metal loading, etc. – have on the properties of the final catalyst (metal dispersion, sintering resistance, metal–support interaction) has been extensively studied and these variables are well established for most conventional supports. Since the use of carbon as a catalyst support is continuously growing, it is not surprising to find in the recent literature reports on the effect of some of the variables mentioned above[60–65] in an attempt to relate metal dispersion with the properties of the support.

One of the best studied is the platinum–carbon system, not only from the point of view of the effect of the type (e.g. pore structure) of carbon used as the support but also the effect of the chemical properties of the surface of the carbon. In fact, there is a growing appreciation of the need to understand the surface chemistry of carbons for achieving reproducible and optimum catalyst performance.[19,66–77] For example, modifying catalyst–support interactions to enhance precursor adsorption should enhance catalyst dispersion on drying and/or reduction. However, what has been neglected in most studies is the fact that it is not sufficient just to create these adsorption (or catalyst anchoring) sites on the support surface; they must also be accessible to the metal precursor.

Earlier studies[62,63] on a platinum–carbon system in which the support was a graphitized carbon black (V3G, Cabot) subjected to various levels of burn-off in air had shown differences in the Pt dispersion as a function of burn-off. The increase in the Pt dispersion with the extent of gasification of the support was related to (i) an increase in surface heterogeneity, which in turn increased the potential energy barrier for the diffusion of platinum species across the carbon surface during sample preparation (e.g. reduction), and (ii) an increase in the surface area of the support, the number of surface platinum atoms per unit area of support being almost constant and independent of carbon burn-off. Ehrburger *et al.*[65,78] analysing the dispersion of small particles on carbon surfaces with other metal–V3G systems, pointed out the importance that the active surface area may have in the dispersion of the metal. This uncertainty about the factor conditioning the Pt dispersion was due to a lack of knowledge of the surface chemistry of the carbon support. In fact, during the gasification of V3G and subsequent cooling to room temperature in air, the chemical nature of the carbon support changes progressively with burn-off.

As mentioned in the previous section, during the preparation of a supported catalyst using an impregnation method, the adsorption behaviour of the carbon cannot be interpreted on the basis of surface area and pore-size distribution alone. Carbons having equal surface area but prepared by different methods or given different treatments show markedly different adsorption characteristics. This is what gives to carbon surfaces the unparalleled flexibility for the preparation of supported catalysts. Thus, for a given carbon, one can modify the metal–support interaction by changing the chemistry of the carbon surface, leading to an enhancement of the metal precursor adsorption and to a subsequent increase in dispersion after reduction. However, it is not sufficient to create adsorption sites on the carbon surface where the metal precursor could have better anchoring, since first the sites have to be accessible to the metal precursor.

There are some reports in the literature discussing metal precursor adsorption and metal dispersion on carbon supports treated in different ways to modify their surface chemistry. In most cases, however, the interpretation of results is not easy because the carbon treatments introduced changes not only in the chemistry of the surface but also in the pore structure, thus making difficult the distinction between the effects due to the two variables.[62,64,67,79–84]

The metal is usually deposited on the carbon surface by processes which apply to other supports: adsorption from solution, impregnation, precipitation and vapour-phase deposition. In some cases the catalyst or its precursor is incorporated by impregnation into the carbonaceous precursor before carbonization.[85]

In the adsorption from solution and impregnation processes, the carbon is contacted with a solution of the metal precursor either in excess of solvent or incipient wetness conditions. After drying, the metal precursor is reduced to the metallic state. Catalyst distribution throughout the support porosity is an important result of the impregnation process and it has been discussed in detail elsewhere.[2,3,86] Depending on the extent of adsorption of the metal precursor on the support, its distribution through the surface will be more or less uniform; if adsorption is important (as in the case of chloroplatinic acid on alumina) the metal precursor will be retained in the external surface and if there is not enough time for redistribution an uneven distribution of the precursor will result. The metal precursor can be distributed more uniformly on the support by appropriately using the competitive adsorption of the solvent and solute by the support surface[2,87]; thus, Machek and co-workers,[60,61,87,88] when impregnating γ-alumina and activated carbon with aqueous and organic solutions of chloroplatinic acid, found that the Pt distribution through the support pellets was strongly influenced by the nature of the solvent used for impregnation. Chloroplatinic acid penetrated further into the pores of alumina when using water as the solvent than when using acetone. In the case of carbon supports the opposite situation occurred: the Pt distribution through the pellets was uniform when using ketones as the solvent.

Let us consider the case of carbon supports. Carbon is essentially non-hydrophilic in nature and consequently it has a very low affinity for solvents of polar character, such as water. Because of this low affinity of water for the carbon surface, the metal precursor (chloroplatinic acid) will be mostly located

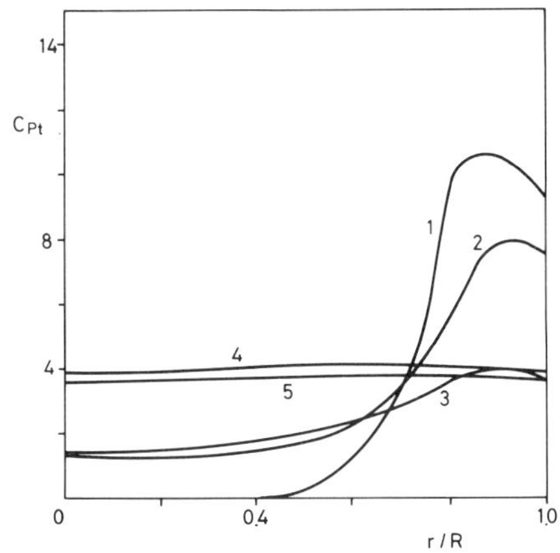

Fig. 10.2 Platinum concentration profiles through grains of catalysts prepared by impregnating activated carbon with different solutions of chloroplatinic acid; $r/R = 0$, grain centre; $r/R = 1$, grain surface. Impregnation from (1) water, (2) methanol, (3) ethanol, (4) acetone and (5) methyl ethyl ketone. Redrawn and adapted from Machek *et al.*[60]

at the external surface of the pellet, with a non-uniform distribution. When the solvent has a larger interaction with the carbon surface, as would be the case for less polar molecules such as ketones, the chloroplatinic acid will penetrate with the solvent into the pores of the carbon, resulting in a more homogeneous distribution through the carbon particles. An intermediate situation will occur with solvents of intermediate polarity between water and ketones (Figure 10.2). Of course, the mean particle size of Pt after reduction in hydrogen of the impregnated catalyst will then be a function of the surface nature of the support and the nature of the solvent used for impregnation.

Confining our attention to metal impregnation from polar solutions, since this is the most common method of catalyst preparation, it can be deduced that appropriately modifying the chemical nature of the carbon surface can condition the distribution of the metal precursor through the carbon particles and, consequently, the metal dispersion. Before going into details it would be convenient to refer to some data found in the literature.

The role played by the conditions of the preparation step of carbon-supported metal catalysts is clearly seen in a study by Palmer and Vannice.[89] A relatively high dispersion (~50%) was obtained when impregnating a conductive carbon black with chloroplatinic acid using the incipient wetness technique, but a very low dispersion was obtained when using a cationic metal precursor, tetraammineplatinum(II) nitrate. Ehrburger *et al.*[62] obtained a low Pt dispersion (16%) when impregnating a graphitized carbon black (V3G) with an organic solution (benzene plus absolute ethanol) of chloroplatinic acid, but the dispersion increased to 55% when the carbon was progressively gasified in air to 49% burn-off; when the 49% burn-off carbon was impregnated with an aqueous solution of the same metal precursor, the Pt dispersion was found to be only 24%. In a first approximation these results could be understood from the point of view of competitive adsorption. The benzene–ethanol mixture will be more strongly forced to the interior of the carbon particles than the water molecules, leaving a more uniform distribution of chloroplatinic acid on the carbon porosity and, consequently, a larger metal dispersion.

There are more results in the literature pointing in the same direction. Thus, Richard and Gallezot[67] showed that high Pt dispersion can be obtained by impregnation with $[Pt(NH_3)_4]^{+2}$ of previously oxidized high surface area graphite and carbon black. Similarly, the adsorption of a cationic silver complex was shown by Ehrburger *et al.*[90] to increase with increasing V3G oxidation. It seems as if oxidation allows a better anchoring of the metal precursor to the oxygen functional groups.

The role played by the nature of the metal precursor, the solvent used for impregnation and the chemical nature of carbon support has also been shown by Rodríguez-Reinoso and co-workers for carbon-supported Pt catalysts.[64,81,91] A recent paper[74] from this group merits a more detailed description. In one attempt to elucidate the role of oxygen surface groups of the carbon supports in the dispersion of Pt, they carried out a systematic study of such a role, maintaining constant all other variables concerned with the support and catalyst preparation. A furnace carbon black (Columbian Chemicals) was heat treated in hydrogen (1223 K, 12 h) to reduce oxygen surface complexes and subsequently oxidized with solutions of H_2O_2 of different strength. The supports were impregnated with an aqueous solution of chloroplatinic acid and the catalysts were characterized by hydrogen chemisorption and transmission electron microscopy (TEM). The N_2 adsorption isotherms for all supports were virtually coincident, showing that the porous texture did not change after the heat treatment in H_2 or the oxidizing treatment with H_2O_2. The nature and amount of oxygen surface groups of the supports were studied by Fourier transform infrared (FTIR) spectrometry and temperature-programmed desorption (TPD). The oxidizing treatment with H_2O_2 introduced carboxylic and phenolic groups, the amount increasing with the oxidation treatment.

Figure 10.3 clearly shows that the Pt dispersion increases with increasing amount of surface groups and this shows that the oxygen groups of the carbon

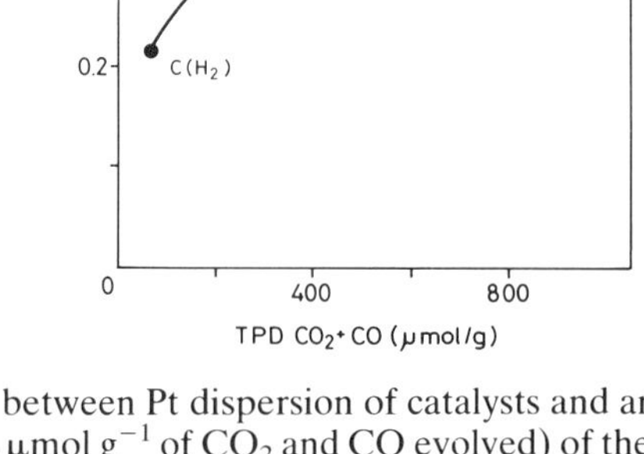

Fig. 10.3 Relationship between Pt dispersion of catalysts and amount of oxygen surface complexes (measured as μmol g^{-1} of CO_2 and CO evolved) of the supports. Redrawn and adapted from Prado-Burguete *et al.*[74]

support play a major role in the final Pt dispersion since the supports had the same porosity and the catalysts were prepared using the same experimental procedure.

The final state of dispersion of the catalysts will be a function of the interactions of the metal precursor with the carbon support at the different stages of the preparation, since all other variables are constant for this series of catalysts. Following the type of reasoning described above, the dependence between Pt dispersion and amount of surface groups of the support can be explained in both the impregnation and reduction steps of the preparation. During impregnation the solvent–carbon interactions and the charge of the metal precursor play an important role. The results[74] indicate that the solvent–carbon interaction is more important because the acidic nature of the treated carbons is not favourable to the anionic character of the metal precursor. Since $[PtCl_6]^{-2}$ is in aqueous solution the decrease in hydrophobicity of the carbon on treatment with H_2O_2 is due to the introduction of acidic groups in the carbon. These groups will increase the wettability of the carbon, facilitating the access of the aqueous solution to the internal pore structure and this, in turn, leads to a more homogeneous distribution of the precursor than in non-oxidized carbon. Of course, carboxylic groups (evolving as CO_2 in TPD) are expected to be most influential in this step of catalyst preparation. These results can be related to those of Machek and co-workers[60,61,87,88] for the adsorption of hexachloroplatinic acid on Al_2O_3 and activated carbon. The distribution of the precursor was more homogeneous for Al_2O_3 when an aqueous solution was used and for activated carbon when a non-polar solvent was used.

On the other hand, once the metal precursor is distributed on the carbon support, it is important to maintain it during the reduction step. The metal and/or its precursor might interact with the support through the surface groups hindering redistribution and agglomeration of the metal (Flynn and Wanke[92] pointed out that sintering of a Pt catalyst may start at temperatures as low as 573 K). The oxygen surface groups may act as chemical anchorage centres for H_2PtCl_6, probably through the formation of some type of Pt–O complex on the surface.

The work of Prado-Burguete *et al.*[74] also makes it possible to relate the Pt dispersion with a given type of oxygen surface group since the more stable the group, the more effective is the anchorage of the platinum and the larger the metal dispersion after reduction. Figure 10.4 shows the plots for metal dispersion as a

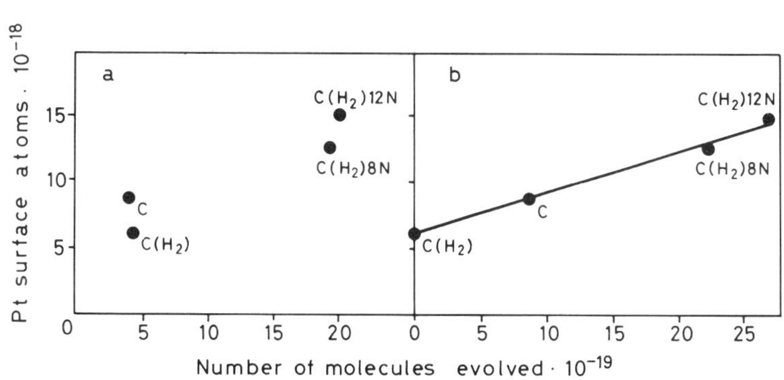

Fig. 10.4 Relationship between the Pt surface atoms and the number of molecules evolved: (a) CO_2; (b) CO. Redrawn and adapted from Prado-Burguete *et al.*[74]

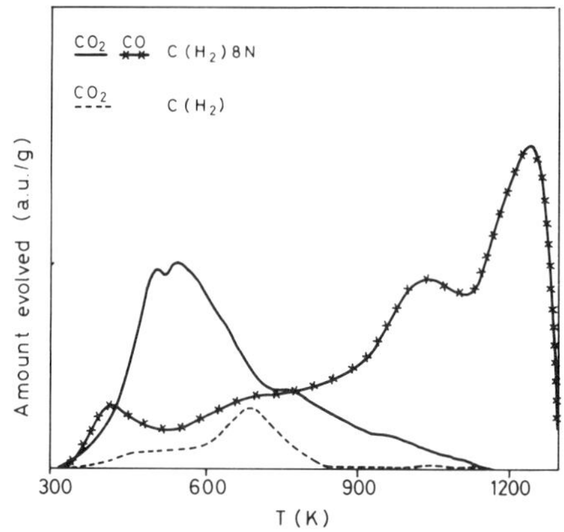

Fig. 10.5 Temperature-programmed desorption profiles for C(H_2) and C(H_2)8N supports. Redrawn and adapted from Prado-Burguete *et al.*[74]

function of the amount of surface complexes evolved during TPD.[74] It is clearly seen that a much better linear correlation is obtained when considering CO complexes and this means that the less acidic groups are more responsible for the anchorage. This is to be expected since only CO complexes are stable at the reduction temperature (623 K), as shown in Figure 10.5.

It may then be deduced that the increase in oxygen surface groups of the carbon support affects the Pt dispersion in both the impregnation and the reduction steps. A decrease in hydrophobicity, caused mainly by carboxylic groups (evolving as CO_2 in TPD), would enhance the diffusion of the metal precursor toward the internal porosity during impregnation with aqueous solution, hence giving a better distribution. An increase in the platinum–carbon interaction, caused mainly through the CO complexes which act as anchorage centres, would increase the Pt dispersion during the reduction step.

A further test of the carbon–platinum interaction could be to follow the sintering behaviour of the catalysts in hydrogen at different temperatures for different periods of time. The change in dispersion for two catalysts, C(H_2) with a small content of surface groups and C(H_2)12N with a large content of surface groups, is shown in Figure 10.6.[74] The behaviour is opposite to that expected on

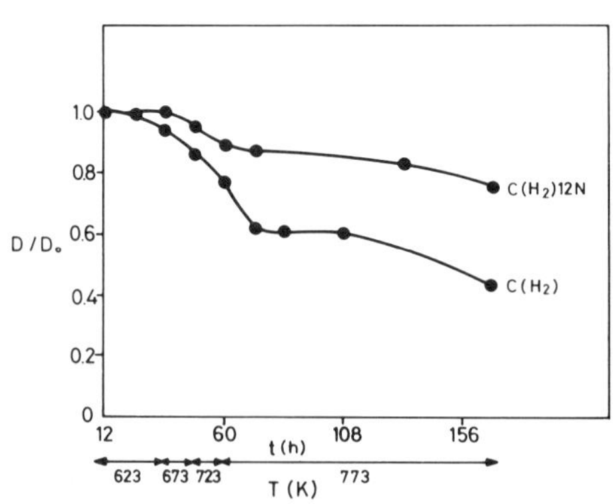

Fig. 10.6 Effect of different sintering treatments on the dispersion of the catalysts C(H_2) and C(H_2)12N. Redrawn and adapted from Prado-Burguete *et al.*[74]

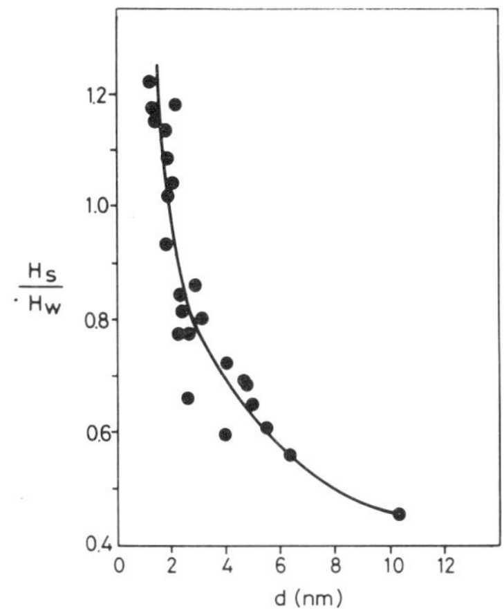

Fig. 10.7 Effect of Pt particle size on weak (H_w) and strong (H_s) hydrogen chemisorption on carbon-supported catalysts[93]

the basis only of initial Pt particle size; sintering should be easier for smaller particles, that is, for the $C(H_2)12N$ catalyst (*see* Figure 10.3). However, the increase in particle size with temperature and length of treatment is much larger for catalyst $C(H_2)$, 136.2%, than for $C(H_2)12N$, 33.3%. This behaviour, which can be explained only on the basis of carbon–platinum interaction, indicates that the oxidized support enhances the thermal stability of platinum particles.

In a further study of this type of carbon-supported Pt catalysts[93] using a wide range of carbon supports, it was shown that in a series of catalysts with similar chemical surface nature, the Pt dispersion was related to the extent of the external surface area; in carbons with different chemical surface composition the increase in Pt dispersion with increasing amount of oxygen surface groups was very large for small changes in porosity.

There is another important feature of carbon-supported Pt catalysts – the large proportion of weak hydrogen chemisorption.[93] Although there are many reports about weak hydrogen chemisorption on platinum black and supported platinum,[94–99] the amount of chemisorbed hydrogen that could be desorbed on evacuating at 298 K for carbon-supported catalysts[93] was significant, being in some cases as high as 70% of the H_2 initially adsorbed at 298 K; this indicates that hydrogen is in general more loosely bound in the platinum–carbon system since figures lower than 50% are found in the literature for Pt supported on other support.[95,96,99] Further, it was found that for the platinum–carbon system there was a relationship between the mean Pt particle size and the weak hydrogen chemisorption, as shown in Figure 10.7,[93] where the strong/weak chemisorption ratio is plotted as a function of Pt particle size. The plot indicates that a greater portion of hydrogen is strongly chemisorbed on smaller rather than on larger Pt particles. This information has a direct practical application since the weakly chemisorbed hydrogen can provide the number of active centres being involved in low-temperature hydrogenation,[100] whereas the strongly chemisorbed hydrogen may provide the number of centres being active for reactions such as hydrogenolysis.[101]

The deactivation of Pt catalysts has received considerable attention, especially in hydrogenation reactions,[102–105] but little study has been made of the deactivation phenomena of Pt particles as a function of the support. In one such study, Ihm *et al.*[106] have shown that during the hydrogenation of cyclopropane the accumulation of carbonaceous materials was large on platinum catalysts dispersed on acidic supports (γ-Al_2O_3, SiO_2–Al_2O_3 and zeolite), whereas no detectable deactivation was observed when dispersed on activated carbon.

Case 2. Iron catalysts

Although iron catalysts are widely used in research and industry, frequently supported on a high surface area material to increase its dispersion and its stability, before 1980 not much work was devoted to iron dispersed on carbon carriers. It is only in the last 10 years that a considerable amount of work has been devoted to the use of carbon-supported iron, especially for the Fischer–Tropsch synthesis.

There are many industrially important reactions centred around synthesis gas, a mixture of carbon monoxide and hydrogen obtained by steam reforming of a variety of hydrocarbon feedstocks ranging from natural gas to naphtha or by coal gasification. The Fischer–Tropsch synthesis is generally taken to mean the formation of a product consisting predominantly of linear hydrocarbons higher than methane. It was used in Germany during World War II to produce fuels and is now used in South Africa to produce, primarily, gasoline.[107] Given the actual situation of the oil industry, it is more attractive from the economic point of view to produce short-chain alkenes, as these are the most valuable bulk hydrocarbons.

The high activity and relatively low cost of traditional Fischer–Tropsch catalysts have given them a pre-eminent position in catalysis research pertaining to CO hydrogenation.[107,108] However, one limitation with these catalysts is that they are poorly dispersed and consequently the preparation of well-dispersed metal particles was considered a desirable goal since they might have improved activity, selectivity and catalytic maintenace. However, Vannice and Garten[109] found evidence indicating that the support could also alter the catalytic properties of the metal component in the CO hydrogenation reaction. Vannice[110,111] carried out a detailed study of the use of alumina and silica supports for iron catalysts in CO hydrogenation, and later[112] considered carbon an interesting support since carbon is an electrical conductor and could conceivably facilitate electron transfer to or from metal crystallites in contact with its surface; indeed, such effects had been attributed to certain carbon-supported systems.[113,114] In fact, carbon is an interesting support for iron in the Fischer–Tropsch synthesis, providing an inert, non-interacting surface. This could alleviate the difficulty often encountered with alumina and silica supports; on a support such as Al_2O_3, iron cations strongly interact with the support surface, which results in the oxide precursor being difficult to reduce to iron in the zero-valent state and it not being possible to obtain small particles of iron. On the other hand, high surface area carbon has certain properties that make it an attractive support for iron. The high surface area provided by the small pores may stabilize highly dispersed particles owing to the presence of physical barriers which suppress surface migration and the resultant agglomeration. Finally, carbon can be made virtually free of sulphur and hydroxyl and carboxyl groups.

A wide range of carbons used for the preparation of carbon-supported iron catalysts can be found in the recent literature, from non-porous graphitized carbon blacks (e.g. refs 115–117) to low and high surface area carbon blacks (e.g. refs 118–120), activated carbons of different types (e.g. refs 30 and 121–123), carbons from polymers (e.g. refs 70 and 124), glassy carbons (e.g. ref. 112), etc. The preparation methods include the incipient wetness technique with iron nitrate or carbonyl clusters, adsorption of iron carbonyls and oxidative decomposition of iron carbonyls in a gas stream; the iron loading ranges from 1 to 10%.

Jung and co-workers[115,116] prepared iron–carbon catalysts by an aqueous impregnation (of iron nitrate) technique on low and high surface area carbon blacks. The low surface area carbon V3G led to catalysts with large iron particles, which showed ferromagnetic behaviour and low activity for CO hydrogenation; the deactivation of these large iron particles was fast and methane was the main product. However, iron catalysts supported on the high surface area carbon were constituted of very small supported iron crystallites, despite the fact that iron is a metal difficult to prepare in such a state. Such results had never been reported before except for MgO-supported iron,[125,126] but two additional advantages are obtained by the use of carbon supports: they are very inert and essentially all the iron can be reduced to the metallic iron. The activity for CO hydrogenation was high (equal to or higher than that of Fe–Al_2O_3) and the small iron crystallites provided a high alkene to alkane ratio (10 times higher than that with Fe–Al_2O_3). However, the authors found that the specific activity for CO hydrogenation decreased with decreasing crystallite size for the iron–carbon system.

Excellent activity maintenace represented another improvement in catalytic behaviour exhibited by catalysts supported on high surface area carbon, as shown in Figure 10.8, where the results[115] are compared with those for Fe–V3G and Fe–Al_2O_3. The methanation activity was constant during 45 hrs of continuous operation for the highly dispersed catalyst; such behaviour is unusual for iron catalysts because the expected behaviour, i.e. slow deactivation after a certain period on-stream,[127,128] was exhibited by the Fe–V3G and Fe–Al_2O_3 catalysts, both of which contained large iron crystallites. Further, the highly dispersed catalyst did not sinter and the product distribution was constant during the 45 hr period.

This highly dispersed carbon-supported iron catalyst was studied using *in situ* Mössbauer spectrometry.[116] The catalyst, after impregnation and drying,

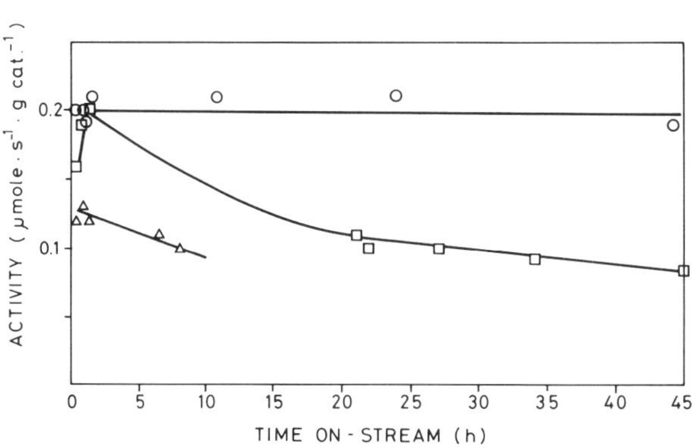

Fig. 10.8 Rate of CH_4 formation as a function of time on-stream. $P = 101$ kPa; $T = 508$ K; $H_2/CO = 3$. (○) 5.0% Fe–C; (△) 4.5% Fe–V3G; (□) 10% Fe–Al_2O_3. Redrawn and adapted from Jung *et al.*[115]

contained iron as Fe^{3+} in highly-dispersed iron(III) oxide. After reduction in hydrogen at 740 K, about 75–80% of the iron was reduced to α-Fe whereas the remaining iron was present as Fe^{2+}; the α-Fe particles exhibited the behaviour of very small particles (2 nm). Exposure of the reduced catalyst to air at 298 K resulted in the oxidation of all iron to Fe^{3+}, whereas the treatment of the reduced catalyst with synthesis gas ($H_2/CO=3$) converted α-Fe into iron carbides whereas the Fe^{2+} remained unaffected. Hence, the results show a high degree of reduction of the iron oxide in the carbon-supported catalyst which cannot occur in silica- or alumina-supported iron catalysts. This behaviour can be explained by the weaker chemical interaction between the iron oxide and the carbon surface compared with that on silica or alumina surfaces.

Sommen *et al.*[129] also observed high alkene selectivities for Fe–carbon catalysts but they found a strong tendency for these catalysts to deactivate owing to the formation of carbon deposits at an H_2/CO ratio of 1. The question then is whether iron–carbon catalysts are stable under typical synthesis conditions, i.e. at an H_2/CO ratio of 2. In this sense, Rodríguez-Reinoso *et al.*[117] found for iron–carbon systems that the alkene/alkane ratio increased when the H_2/CO ratio in the stream was changed from 3 to 2. Jones *et al.*[121] using an H_2/CO ratio of 2, found that both the initial and steady-state specific activities and the alkene/alkane ratio decreased with decreasing iron crystallite size. It is not known whether this trend, contrary to that found by Jung and co-workers,[115,116] is due to the different reaction conditions, but nevertheless both sets of data indicate that crystallite size does affect alkene selectivity. What is also clear is that iron–carbon catalysts are more selective for alkenes than are Fe–Al_2O_3 and Fe–SiO_2 catalysts.

An interesting approach to increase the activity and selectivity would be the use of iron carbonyls supported on high surface area carbon supports, preferably with no reactive surface oxygen that can entrap and stabilize small metal clusters or particles. The use of $Fe(CO)_5$ and $Fe_3(CO)_{12}$ has been reported in the last few years for producing small iron particles on different supports.[30,70,117,122,124,130–134] The advantages of this approach are clear: accessibility to the internal porosity in the liquid or gas phase, production of a high iron dispersion and, especially, zero-valent Fe is initially present in the metal precursor.

Rodríguez-Reinoso *et al.*[117] compared catalysts supported on high surface area carbon (prepared from the copolymer Saran) prepared by impregnation with $Fe(NO_3)_3$ and $Fe(CO)_5$. They found that $Fe(CO)_5$ produced catalysts with a smaller iron particle size, which in the CO hydrogenation reaction produced less methane and a much higher alkene/alkane ratio. Further, the alkene/alkane ratio for a catalyst S–5.4% Fe(P), prepared with $Fe(CO)_5$, was not only higher than with the corresponding S–5.2% Fe prepared with $Fe(NO_3)_3$, but in addition increased with reaction time, as shown in Figure 10.9, to a constant value (after 5–6 h on-stream) approximately seven times that given by the S–5.2% Fe catalyst. Something similar was shown later by Rodríguez-Reinoso *et al.*[30] when using activated carbon as a support. These results seemed to be important because for other catalysts described in the literature (e.g. ref. 135) and prepared from $Fe(CO)_5$ the alkene/alkane ratio decreased because of metal particle sintering. The sintering took place at the temperature of the reaction because the precursor was not reduced at high temperature (673 K in hydrogen, as in the work of Rodríguez-Reinoso *et al.*). It is also interesting that Rodríguez-Reinoso *et al.*[117] reported that the catalytic activity decreased with reaction time

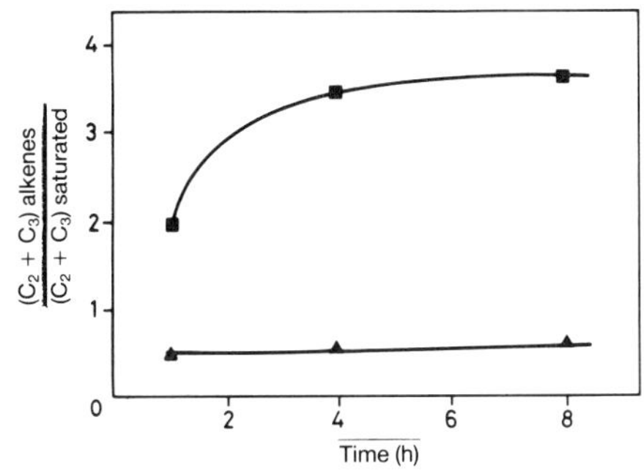

Fig. 10.9 Alkene/alkane ratio during reaction. $H_2/CO = 2$. (▲) S–5.2% Fe; (■) S–5.4% Fe(P). Redrawn and adapted from Rodríguez-Reinoso *et al.*[117]

but the decrease was especially marked for catalysts prepared from $Fe(CO)_5$, with the smallest particle size, compared with those prepared with $Fe(NO_3)_3$ on the same Saran carbon support, as shown in Figure 10.10 where a catalyst supported on a graphitized carbon black (V3G), with a much larger metal particle size, is included for comparison.

It is also interesting that Rodríguez-Reinoso *et al.*,[30] when studying iron catalysts [prepared using $Fe(CO)_5$ and an activated carbon] with different degrees of dispersion, reported that the specific activity for the CO hydrogenation decreased with decreasing iron metal particle size, as described by Jung and co-workers.[115,116] However, the large alkene/alkane ratios were maintained even for the large metal particles. They concluded that the selectivity for the formation of alkenes could not be attributed solely to the small iron particle size but, additionally, to some other characteristics of catalysts prepared using $Fe(CO)_5$ as the precursor, such as the different morphology of the iron crystallites with respect to those prepared from $Fe(NO_3)_3$.

Since the advantage of using carbon as support for iron catalysts is clear, it would be interesting to see how the structure, stability and chemical nature of the metal particles are influenced by the porosity and/or the surface chemical nature of the carbon. To start with, large differences have been found between iron particles dispersed on graphitic supports and those dispersed on porous high surface area carbons.[136] Fe particles [from $Fe(NO_3)_3$] supported on graphitized

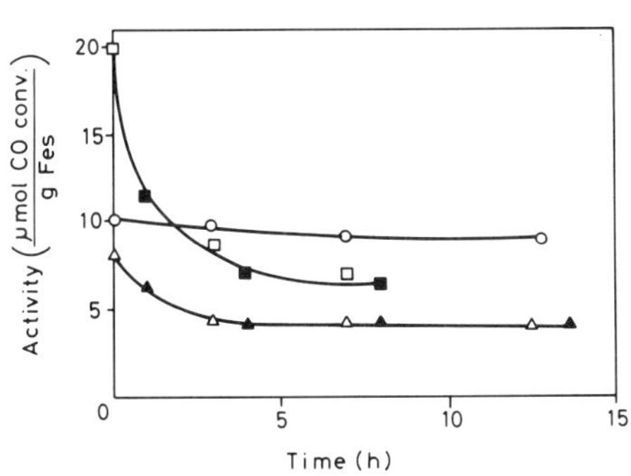

Fig. 10.10 Activity during CO hydrogenation for iron catalysts. Open symbols, $H_2/CO = 3$; closed symbols, $H_2/CO = 2$. (○) V3G–4.8% Fe; (△, ▲) S–5.2% Fe; (□, ■) S–5.4% Fe(P). Redrawn and adapted from Rodríguez-Reinoso *et al.*[117]

carbon black (V3G) exist in phases which are essentially those expected for bulk Fe, with little apparent interaction with the support and sintering rapidly during heating,[116,130] this not being the case when the metal precursor is $Fe(CO)_5$. However, as shown above, very high metal dispersions can be obtained when porous carbon supports are used and it seems there is evidence which indicates that the support can influence the morphology and chemistry of the particles.

The effect of the porosity of the carbon support on the iron dispersion and catalytic activity and selectivity for the CO hydrogenation reaction has recently been studied in detail for different types of metal precursors. $Fe_3(CO)_{12}$, a metal precursor used by Vannice and co-workers (e.g. refs 137–140), was selected in one of the studies,[141] whereas $Fe(NO_3)_3$ and $Fe(CO)_5$ were selected by Rodríguez-Reinoso *et al.*,[142] similar types of activated carbons being chosen as supports. In the first case,[141] the results obtained show that the carbon microporosity affects the methanation kinetics. Thus, although the conversion to hydrocarbons was high in all cases, it was higher in the catalyst supported on the more microporous carbon, with a large turnover. However, this catalyst showed a high methane make and lower selectivity for low molecular weight alkenes; the selectivity increased with the development of porosity in the support.

Similarly, in the work of Rodríguez-Reinoso *et al.*[142] three activated carbons with increasing degrees of activation were used as supports. When $Fe(NO_3)_3$ in aqueous solution was the metal precursor the iron loading increased linearly with activation of the carbon support, whereas for $Fe(CO)_5$ in pentane the iron loading remained constant above 30% activation of the carbon. This different behaviour can be related to the evolution of both porosity and chemical nature of the carbon surface during activation; the development of porosity is paralleled by an increase in the number of oxygen surface groups (especially carboxyl). When the precursor is iron nitrate in aqueous solution, the oxygen surface groups reduce the hydrophobicity of the carbon and the carboxyl groups can exchange Fe(III) ions; as a consequence, both the access of the solution and the retention of iron are increased with activation of the carbon support. The situation is different for the iron pentacarbonyl–pentane system; the solvent, being non-polar, will be more accessible than water to the internal porosity so that a larger amount of solute can be deposited compared with the iron nitrate–water system in carbons with narrow microporosity. For highly activated carbons, the wider porosity and the number of oxygen surface groups make the surface almost equally accessible to both precursors and the iron loading is similar. The two parameters also affect the iron particle size, particle sintering, catalytic activity and selectivity. Thus, iron dispersion increased with the degree of activation of the carbon support when iron nitrate was the metal precursor but was high and uniform for iron pentacarbonyl, indicating that the porosity of the carbon support affects the iron dispersion for the nitrate but not for the pentacarbonyl.

The effect of the chemical nature alone of the carbon support on the supported iron particles is not so well understood. A useful approach was recently adopted by Vannice and co-workers[136,143] using a high surface area carbon black (CSX-203, Cabot) to which three different pretreatments were applied prior to impregnation with $Fe_3(CO)_{12}$: (i) simply evacuation; (ii) treatment in H_2 for 12 hrs at 1225 K, cooling and exposing to air; and (iii) treatment in H_2 for 12 hrs at 1255 K, cooling and keeping under vacuum (this sample was never exposed to air between pre-treatment and impregnation). The carbonyl-derived iron particles placed on the three carbons exhibited differing structures and chemical behaviours as detected by TEM, Mössbauer spectroscopy, adsorption and CO hydrogenation kinetic measurements. Since there are no significant changes in the porosity of the three carbons, the only real difference should be in the number of unsaturated active sites, and the authors concluded that on carbon that was heat treated in H_2 and never exposed to air most of the Fe particles had an unusual structure and resisted sintering as a result of an interaction with unsaturated active sites of the carbon; in contrast, particles which form on similarly treated but air-exposed carbon sinter more rapidly whereas carbon not treated in H_2 yields larger particles reminiscent of those found on graphitic supports because in the absence of the unsaturated active sites there is nothing to anchor the iron and it sinters rapidly. These are very interesting results, but more experimental work is needed to learn about the effect of carbon surface groups on the accessibility and stabilization of metal complexes and the subsequent metal dispersion following reduction. A recent study[70] was directed toward the ellucidation of the role of nitrogen and oxygen surface groups (introduced by reaction of the porous carbon with NH_3 and HNO_3, respectively) on the behaviour of iron catalysts prepared from $Fe(NO_3)_3$ and $Fe(CO)_5$ as precursors. It was found that the surface groups may affect the iron particle size and their reducibility, although not much change was observed in the activity and selectivity for the CO hydrogenation reaction; nitrogen surface groups, however, clearly increased the catalytic activity for the hydrogenolysis of butane, especially in catalysts prepared from $Fe(CO)_5$, with a smaller iron metal particle size.

In addition to favouring the preparation of highly-dispersed iron catalysts, the use of iron carbonyls allows the preparation of bimetallic and even promoted catalysts. This is important from the point of view of trying to solve one of the basic problems associated with the Fischer–Tropsch synthesis reaction, namely the lack of control of product selectivity. Supported alloy catalysts offer one method to alter product selectivities, especially regarding alkene/alkane ratios and oxygenate formation, compared with single-component systems, since electronic and crystal structure effects can change the relative rates of parallel reactions and perhaps suppress undesirable reactions. One way of obtaining such supported bimetallic catalysts is by utilizing organometallic mixed-metal carbonyl clusters as the metal precursor.[137,144,145] In this way, a variety of stoichiometric metal compositions can be obtained which would be difficult to guarantee by co-impregnation techniques. Further, the mixed-metal carbonyl cluster should be activated just by heating to a temperature that would decompose the cluster to yield reduced metal and CO, the temperature usually being below 475 K. Hence it is possible to obtain reduced metal under much less severe reduction conditions than those used for conventional metal salt precursors. Vannice and co-workers[137,146] used this approach to prepare Fe–Ru and Fe–Co among other carbon-supported bimetallic catalysts and, what is more important, to introduce a variety of promoters. It should be remembered that carbon is an ideal support for this type of study: since, in contrast to Al_2O_3, SiO_2 or MgO, the carbon–active phase interaction is very weak, the behaviour of the catalyst will be basically governed by the chemical nature of the active phase and carbon then is ideal for studying the effect of adding other metals or promoters to a given metal.

It is known that the addition to iron of promoters such as alkali metals, especially potassium[147,148] and, more recently, manganese[119,147] can enhance the selectivity to low molecular weight alkenes in the CO hydrogenation reaction.

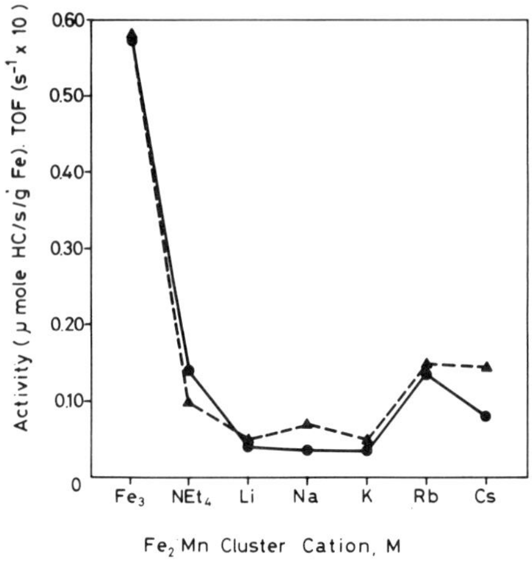

Fig. 10.11 Activity (▲) and turnover frequency (●) for hydrocarbon formation at 548 K, 0.1 MPa and $H_2/CO = 3$ of carbon-supported clusters derived from $Fe_3(CO)_{12}$ and M–$Fe_2Mn(CO)_{13}$. Redrawn and adapted from Martín-Martínez *et al.*[123]

Studies of carbon-supported Mn–Fe catalysts prepared from stoichiometric mixed-metal carbonyl clusters showed them to be unique in that they had exceptionally high selectivities to light alkenes and were also very active, and the addition of one potassium atom to a cluster improved the performance further.[119,138] In a further step, recent studies[123,149] concerned the effect of all alkali metals as promoters on Mn-promoted Fe catalysts which were prepared by exchange of the appropriate alkali metal cation with the tetraethylammonium cation in $N(C_2H_5)_4[Fe_2Mn(CO)_{12}]$ clusters; the supports were activated carbons prepared from olive stones. The activity and the turnover frequency for hydrocarbon formation at 548 K, 0.1 MPa and $H_2/CO=3$ are plotted in Figure 10.11. The values are highest on the unpromoted Fe crystallites, the addition of Mn decreasing them and the further addition of the alkali metals, with the exception of Rb, reducing the values even more. However, the addition of Mn increases the alkene/alkane ratio (see Figure 10.12), the increase being greater with the further addition of alkali metals except Cs), especially K. An interesting aspect of the kinetic behaviour of this family of catalysts is the correlation between the activation energy for hydrocarbon formation and the alkene/alkane ratio, as shown in Figure 10.12. Of course, an additional advantage of these mixed-metal carbonyl clusters is that they require no pretreatment such as a reduction step to be active catalysts.[123]

Carbon-supported metal compounds

Carbon, especially in the form of activated carbon, is also used as a support for compounds such as sulphides, oxides and halides.[22,150–154] Of the different types of carbon-supported metal compound catalysts, this section will be devoted to sulphide catalysts because they are a very good example of the important role of the carbon surface as a catalyst support. In the last 10 years there has been

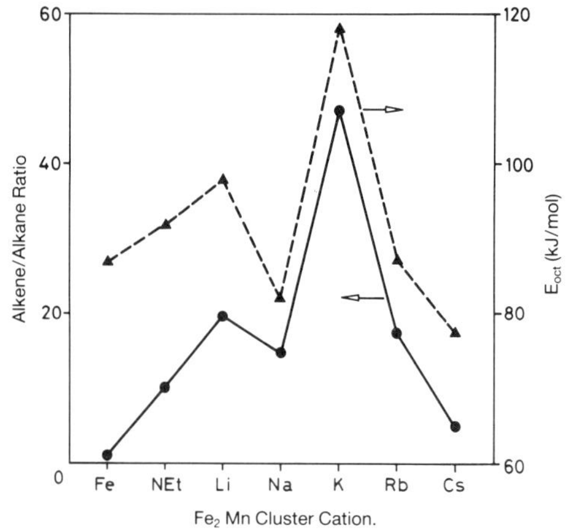

Fig. 10.12 Activation energy for hydrocarbon formation at 0.1 MPa (▲) and alkene/alkane ratio at $H_2/CO = 3$ (●). Redrawn and adapted from Martín-Martínez *et al.*[149]

growing interest in the study of carbon-supported sulphides because of the great potential they offer in industrial applications for hydrotreating processes. Carbon-supported sulphided catalysts, however, have been known for some time and used in the selective hydrogenation of nitro compounds.[22]

Sulphide catalysts are of great current industrial interest because they are widely used in petroleum refining for hydrotreating applications such as hydrodesulphurization (HDS) and hydronitrogenation (HDN) (more than 1 billion tons of feed treated annually). With the ever-increasing emphasis on preservation of the natural environment, the importance of hydrotreating will increase even further in the coming decade. Currently, sulphide catalysts applied in industry are derived from oxides of an element from Group 6 (Mo or W) and an element such as Co or Ni, supported on alumina, sulphided in operation. It has been assumed that the catalytic activity is connected with Mo or W, Co or Ni acting as promoters,[10,155–165] the term promoter being related to the fact that the activity of sulphided alumina-supported Mo or W is largely enhanced by the addition of Co or Ni sulphide, while sulphided Co or Ni is considered to be inactive.

Intensive research has shown that there are various degrees of chemical interaction between the alumina and the metal oxides, some of the species formed being very stable and resisting complete sulphidation. This means that after sulphidation the catalyst probably contains sulphides and oxides. In 1976, de Beer and co-workers[13–15] found for the Co–Mo system that there is no need for the exclusive use of alumina support in hydrodesulphurization catalysts; by appropriate changes in the preparation method high activity levels could be obtained with samples based on a less reactive support such as silica. They concluded that even carbon could be successfully used as a support for sulphide catalysts.

The reason for the viability of carbon as a support for sulphides lies in the fact that being more inert than either alumina or silica, the final catalyst after sulphidation will be formed by almost exclusively sulphides of the transition metals, this implying that the metals can be more effectively used. Of course, a

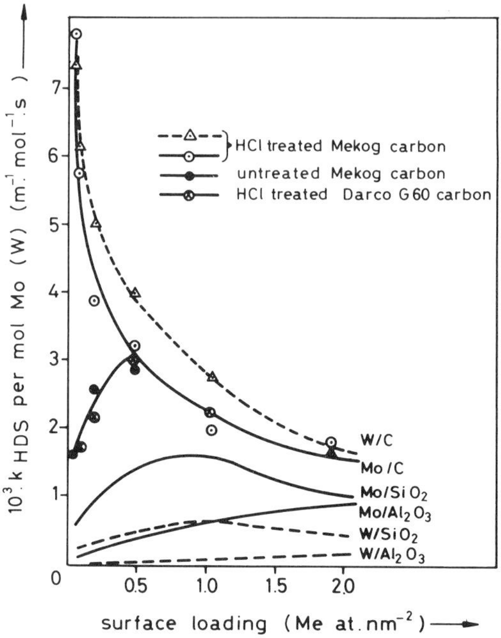

Fig. 10.13 Thiophene hydrodesulphurization reaction rate constant per mole of Mo or W as a function of surface loading. Redrawn and adapted from Duchet *et al.*[19]

further advantage of carbon is that the metals are readily recoverable from spent catalyst by burning away the carbon support. However, among the vast number of papers and patents on hydrodesulphurization that appeared before the early 1980s, only a few[164–174] dealt with carbon-supported sulphide catalysts. It is in the last 10 years that a relatively large number of papers and patents have been published on the subject, some of which will now be discussed.

Duchet *et al.*[19] carried out an interesting piece of work, comparing the catalytic properties for thiophene hydrodesulphurization of sulphided Mo and W supported on carbon, alumina and silica. A summary of their results is shown in Figure 10.13 where the HDS reaction rate constant is plotted as a function of surface loading. The thiophene HDS activity per mole of metal increases in the order γ-$Al_2O_3 < SiO_2 <$ carbon, and the activity is a function of surface loading. With Al_2O_3 and SiO_2, a high efficiency of the supported sulphide phase can be obtained only when the Mo or W species in the oxidic precursor are well dispersed and are easily and quantitatively converted into the sulphide form. At low surface loading the interaction between alumina and the oxidic species is very strong and decreases slowly with increasing metal content. This means that it is difficult to convert the oxidic metal species into the sulphide phase and the HDS activity is low, increasing with increasing metal content. In the case of SiO_2, because it has a weaker interaction with Mo or W oxides than γ-Al_2O_3, the requirement for high efficiency of the HDS reaction can be met at lower surface loading and the plots exhibit maximum HDS activity at about 1 atoms nm^{-2} loading; below such a loading the oxides are well dispersed but the interaction with the SiO_2 is too strong to allow complete sulphidation and above such a loading the limiting factor is the decreasing dispersion resulting in larger sulphide crystals. For carbon supports the situation is different. For the Darco carbon and for the untreated Mekog carbon, maximum efficiency as reached at lower surface loading than with γ-Al_2O_3 and SiO_2, reflecting that there is only a weak interaction between the oxidic metal species and the carbon supports. Further, when the Mekog carbon is previously treated with HCl the interaction between the support and the oxidic species is further decreased, the HDS activity becoming very high at very low surface loading. The authors claimed a maximum below 0.03 atoms nm^{-2} surface loading, this being indicative of the fact that this treated carbon had to be an almost inert support material.

It should also be mentioned that in the above work[19] it was shown that cobalt and nickel sulphides, when supported on carbon, had an HDS activity which was higher (Co) or similar (Ni) to that measured for molybdenum and tungsten sulphides. This was thought to show that contrary to what was generally assumed in previous models, Co or Ni present in Co (Ni)–Mo (W) sulphide catalysts had the potential to act not only as promoters for the molybdenum or tungsten sulphide phase but also as an additional active phase. This was confirmed in later work.[175–177]

Vissers *et al.*[178] attempted, by means of combined oxygen chemisorption, X-ray photoelectron spectrometry (XPS) and thiophene HDS activity measurements, to explain the observed activity differences between activated carbon-supported and alumina-supported molybdenum sulphide catalysts. They concluded that the observed superior activity of the Mo–C catalysts (especially at low metal loadings) could be attributed to the presence at the carbon surface of a molybdenum sulphide phase which had both a higher fraction of catalytically active surface area and a higher HDS activity per active site compared with the molybdenum sulphide phase present on the alumina support surface. This, of course, suggests that at low Mo loadings the molybdenum sulphide phases present on carbon and alumina are different. However, at high Mo loadings the properties of the molybdenum sulphide present on alumina tended toward those of the Mo–Co system, emphasizing that at very high metal loadings the carbon- and alumina-supported sulphide phases were essentially the same. In a more recent investigation, Vissers *et al.*[179] studied the sulphidation process of Mo–C catalysts and compared it with the Mo–Al_2O_3 system. The results for the carbon-supported catalysts showed that in the oxide state the Mo phase was highly dispersed as isolated or polymerized monolayer species at Mo loadings below 3 wt% and as very small three-dimensional particles at higher loadings. On sulphiding some particle growth took place even at low Mo loadings, indicating a weak interaction with the carbon surface, in contrast to the alumina-supported system. Nevertheless, the carbon surface stabilizes the small sulphide particles sufficiently since no bulky sulphide particles are observed (the size of the sulphide particles remained below 4.6 nm even in the sample with the highest Mo loading, 14.1 wt%). However, in the Mo–Al_2O_3 system the sulphide phase is present as an MoS_2-like two-dimensional single slab structure; on the other hand, in the Mo–Al_2O_3 system the larger interaction of the oxide form with the support results in a need for increasing the sulphidation temperature (this effect decreases with increasing Mo loading, as shown above). Clearly, these strong interactions are not present in the carbon-supported catalysts so that the formation of MoS_2 takes place at lower temperatures. It may be concluded that different molybdenum sulphide structures are formed on alumina and carbon supports due to differences in interaction between the MoS_2 phase and the support surface, a single slab monolayer strongly interacting with the support on alumina and small three-dimensional particles essentially free of interaction with the support on carbon. These results, together with those mentioned above for Co and Ni supported on carbon[19] giving more (Co) or similar (Ni) activity than the corresponding Mo catalysts, seem clearly to indicate that the application of a material as inert as carbon offers the advantage that all transition metal compounds present in the precursor state will be quantitatively converted into their active sulphide form, something which is not possible with the conventional industrial alumina support.

The above results for simple, monometallic systems have been extended to bimetallic catalysts similar to those used in industry, especially the Co–Mo sulphide catalyst,[19,21,180–183] and in recent years to other systems such as Fe,[20,184–186] Fe–Mo,[184] Ni–W,[19] Ni–Mo[19] and many other metals.[187–189] In all cases, the thiophene HDS activity was found to be larger when the active phase was supported on carbon.

For the case of the most studied system, the sulphided Co–Mo catalyst, as in the case of the sulphided Mo system[179] described above, the essential difference between the sulphided Co–Mo–C and Co–Mo–Al_2O_3 systems is that in the case of the alumina-supported catalyst there is an interaction of the active phase with the support, probably through Mo–O–Al linkages, whereas in the case of the carbon-supported catalyst the Co–Mo component is fully sulphided with a very weak interaction with the support. This, again, should be taken as the catalysts having different HDS activities essentially because they contain different active phases. According to Topsoe and Clausen,[190] the former is a Co–Mo–S type I phase whereas in carbon there is a Co–Mo–S type II phase. In order to study the real support effect, van Veen *et al.*[182] prepared sulphided Co–Mo catalysts supported on alumina, silica and activated carbon in such a way as to ensure that the same phase (Co–Mo–S type II) and the same dispersion were present in the three catalysts. Figure 10.14 shows the thiophene HDS activities, expressed as pseudo-first-order rate constants, of Co–Mo–S type II as a function of Mo loading at fixed Co/Mo ratio. It is seen that whereas the change from alumina to

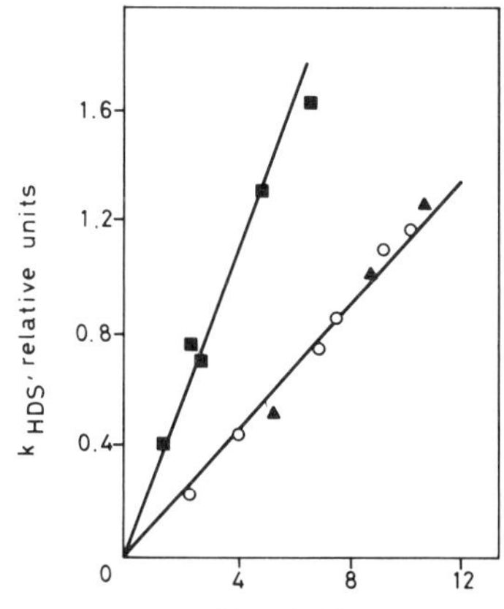

Fig. 10.14 Thiophene HDS activity of Co–Mo–S II versus Mo loading, Co/Mo = 0.33 atom/atom. Supports: (○) γ–Al_2O_3; (▲) SiO_2; (■) carbon. Catalysts were pre-sulphided at 350°C in 1:9 (v/v) H_2S–H_2 for 1 h. Test conditions: 0.2–0.3 g catalysts, H_2 flow rate 55 ml min^{-1}, [thiophene] = 6 vol.%, $T = 350°C$, $P = 0.1$ MPa. Redrawn and adapted from van Veen *et al.*[182]

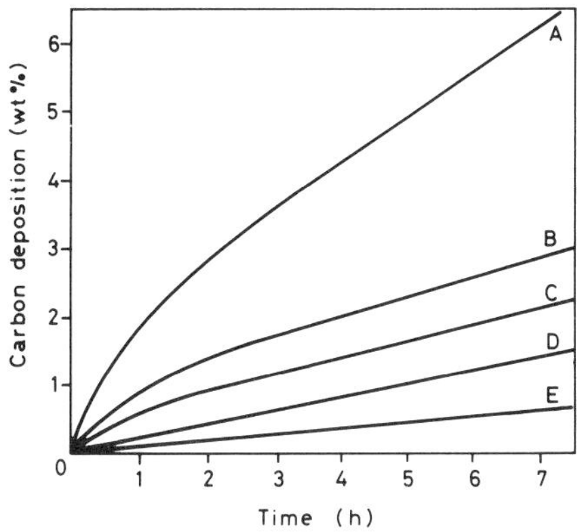

Fig. 10.15 Deposition of carbon from propylene cracking over catalyst supports (673 K, 100 cm^3 min^{-1}, atmospheric pressure). (A) Active carbon (Norit); (B) carbon black composite (Ketjen); (C) carbon black composite (Monarch); (D) γ-Al_2O_3 (Ketjen); (E) γ-Al_2O_3 (Gulf). Redrawn and adapted from de Beer *et al.*[20]

silica does not affect the catalytic activity of Co–Mo–S type II, the carbon support induces an appreciable higher activity. Hence, it appears that even when the active phase and its dispersion are invariant, the support influences the specific activity.

On top of carbon-supported catalysts having higher HDS and HDN activities there is a further advantage of carbon as a support for sulphide catalysts, viz. a lower coking propensity. The side-reactions taking place during the thiophene HDS reaction leading to cracking of hydrocarbons and coke deposition on the surface of the catalyst will reduce the life of the catalyst. This aspect has been studied by Scaroni and co-workers[20,21] using Fe, Mo and Co–Mo catalysts supported on alumina and carbon and following the thiophene HDS reaction and coke deposition during propylene cracking and anthracene carbonization. Some of their results may be summarized as follows. For propylene cracking, Figure 10.15 shows that the extent of carbon deposition on the blank supports after a reaction time of 7.5 h is higher for the carbons than for the aluminas and increases in order of increasing surface area[20]; in the absence of a metallic component the cracking of propylene appears to be related more to the amount of accessible surface than to any particular surface property. The addition of metals causes an increase in the rates and amounts of carbon deposition from propylene but the increase is much higher in the alumina-supported catalyst. Thus, Figure 10.16 shows the coke deposition as a function of Mo loading for an alumina and a carbon black composite having about the same porosity[20]; the latter induces a much smaller carbon deposition and, in addition, is essentially insensitive to the metal concentration. Similar results were obtained in the investigation of the liquid-phase carbonization of anthracenes over the same supports.[21] Figure 10.17 shows the comparative coke deposition after 1 h of anthracene coking for Co–Mo catalysts supported on alumina and carbon black as a function of temperature. It is clear that the Co–Mo–Al_2O_3 catalyst has a significantly greater coking propensity than the Co–Mo–carbon black catalyst. Again, the reason for this behaviour can be attributed to the lower surface acidity of carbon compared with alumina. The results of Zdrazil[170] indicate that

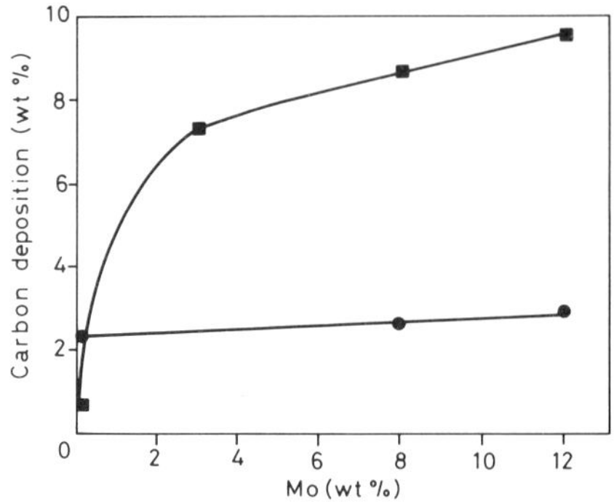

Fig. 10.16 Effect of molybdenum loading on carbon deposition for carbon and alumina supports (673 K, 7.5 h, 100 $cm^3 min^{-1}$ propylene, atmospheric pressure). (●) carbon black composite (Monarch); (■) γ-Al_2O_3 (Gulf). Redrawn and adapted from de Beer *et al.*[20]

the acidity of MoS_2–C is similar to that of MoS_2 alone, whereas that of MoS_2–Al_2O_3 is much higher. The impregnation of metals on alumina increases the surface acidity, which raises the coking propensity. In the case of carbon, as the acidity is initially much lower and the interaction with the active phase weaker, the effect of metal addition on coking propensity will be small. However, it has recently been shown[191] that carbon-supported molybdenum sulphide catalysts are poisoned by phosphorus (added as phosphoric acid to the carrier prior to the molybdenum component), this not being the case for similar catalysts supported on γ-Al_2O_3. In this case the strong interaction of phosphate with the alumina support prevents the reduction to phosphine (the poison) during sulphiding, a reduction which is easy in carbon-supported catalysts.

The advantages of carbon as a support for mono- and bimetallic sulphide catalysts have been extended to reactions other than hydrotreating petroleum feedstocks. This is the case for the purification of liquids derived from coal. Thus, Groot *et al.*[184] found that carbon-supported sulphide catalysts are especially attractive for application in a coal liquefaction process, this being confirmed by the independent work of Scaroni *et al.*[20,21]

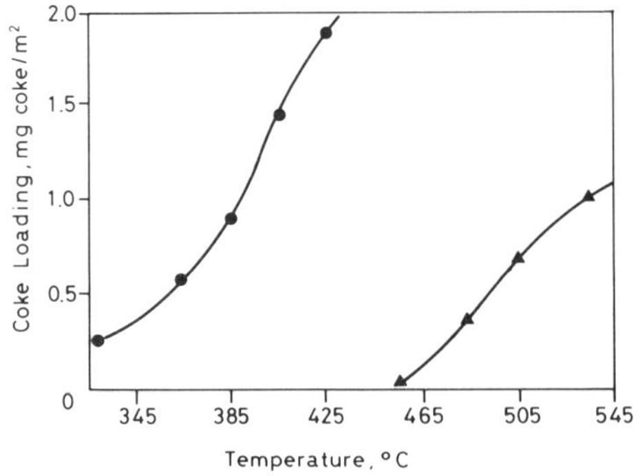

Fig. 10.17 Comparative coke loadings after 1 h on the Co–Mo–Al_2O_3 and Co–Mo–carbon black catalysts: (●) Co–Mo–Al_2O_3; (▲) Co–Mo–carbon black. Redrawn and adapted from Scaroni *et al.*[21]

All the studies described coincide with the fact that hydrotreating activity is higher in carbon-supported catalysts than in alumina-supported systems, coupled with lower coking propensities in the former. However, one point which is not clear is the influence of the physical (surface area, porosity) and chemical (e.g. surface group) properties of the carbon. The reason lies in the fact that the carbons used ranged from carbon blacks to activated carbons, with very different textural and chemical compositions and contents of impurities. Further, the comparison between different studies is difficult because the catalysts are prepared using different experimental procedures and have different metal loadings and the reactions are carried out under different experimental conditions. However, some comparison is possible. Thus, when comparing sulphided Mo catalysts supported on a graphitized carbon black[172] and activated carbon,[174] it is deduced that the former is more selective for thiophene HDS. It has already been shown in Figure 10.13 that the washing of activated carbon with HCl increased considerably the HDS activity at low Mo or W loadings. Groot *et al.*[185] found, for sulphided iron oxide supported on different carbons, that the thiophene HDS is higher for activated carbon supports than for carbon blacks, the washing of the activated carbon with HCl leading also to an increase in the HDS activity. Vissers *et al.*[192] found also that the thiophene HDS activity for molybdenum sulphide catalysts supported on activated carbon was higher than when supported on different carbon black composites; they also found that there was a correlation between the HDS activity of the composite-supported catalysts and the surface area of the support.

Most carbons used in published studies have large surface areas and pore volumes and this, coupled with the general inertness of the surface, provides the basis for their good HDS results. There is, however, one point which is very important but has not been much studied, viz. the role that the carbon surface functionality might play with respect to the dispersion of the oxidic precursor and the dispersion on the catalytically active sulphide phase obtained after sulphidation. Oxygen surface groups would be most important in this respect,[74] although not exclusively so.

Derbyshire *et al.*[66] used Ambersorb XE-340 (produced by carbonization of sulphonated styrene–divinylbenzene copolymer) and a carbon black composite as supports for molybdenum sulphide catalysts. The carbons were also oxidized and nitrided to introduce oxygen and nitrogen surface groups. Pre-oxidation of the polymer-derived carbon lowered the catalytic activity for coal asphaltene hydrogenation whereas pre-oxidation of the carbon black composite support increased it. They concluded that the modified catalytic activity cannot be unequivocally attributed to surface oxygen; it is the structure of the carbon which is ultimately the controlling factor. However, pre-nitriding of the support was found to have a distinct influence in enhancing catalyst activity; they argued that the presence of nitrogen-containing surface groups may provide preferential sites for the adsorption of Mo species.

Vissers *et al.*[192] recognized that molybdenum sulphide was better dispersed on an activated carbon support than on the more inert carbon black composite support. A number of oxidative treatments applied to the latter caused an increase in the concentration of oxygen functional groups and this was found to result in a catalytic activity for the thiophene HDS that was higher on the catalysts supported on the oxidized support, although still lower than with the activated carbon. In another study[68] they used four carbon blacks differing in surface area, pH and surface properties (oxygen functionality) and examined the dispersion of the molybdate precursor and the sulphided phase. The molybdate was highly dispersed in all cases but after sulphidation particle growth took place, suggesting that the interaction was insufficient to maintain a high degree of dispersion. However, they found that dispersion was best for carbon blacks with high oxygen functionality.

These results can be compared with those described in another section of this chapter for the dispersion of platinum on carbon.[74] Oxygen surface groups are needed to produce the anchoring of the metal precursor to the support surface. For the case of molybdenum sulphide, this has recently been proved in our laboratories.

Although it has been clearly shown that carbon-supported sulphide catalysts present advantages over industrial alumina-supported catalysts, most carbon materials used have an important drawback in practical applications in that the large surface area is mainly contained within micropores and these would be inaccessible to large sulphur-containing molecules present in industrial feedstocks and, on the other hand, would cause diffusion limitations. Another possible drawback of carbon supports is the poor mechanical strength (causing crushing when packed in an industrial reactor). If the carbon was made to have a lower surface area but with an increased mesopore volume, then the mechanical properties would be poorer and the bulk density too low.

One way to circumvent these drawbacks is to use carbon black composites of the type originally proposed by Schmitt *et al.*[59] These have been used in some studies mentioned above and are now being used with many different variations in the preparation procedure and even using activated carbons instead of carbon blacks in the preparation. Another approach has recently been presented by Vissers *et al.*,[193] based on the covering (deposition by pyrolysis of hydrocarbons at high temperature) of the Al_2O_3 surface with a thin layer of carbon prior to impregnation of the metal precursor. In this way, the favourable carbon surface properties (quantitative conversion of the precursor into the active sulphide form) are combined with the optimum textural and mechanical properties of the alumina support. As expected, oxidation of the carbon surface prior to the introduction of the metal precursor leads to a further improvement of the catalytic activity. More recently, a novel type of carbon-supported catalyst has been introduced[194,195] with metal particles supported on carbon, which in turn is situated on alumina; starting off with metal oxide–alumina catalyst, the carbon is catalytically deposited by CO treatment between the *in situ* reduced metal particle and the alumina (the authors call this a 'tulip-field' catalyst). these catalysts combine the favourable mechanical characteristics of the alumina support with the beneficial effect of the carbon support on catalytic activity.

10.4 Concluding remarks

The examples used here have clearly shown the unparalleled flexibility of carbon (in its different forms, especially carbon black and activated carbon) for the preparation of catalysts, in spite of some specific problems that can be found in its use for some reactions. This unique flexibility arises from the porous structure and the chemical nature of the surface, both of which may be appropriately tailored. In some cases, the high surface area of carbon provided by the small pores stabilizes highly dispersed metal particles owing to the presence of physical barriers which suppress surface migration and the resultant agglomeration. In many cases, the inertness of the carbon surface alleviates the difficulty often encountered when attempting the reduction of oxided metal precursors to the metal in the zero-valent state or the quantitative conversion of such precursors to a different form such as sulphide. On the other hand, the flexibility of the carbon surface, which may be functionalized, is an especially attractive feature of this support for the preparation of catalysts to suit many specific needs. Finally, it has been shown that many carbon-supported catalysts have the additional advantage of exhibiting a lower deactivation propensity than the corresponding catalysts supported on more conventional oxide supports.

10.5 Acknowledgements

Partial support from the CICYT (project PB86/279) is gratefully acknowledged.

10.6 References

1. MILLER, A.W., ATKINSON, W., BARBER, M. and SWIFT, P. 1971: *Journal of Catalysis* **22**, 140.
2. MOSS, R.L. 1976: in Anderson, R.B. and Dawson, P.T. (eds) *Experimental Methods in Catalytic Research*. New York: Academic Press, Vol. II, p. 43.
3. SATTERFIELD, C.N. 1980: *Hetrogeneous Catalysis in Practice*. New York: McGraw-Hill.
4. HONALLA, M. and DELMON, B. 1971: *Journal of Physical Chemistry* **84**, 2194.
5. BOUDART, M. and DJEGA-MARIADASSOU, G. 1984: *Kinetics of Heterogeneous Catalytic Reactions*. Princeton, NJ: Princeton University Press.
6. WACHS, I.E., SALCH, R.Y., CHAN, S.S. and CHERSICH, C.C. 1985: *Applied Catalysis* **15**, 339.
7. MARCELIN, G. and LESTER, J.E. 1985: *Journal of Catalysis* **93**, 270.
8. SPENCER, M.S. 1985: *Journal of Catalysis* **93**, 216.
9. ROMANOWSKI, W. 1987: *Highly Dispersed Metals*. Chichester: Ellis Horwood.
10. WEISSER, I. and LANDA, S. 1973: *Sulphide Catalysts. Their Properties and Applications*. New York: Pergamon Press.
11. GRANGE, P. 1980: *Catalysis Reviews. Science and Engineering* **21**, 135.
12. RATNASAMY, P. and SIVASANKER, S. 1980: *Catalysis Reviews. Science and Engineering* **22**, 401.
13. De BEER, V.H.J., van der AALST, M.J.M., MACHIELS, C.J. and SCHUIT, G.C.A. 1976: *Journal of Catalysis* **43**, 78.
14. De BEER, V.H.J. and SCHUIT, G.C.A. 1976: in Delmon, B., Grange, P. Jacobs, P.A. and Poncelet, G. (eds) *Preparation of Catalysts*. Amsterdam: Elsevier, p. 343.
15. De BEER, V.H.J., van SINT FIET, T.H.M., van der STEEN, G.H.A.M., ZWAGA, A.C. and SCHUIT, G.C.A. 1974: *Journal of Catalysis* **35**, 297.
16. LYCOURGHIOTIS, A., VATTIS, D. and ARONI, P. 1980: *Zeitschrift für Physikalische Chemie. Neue Folge* **121**, 257.
17. LYCOURGHIOTIS, A., TSIATSIOS, A. and KATSAMOS, N.A. 1981: *Zeitschrift für Physikalische Chemie. Neue Folge* **126**, 95.

18. LO JACONO, M., SCHIAVELLO, M., de BEER, V.H.J. and MINELLI, J. 1977: *Journal of Physical Chemistry* **81**, 1583.
19. DUCHET, J.C., van OERS, E.M., de BEER, V.H.J. and PRINS, R. 1983: *Journal of Catalysis* **80**, 386.
20. De BEER, V.H.J., DERBYSHIRE, F.H., GROOT, C.K., PRINS, R., SCARONI, A.W. and SOLAR, J.M. 1984: *Fuel* **63**, 1095.
21. SCARONI, A.V., JENKINS, R.G. and WALKER, P.L. Jr, 1985: *Applied Catalysis* **14**, 173.
22. BIRD, A.J. 1987: in Stiles, A.B. (ed.) *Catalyst Supports and Supported Catalysts.* Guildford: Butterworth, p. 107.
23. TRIMM, D.L. 1981: in *Catalysis (Specialist Periodical Report),* London: Chemical Society, Vol. 4, p. 210.
24. KLEIN, G.J. and HENNING, K.D. 1984: *Fuel* **63**, 1064.
25. JUNTGEN, H., RICHTER, E. and KUHL, H. 1988: *Fuel* **67**, 775.
26. BOEHM, H.P., MAIR, G., STOEHR, T., DE RINCÓN, A.R. and TERECZKI, B. 1984: *Fuel* **63**, 1061.
27. JUNTGEN, H. and KUHL, H. 1989: in Thrower, P.A. (ed) *Chemistry and Physics of Carbon.* New York: Marcel Dekker, Vol. 22, p. 145.
28. JUNTGEN, H. 1986: *Fuel* **65**, 1436.
29. LINARES-SOLANO, A., LÓPEZ-GONZÁLEZ, J.D., RODRÍGUEZ-REINOSO, F., SALINAS-MARTÍNEZ DE LECEA, C. and SEPÚLVEDA, A. 1988: in McEnaney, B. and Mays, T.J. (eds) *Proceedings of Carbon '88.* Bristol: IOP Publishing, p. 80.
30. RODRÍGUEZ-REINOSO, F., RODRÍGUEZ-RAMOS, I., GUERRERO-RUIZ, A. and LÓPEZ-GONZÁLEZ, J.D. 1986: *Applied Catalysis* **21**, 251.
31. RODRÍGUEZ-REINOSO, F., RODRÍGUEZ-RAMOS, I. MORENO-CASTILLA, C., GUERRERO-RUIZ, A. and LÓPEZ-GONZÁLEZ, J.D. 1986: *Journal of Catalysis* **99**, 171.
32. SMISEK, M. and CERNY, S. 1970: *Active Carbon Manufacture, Properties and Applications.* New York: Elsevier.
33. MATTSON, J.S. and MARK, H.B. Jr, 1971: *Activated Carbon.* New York: Marcel Dekker.
34. *Activated Carbon, a Fascinating Material.* 1984: Amersfoort: Norit.
35. WIGMAM, T. 1986: In Figuereido, J.L. and Moulijn, J.M. (eds) *Carbon and Coal Gasification.* Dordrecht: Martinus Nijhoff, p. 559.
36. RODRÍGUEZ-REINOSO, F. 1986: in Figuereido, J.L. and Moulijn, J.M. (eds) *Carbon and Coal Gasification.* Dordrecht: Martinus Nijhoff, p. 601.
37. RODRÍGUEZ-REINOSO, F. and LINARES-SOLANO, A. 1988: in Thrower, P.A. (ed.) *Chemistry and Physics of Carbon.* New York: Marcel Dekker, Vol. 21, p. 1.
38. BANSAL, R.C., DONNET, J.B. and STOECKLI, F. 1988: *Active Carbon.* New York: Marcel Dekker.
39. SING, K.S.W., EVERETT, D.H., HAUL, R.A.W., MOSCOU, L., PIEROTTI, R.A., ROUQUEROL, J. and SIEMIENIEWSKA, T. 1985: *Pure and Applied Chemistry* **57**, 603.
40. RODRÍGUEZ-REINOSO, F. 1989: *Pure and Applied Chemistry* **61**, 1859.
41. FREEMAN, J.J., GIMBLETT, F.G.R., ROBERTS, R.A. and SING, K.S.W. 1988: *Carbon* **26**, 7.
42. DUBININ, M.M. and FEDOSEEV, D.V. 1982: *Izvestiia Akademiia Nauk SSSR: Seriia Khimischeskaia*, p. 246.
43. CATURLA, F., GUTIÉRREZ, A., MOLINA-SABIO, M. and RODRÍGUEZ-REINOSO, F. 1988: In McEnaney, B. and Mays, T.J. (eds) *Proceedings of Carbon '88.* Bristol: IOP Publishing, p. 4.
44. *US Patent* 4532227 1985.
45. ILEY, M., MARSH, H. and RODRÍGUEZ-REINOSO, F. 1973: *Carbon* **11**, 633.

46. LÓPEZ-GONZÁLEZ, J.D., MARTINEZ-VILCHEZ, F. and RODRÍGUEZ-REINOSO, F. 1980: *Carbon* **18**, 413.
47. RODRÍGUEZ-REINOSO, F., LÓPEZ-GONZÁLEZ, J.D. and BERENGUER, C. 1982: *Carbon* **20**, 513.
48. RODRÍGUEZ-REINOSO, F., LÓPEZ-GONZÁLEZ, J.D. and BERENGUER, C. 1984: *Carbon* **22**, 13.
49. RODRÍGUEZ-REINOSO, F., MARTÍN-MARTÍNEZ, J.M., MOLINA-SABIO, M., PÉREZ-LLEDÓ, I. and PRADO-BURGUETE, C. 1985: *Carbon* **23**, 19.
50. RODRÍGUEZ-REINOSO, F., MARTÍN-MARTÍNEZ, J.M. MOLINA-SABIO, M., TORREGROSA-MACIÁ, R. and GARRIDO-SEGOVIA, J. 1985: *Journal of Colloid Interface Science* **106**, 315.
51. GARRIDO-SEGOVIA, J., LINARES-SOLANO, A., MARTÍN-MARTÍNEZ, J.M., MOLINA-SABIO, M., RODRÍGUEZ-REINOSO, F. and TORREGROSA, R. 1987: *Langmuir* **3**, 76.
52. RODRÍGUEZ-REINOSO, F., GARRIDO, J., MARTÍN-MARTÍNEZ, J.M., MOLINA-SABIO, M. and TORREGROSA, R. 1989: *Carbon* **27**, 23.
53. YERMAKOV, Y.I., SUROVIKM, V.F., PLAKSIN, G.V., SAMIKOLENOV, V.A., LIKHOLOBOV, V.A., CHUVILIN, L.V. and BOGDANOV, S.V. 1987: *Reaction Kinetics and Catalysis Letters* **33**, 435.
54. DALY, F.P. and BRINEN, J.S. 1987: *Applied Catalysis* **30**, 91.
55. VENTER, J.J., CHEN, A.A., PHILLIPS, J. and VANNICE, M.A. 1989: *Journal of Catalysis* **119**, 451.
56. WAGNER, W., CHEN, A.A. and VANNICE, M.A. 1989: *Catalysis Letter* **2**, 413.
57. MULAY, L.N., RAO, A.V.P., WALKER, P.L. Jr, YOON, K.J. and VANNICE, M.A. 1985: *Carbon* **23**, 501.
58. CHRISTENSEN, H., MORUP, S. and NIEMANTSVERDRIET, J.W. 1985: *Journal of Physical Chemistry* **89**, 4898.
59. SCHMITT, J.L. Jr, WALKER, P.L. Jr and CASTELLION, G.A. 1976: *US Patent* 3978000.
60. MACHEK, V., HANIKA, J., SPORKA, K., RUZICKA, V. and KUNZ, J. 1981: *Collection of Czechoslovak Chemical Communications* **46**, 3270.
61. HANIKA, J., MACHEK, V., NEMEC, V., RUZICKA, V. and KUNZ, J. 1982: *Journal of Catalysis* **77**, 248.
62. EHRBURGER, P., MAHAJAN, O.P. and WALKER, P.L. Jr 1976: *Journal of Catalysis* **43**, 61.
63. LINARES-SOLANO, A., RODRÍGUEZ-REINOSO, F., SALINAS-MARTÍNEZ DE LECEA, C., MAHAJAN, O.P. and WALKER, P.L. Jr 1982: *Carbon* **20**, 177.
64. FERNÁNDEZ-MORALES, I., GUERRERO-RUIZ, A., LÓPEZ-GARZÓN, F.J., RODRÍGUEZ-RAMOS, I. and MORENO-CASTILLA, C. 1985: *Applied Catalysis* **14**, 159.
65. EHRBURGER, P. 1984: *Advanced Colloid Interface Science* **21**, 275.
66. DERBYSHIRE, F.J., DE BEER, V.H.J., ABOTSI, G.M.K., SCARONI, A.W., SOLAR, J.M. and SKROVANEK, D.J. 1986: *Applied Catalysis* **27**, 117.
67. RICHARD, D. and GALLEZOT, P. 1987: in Delmon, B., Grange, P., Jacobs, P.A. and Poncelet, G. (eds) *Preparation of Catalysts IV*. Amsterdam: Elsevier, p. 71.
68. VISSERS, J.P.R., BOUWENS, S.M.A.M., de BEER V.H.J. and PRINS, R. 1987: *Carbon* **25**, 485.
69. CHEN, A.A., VANNICE, M.A. and PHILLIPS, J. 1987: *Journal of Physical Chemistry* **91**, 6257.
70. GUERRERO-RUIZ, A., RODRÍGUEZ-RAMOS, I., RODRÍGUEZ-REINOSO, F., MORENO-CASTILLA, C. and LÓPEZ-GONZÁLEZ, J.D. 1988: *Carbon* **26**, 417.
71. BAKER, R.T.K. 1982: in Imelik, B., Naccache, C., Coudurier, G., Praliaud, H.,

Meriandeau, P., Gallezot, P., Martin, G.A. and Vedrine, J.C. (eds) *Metal–Support and Metal–Additive Effects in Catalysis*. Amsterdam: Elsevier, p. 37.
72. Del ANGEL, G., COQ, B. and FIGUERAS, F. 1982: in Imelik, B., Naccache, C., Coudurier, G., Praliaud, H., Meriandeau, P., Gallezot, P., Martin, G.A. and Vedrine, J.C. (eds) *Metal–Support and Metal–Additive Effects in Catalysis*. Amsterdam: Elsevier, p. 85.
73. PAJONK, G.M. and TEICHNER, S.J. 1986: in Cerveny, L. (ed.) *Catalytic Hydrogenation*. Amsterdam: Elsevier, p. 257.
74. PRADO-BURGUETE, C., LINARES-SOLANO, A., RODRÍGUEZ-REINOSO, F. and SALINAS-MARTÍNEZ DE LECEA, C. 1989: *Journal of Catalysis* **115**, 98.
75. FOUILLOUX, P. 1988: in Guisnet, M., Barrault, J., Bouchoule, C., Duprez, D., Montassier, C. and Perot, G. (eds) *Heterogeneous Catalysis and Fine Chemicals*. Amsterdam: Elsevier, p. 123.
76. ORITO, Y., IMAI, S. and NIWA, S. 1982: *Journal of the Chemical Society of Japan*, p. 137.
77. BLASER, H.U., JALETT, H.P., MONTI, D.M., REBER, J.F. and WEHRLI, J.T. 1988: in Guisnet, M., Barrault, J., Bouchoule, C., Duprez, D., Montassier, C. and Perot, G. (eds) *Heterogeneous Catalysis and Fine Chemicals*. Amsterdam: Elsevier, p. 153.
78. EHRBURGER, P., MONGILARDI, A. and LAHAYE, J. 1983: *Journal of Colloid Interface Science* **91**, 151.
79. GALLEZOT, P., LECLERQ, C., MUTIN, I., NICOT, C. and RICHARD, D. 1985: *Journal de Microscopie et de Spectroscopie Electroniques* **10**, 479.
80. RICHARD, D., OCKELFORD, J., GIROIR-FENDLER, A. and GALLEZOT, P. 1989: *Catalysis Letter* **3**, 53.
81. CASTILLA, C., GUERRERO-RUIZ, A. and LÓPEZ-GONZÁLEZ, J.D. 1987: *Journal of Catalysis* **107**, 1.
82. HEAL, G.R. and MKAYULA, L.L. 1988: *Carbon* **26**, 815.
83. ABOTSI, G.M.K., SCARONI, A.W. and DERBYSHIRE, F.J. 1988: *Applied Catalysis* **37**, 93.
84. LEON, C.A. and RODOVIC, L.R. 1989: in Pennsylvania State University (eds) *Proceedings of the XIXth Biennial Conference on Carbon*. American Carbon Soceity, p. 58.
85. FIGUEIREDO, J.L., FERRAZ, M.C.A. and ORFAO, J.J.M. 1983: in Poncelet, G., Grange, P. and Jacobs, P.A. (eds) *Preparation of Catalysts III*. Amsterdam: Elsevier, p. 571.
86. ANDERSON, J.R. 1975: *Structure of Metallic Catalysts*. London: Academic Press, p. 163.
87. MACHEK, V., HANIKA, J., SPORKA, K. and RUZICKA, V. 1981: *Collection of Czechoslovak Chemical Communications* **46**, 1588.
88. MACHEK, V., RUZICKA, V., SOURKOVA, M., KUNZ, J. and JANACEK, L. 1983: *Collection of Czechoslovak Chemical Communications* **48**, 517.
89. PALMER, M.B. and VANNICE, M.A. 1980: *Journal of Chemical Technology and Biotechnology* **30**, 205.
90. EHRBURGER, P., DENTZER, J. and LAHAYE, J. 1987: *Carbon* **25**, 129.
91. RODRÍGUEZ-REINOSO, F., MORENO-CASTILLA, C., GUERRERO-RUIZ, A., RODRÍGUEZ-RAMOS, I. and LÓPEZ-GONZÁLEZ, J.D. 1985: *Applied Catalysis* **15**, 293.
92. FLYNN, P.C. and WANKE, S.W. 1975: *Journal of Catalysis* **37**, 423.
93. PRADO-BURGUETE, C., LINARES-SOLANO, A., RODRÍGUEZ-REINOSO, F. and SALINAS-MARTÍNEZ DE LECEA, C. 1991: *Journal of Catalysis* **128**, 397.
94. YATES, D.J.C. and SINFELT, J.M. 1967: *Journal of Catalysis* **8**, 348.
95. CLEWLEY, J.D., LYNCH, J.F. and FLANAGAN, T.B. 1975: *Journal of Catalysis* **36**, 291.

96. PALMER, M.B. Jr and VANNICE, M.A. 1980: *Journal of Chemical Technology and Biotechnology* **30**, 205.
97. CANDY, J.P., FOUILLOUX, P. and PRIMET, M. 1978: *Surface Science* **72**, 167.
98. SCHOLTEN, J.J.F., PIJPERS, A.P. and HUSTINGS, A.M.L. 1985: *Catalysis Reviews. Science and Engineering* **27**, 151.
99. SAYARI, A., WANG, H.T. and GOODWIN, J.G. Jr 1985: *Journal of Catalysis* **93**, 368.
100. BASSET, J.M., DALMAI-IMELIK, G., PRIMET, M. and MUTIN, R. 1975: *Journal of Catalysis* **37**, 22.
101. SOMORJAI, G.A. 1981: *Chemistry in Two Dimensions: Surfaces*. Ithaca, NY: Cornell University Press.
102. DELMON, B. and FROMENT, G.F. 1980: *Catalyst Deactivation*. Amsterdam: Elsevier.
103. HUGHES, R. 1984: *Deactivation of Catalysts*. New York: Academic Press.
104. TRIMM, D.L. 1982: in Figueiredo, J.L. (ed.) *Progress in Catalyst Deactivation*. Dordrecht: Martinus Nijhoff, p. 3.
105. TRIMM, D.L. 1982: in Figueiredo, J.L. (ed.) *Progress in Catalyst Deactivation*. Dordrecht: Martinus Nijhoff, p. 23.
106. IHM, S.K., LEE, D.K. and YEUM, Y.S. 1988: *Korean Journal of Chemical Engineering* **5**, 14.
107. FALBE, J. (ed.) 1980: *New Syntheses with Carbon Monoxide*. New York: Springer.
108. HENRICI-OLIVÉ, G. and OLIVÉ, S. 1984: *The Chemistry of the Catalyzed Hydrogenation of Carbon Monoxide*. New York: Springer.
109. VANNICE, M.A. and GARTEN, R.L. 1979: *Journal of Catalysis* **56**, 236.
110. VANNICE, M.A. 1975: *Journal of Catalysis* **37**, 449.
111. VANNICE, M.A. 1975: *Journal of Catalysis* **50**, 228.
112. VANNICE, M.A., WALKER, P.L. Jr, JUNG, H.-J., MORENO-CASTILLA, C. and MAHAJAN, O.P. 1981: in *Proceedings of the 7th International Congress on Catalysis*. Amsterdam: Elsevier, p. 460.
113. AIKA, K. HORI, H. and OZAKI, A. 1972: *Journal of Catalysis* **27**, 424.
114. SAGERT, N.H. and PONTEAU, R.M.L. 1972: *Canadian Journal of Chemistry* **50**, 3686.
115. JUNG, H.-J., WALKER, P.L. Jr and VANNICE, M.A. 1982: *Journal of Catalysis* **75**, 416.
116. JUNG, H.-J., MULAY, L.N., VANNICE, M.A., STANFIELD, R.M. and DELGASS, W.N. 1982: *Journal of Catalysis* **76**, 208.
117. RODRÍGUEZ-REINOSO, F., LÓPEZ-GONZÁLEZ, J.D., MORENO-CASTILLA, C., GUERRERO-RUIZ, A. and RODRÍGUEZ-RAMOS, I. 1984: *Fuel* **63**, 1089.
118. TAU, L.M. and BENNETT, C.O. 1986: *Journal of Physical Chemistry* **90**, 4825.
119. VENTER, J.J., KAMINSKY, M., GEOFFROY, G.L. and VANNICE, M.A. 1987: *Journal of Catalysis* **103**, 450.
120. NIEMANTSVERDRIET, J.W., van der KRAAN, A.M., DELGASS, W.N. and VANNICE, M.A. 1985: *Journal of Physical Chemistry* **89**, 67.
121. JONES, V.K., NEUBAUER, L.R. and BARTHOLOMEW, C.H. 1986: *Journal of Physical Chemistry* **90**, 4832.
122. PETERS, U., CREB, H., JOCKERS, R. and KLEIN, J. 1987: in Delmon, B., Grange, P., Jacobs, P.A. and Poncelet, G. (eds) *Preparation of Catalysts IV*. Amsterdam: Elsevier, p. 493.
123. MARTÍN-MARTÍNEZ, J.M., RODRÍGUEZ-REINOSO, F. and VANNICE, M.A. 1989: *Applied Catalysis* **51**, 93.
124. RODRÍGUEZ-REINOSO, F., GUERRERO-RUIZ, A., MORENO-CASTILLA, C., RODRÍGUEZ-RAMOS, I. and LÓPEZ-GONZÁLEZ, J.D. 1986: *Applied Catalysis* **23**, 299.
125. TOPSOE, H., TOPSOE, N., BOHLBRO, H. and DUMESIC, J.A. 1981: in

Proceedings of the 7th International Congress on Catalysis. Amsterdam: Elsevier, p. 415.
126. BOUDART, M., DELBOUILLE, A., DUMESIC, J.A., KHAMMOUMA, S. and TOPSOE, H. 1975: *Journal of Catalysis* **37**, 486.
127. RAUPP, G.B. and DELGASS, W.N. 1979: *Journal of Catalysis* **58**, 361.
128. DWYER, D.J. and SOMORJAI, G.A. 1978: *Journal of Catalysis* **52**, 291.
129. SOMMEN, A.P.B., STOOP, F. and van der WIELE, K. 1985: *Applied Catalysis* **14**, 277.
130. PHILLIPS, J., CLAUSEN, B. and DUMESIC, J.A. 1980: *Journal of Physical Chemistry* **84**, 1814.
131. PHILLIPS, J. and DUMESIC, J.A. 1984: *Applied Catalysis* **9**, 1.
132. LAZAR, K., MATESEK, K., MINK, J., DOBOS, S., GUCZI, L.J., VIZI, A., MARKO, L. and REIFF, W.M. 1984: *Journal of Catalysis* **87**, 163.
133. GUGLIELMINOTI, E. and ZECCHINA, A. 1984: *Journal of Molecular Catalysis* **24**, 331.
134. GUERRERO-RUIZ, A., LÓPEZ-GONZÁLEZ, J.D., RODRÍGUEZ-RAMOS, I. and RODRÍGUEZ-REINOSO, F. 1986: *Reaction Kinetics and Catalysis Letters* **31**, 349.
135. HUGES, F., BUSSIERE, P., BASSET, J.M., COMMERENC, D., CHAUSIN, Y., BONMEVIOT, L. and OLIVIER, D. 1981: in Seiyama, T. and Tanabe, K. (eds) *New Horizons in Catalysis*. Amsterdam: Elsevier, p. 418.
136. CHEN, A.A., VANNICE, M.A. and PHILLIPS, J. 1987: *Journal of Physical Chemistry* **91**, 6257.
137. KAMINSKY, M., YOON, K.J., GEOFFROY, G.L. and VANNICE, M.A. 1985: *Journal of Catalysis* **91**, 338.
138. VENTER, J.J., KAMINSKY, M., GEOFFROY, G.L. and VANNICE, M.A. 1987: *Journal of Catalysis* **105**, 155.
139. VENTER, J.J., CHEN, A. and VANNICE, M.A. 1989: *Journal of Catalysis* **117**, 170.
140. VENTER, J.J. and VANNICE, M.A. 1989: *Journal of Physical Chemistry* **93**, 4158.
141. MARTÍN-MARTÍNEZ, J.M. and VANNICE, M.A. 1991: *Industrial and Engineering Chemistry Research* **30**, 2263.
142. RODRÍGUEZ-REINOSO, F., SALINAS-MARTÍNEZ DE LECEA, C., SEPULVEDA, A. and LÓPEZ-GONZÁLEZ, J.D. 1990: *Catalysis Today* **7**, 287.
143. PHILLIPS, J., VANNICE, M.A. and CHEN, A.A. 1989: in *Extended Abstracts of the 19th Biennial Conference on Carbon*. Pennsylvania: Penn. State University, p. 56.
144. ANDERSON, J.R. and MAINWARING, D.E. 1974: *Journal of Catalysis* **35**, 162.
145. GUCZI, L. 1981: *Catalysis Reviews. Science and Engineering* **23**, 349.
146. CHEN, A.A., KAMINSKY, M., GEOFFROY, G.L. and VANNICE, M.A. 1986: *Journal of Physical Chemistry* **90**, 4810.
147. KOLBEL, H. and RALEK, M. 1980: *Catalysis Reviews. Science and Engineering* **21**, 225.
148. DRY, M.E. 1981: in Anderson, J.R. and Boudart, M. (eds), *The Fischer–Tropsch Synthesis. Catalysis. Science and Technology*. Berlin: Springer, p. 159.
149. MARTÍN-MARTÍNEZ, J.M., RODRÍGUEZ-REINOSO, F. and VANNICE, M.A. 1988: in Phillips, M.J. and Ternan, M. *Proceedings of the 9th International Congress on Catalysis*. Toronto: Chemical Society of Canada, Vol. 2, p. 666.
150. BECHADERGUE, D., BLANCHART, M. and CANESSON, P. 1986: *Applied Catalysis* **20**, 179.
151. BECHADERGUE, D., BLANCHART, M. and CANESSON, P. 1988: in Guisnet, M., Barrault, J., Bouchoule, C., Duprez, D., Montassier, C. and Perot, G. *Heterogeneous Catalysis and Fine Chemicals*. Amsterdam: Elsevier, p. 257.
152. KAWAGUCHI, T., NAKAGAWA, J. and WAKAZUGI, T. 1987: *Applied Catalysis* **32**, 23.
153. EHRBURGUER, P. and LAHAYE, J. 1986: in Bacha, J.D., Newman, J.W. and White, J.L. (eds) *Petroleum Derived Carbons*. Washington, DC: ACS Symposium

Series, American Chemical Society, p. 310.
154. ZDRAZIL, M. 1988: *Catalysis Today* **3**, 269.
155. GATES, B.C., KATZER, J.R. and SCHUIT, G.C.A. 1979: *Chemistry of Catalytic Processes*. New York: McGraw-Hill.
156. MASSOTH, F.E. 1975: *Journal of Catalysis* **36**, 164.
157. MASSOTH, F.E. 1977: *Journal of Catalysis* **47**, 316.
158. VOORHOEVE, R.J.H. and STUIVER, J.C.M. 1971: *Journal of Catalysis* **23**, 228.
159. VOORHOEVE, R.J.H. 1971: *Journal of Catalysis* **23**, 236.
160. HAGENBACH, G., COURTY, P. and DELMON, B. 1971: *Journal of Catalysis* **23**, 295.
161. TOPSOE, H., CLAUSEN, B.S., CANDIA, R., WIVEL, C. and MORUP, S. 1981: *Journal of Catalysis* **68**, 433.
162. MASSOTH, F.E. 1978: *Advances in Catalysis* **27**, 265.
163. TOPSOE, H. 1983: in Bonelle, J.P. *et al.* (ed.) *Surface Properties and Catalysis by Non-metals*. Dordrecht: Reidel, p. 329.
164. *French Patent* 981 468 1951.
165. *US Patent* 3725303 1973.
166. *US Patent* 3997473 1976.
167. *Us Patent* 4032435 1977.
168. KRAUS, J. and ZDRAZIL, M. 1977: *Reaction Kinetics and Catalysis Letters* **6**, 475.
169. YAMAGATA, N., OWADA, Y., OKAZAKI, S. and TANABE, K. 1987: *Journal of Catalysis* **47**, 358.
170. ZDRAZIL, M. 1979: *Journal of Catalysis* **58**, 436.
171. TOPSOE, H., CLAUSEN, B.S., BURRIESER, N., CANDIA, R. and MORUP, S. 1979: in Delmon, B., Grange, P., Jacobs, P.A. and Poncelet, G. (eds) *Preparation of Catalysts II*. Amsterdam: Elsevier, p. 479.
172. STEVENS, G.C. and EDMONDS, T. 1979: in Delmon, B., Grange, P., Jacobs, P.A. and Poncelet, G. (eds) *Preparation of Catalysts II*. Amsterdam: Elsevier, p. 507.
173. De BEER, V.H.J., DUCHET, J.C. and PRINS, R. 1981: *Journal of Catalysis* **72**, 369.
174. BRIDGEWATER, A.J., BURCH, R. and MITCHELL, P.C.H. 1982: *Applied Catalysis* **4**, 267.
175. VISSERS, J.P.R., de BEER, V.H.J. and PRINS, R. 1987: *Journal of the Chemical Society, Faraday Transactions 1* **83**, 2145.
176. VAN DER KRAAN, A.M., GRAJE, M.W.J., GERKEMA, E., RAMSELAAR, W.L.T.M. and de BEER, V.H.J. 1989: *Hyperfine Interactions* **46**, 567.
177. KARRONA, M., CENTENO, A., MATRALIS, H.K., GRANGE, P. and DELMON, B. 1989: *Applied Catalysis* **51**, L21.
178. VISSERS, J.P.R., BACHELIER, J., ten DOESCHATE, H.J.M., de BEER, V.H.J. and PRINS, R. 1984: in *Proceedings of the 8th International Congress on Catalysis*. Weinheim: Verlag Chemie, Vol. 2, p. 387.
179. VISSERS, J.P.R., SCHEFFER, B., de BEER, V.H.J., MOULIJN, J.A. and PRINS, R. 1987: *Journal of Catalysis* **105**, 277.
180. DALY, F.P., BRINEN, J.S. and SCHMITT, J.L. 1984: *Applied Catalysis* **11**, 161.
181. REDDY, B.M. and SUBRAHMANYAM, V.S. 1986: *Applied Catalysis* **27**, 1.
182. Van VEEN, J.A.R., GERKEMA, E., van der KRAAN, A.M. and KNOESTER, A. 1987: *Journal. Chemical Society. Chemical Communications*, p. 1684.
183. Van der KRAAN, A.M., CRAJÉ, M.W.J., GERKEMA, E. and RAMSELAAR, W.L.T.M. 1988: *Applied Catalysis* **39**, L7.
184. GROOT, C.K., de BEER, V.H.J., PRINS, R., STOLARSKI, M. and NIEDZWIEDZ, W.S. 1986: *Industrial Engineering Chemistry. Process Design Development* **25**, 522.
185. GROOT, C.K., van der GENDER, P.J.G.D., NIEDZWIEDZ, W.S., van der KRAAN, A.M., de BEER, V.H.J. and PRINS, R. 1988: *Bulletin Societés Chimiques Belges* **97**, 167.
186. RAMSELAAR, W.L.T.M., HADDERS, R.H., GERKEMA, E., de BEER,

V.H.J., van OERS, E.M. and van der KRAAN, A.M. 1989: *Applied Catalysis* **51**, 263.
187. VISSERS, J.P.R., GROOT, C.K., van OERS, E.M., de BEER, V.H.J. and PRINS, R. 1984: *Bulletin Societés Chimiques Belges* **93**, 813.
188. LEDOUX, M.J., MICHAUX, O. and AGOSTINI, G. 1986: *Journal of Catalysis* **102**, 275.
189. ARNOLDY, P., van OERS, E.M., de BEER, V.H.J., MOULIJN, J.A. and PRINS, R. 1989: *Applied Catalysis* **48**, 241.
190. TOPSOE, H. and CLAUSEN, B.S. 1986: *Applied Catalysis* **25**, 273.
191. BONWENS, S.M.M.M., VISSERS, J.P.R., de BEER, V.H.J. and PRINS, R. 1988: *Journal of Catalysis* **112**, 401.
192. VISSERS, J.P.R., LENSING, T.J., de BEER, V.H.J. and PRINS, R. 1987: *Applied Catalysis* **30**, 21.
193. VISSERS, J.P.R., MERCX, F.P.M., BOUWENS, S.M.A.M., de BEER, V.H.J. and PRINS, R. 1988: *Journal of Catalysis* **114**, 291.
194. Van DOORN, J., STANGAARD, P., MOULIJN, J.A. and de BEER, V.H.J. 1988: *Applied Catalysis* **48**, 253.
195. Van DOORN, J., WAKKER, J.P., MOULIJN, J.A. and de BEER, V.H.J. 1989: *Applied Catalysis* **49**, 319.
196. DONNET, J.B. and VOET, A. 1976: *Carbon Black*. New York: Marcel Dekker.

11 Natural Gas Adsorbed on Carbon

Norman D. Parkyns
British Gas plc
Research and Technology Division
Gas Research Centre
Ashby Road
Loughborough
Leicestershire LE11 3QU
United Kingdom

David F. Quinn
Royal Military College of Canada
Department of Chemistry and Chemical Engineering
Kingston
Ontario K7K 5LO
Canada

11.1 Introduction

Natural gas is a fuel of considerable natural abundance, for which the value of reserves at the current rate of use is around 60 years,[1] as opposed to about 30 years for liquid petroleum. It normally has very little sulphur as organosulphur compounds, which, in any case, are easy to remove. As a result, it is an excellent fuel from the environmental point of view and can be burnt in such a way as to easily minimize emissions of NO_x and CO.

Natural gas has, however, a twofold disadvantage compared with liquid fuels: it is relatively expensive to transport from the remote areas where many large reserves are found,[2] and the low energy density (heat of combustion/volume) restricts it currently to use where it is constantly replenished by a piped supply. This last point has largely kept natural gas out of the very large transport market, although current concern over emissions from conventional liquid-fuelled vehicles is causing increased interest to be shown in natural gas-fuelled vehicles (NGV).

There is already a considerable market worldwide in NGVs fuelled by compressed natural gas (CNG) at pressures up to 25 MPa (ca 3600 psig). The size and cost of the cylinders used for storage are, however, considerable deterrents to use on a wide scale and the cost of providing the high-pressure facilities is relatively high. Nonetheless, the worldwide use of NGVs is substantial and increasing steadily. If a method could be found whereby amounts of gas comparable to those in CNG cylinders could be stored at much lower pressures, say 2–4 MPa (ca 300–600 psig), then the competitive position of NGVs *vis-à-vis* their liquid-fuelled counterparts could be greatly improved.

Possible techniques of methane storage

Natural gas is conveniently stored on a large scale in liquid form (LNG) where the density is 0.42 g ml^{-1}, but the cost of liquefaction, the special insulated vessels required and the potential fire hazard are such as to make it unsuitable for use on a small scale, such as in a home.

Three alternatives have been considered at various times: the use of a solvent, the formation of molecular cage compounds (clathrates) and adsorption in a porous medium. Dignam[3] has covered the theoretical aspects of these three potential methods.

The higher paraffinic hydrocarbons are suitable solvents for methane but the solubility is low: in pentane, methane densities of 0.13 g ml^{-1} can be obtained but only at pressures in excess of 17 MPa. The corresponding figures for decane are 0.15 g ml^{-1} at over 35 MPa.[4,5] These figures represent little improvement over those for pure methane in the gas phase at these pressures.

Clathrates of water and methane are very well known and have been extensively investigated.[6] The most methane-rich of these has a methane density of 0.12 g ml^{-1}; again, the equilibrium pressure is very high, 24 MPa at 25 °C.[7] Other systems, e.g. methane–diphenol, give values of 0.09 g ml^{-1} at a pressure of 3.6 MPa.[7]

Adsorbed natural gas (ANG)

Any material with pores of diameter <2 nm (micropores according to the IUPAC definition) is able to adsorb gases above their critical temperature to an extent proportional to the micropore volume. Carbons, by virtue of the very high micropore volumes that are readily obtained, have shown the greatest promise, although several workers have also investigated the use of zeolites and even of pillared clays.[8] The costs of the cheaper zeolites are of the same order as those of highly activated carbons, but the greater potential of the latter for high adsorptive capacity has led to increasing concentration on them as materials of choice for adsorbed natural gas (ANG).

The largest component of natural gas, generally to the extent of >90% v/v, is methane (see Table 11.1). The first member of the alkane (paraffin) series, it has the lowest normal boiling point (111.7 K) and is correspondingly the most difficult to adsorb. The critical temperature (T_c) is also low (190.5 K), so methane behaves approximately as ideal gas at ambient temperatures. As a spherically symmetric molecule, it has neither dipole nor quadrupole moment, unlike common adsorptives such as N_2 and CO_2, which have appreciable quadrupole moments. As Dignam[3] has pointed out, the type of van der Waals forces leading to adsorption is solely that of London dispersion interactions and, as a result, methane tends to show no preference for specific, e.g. polar, adsorption sites.

This last is an important point because real natural gases (Table 11.1) do contain significant quantities of polar molecules, CO_2, H_2O, S-containing odorants, in addition to higher hydrocarbons, and some consideration must be given in any practical system for ANG to the possible preferential adsorption of these minor components. The essential technical problem of ANG is, however, maximizing the methane uptake per unit volume of adsorbent at a particular temperature and pressure. It is important to stress that the key factor is methane density, that is, uptake/volume rather than uptake/mass. The significance of this factor will be elaborated upon later.

Table 11.1 Some compositions of typical natural gases[a]

Component/ % v/v	Source: Leman Bank (UK)	Source: Frigg (Norway)	Source: Groningen (L) (Netherlands)
Methane	94.81	95.72	81.50
Ethane	3.00	3.37	2.90
Propane	0.55	0.03	0.40
Butanes	0.19	0.01	0.15
$>C_4$	0.14	0.02	0.10
Carbon dioxide	0.04	0.31	1.00
Nitrogen	1.22	0.53	14.00

[a] Content of $>C_4$ compounds may be regulated for pipeline transmission purposes. Level of sulphur-containing compounds ≈5–10 ppm (v/v). Pipeline gas is dried to a water content of 75–180 ppm (v/v).

Potential commercial and large-scale applications of ANG

A number of possible applications for ANG may be envisaged; for convenience, they may be grouped as follows:

- fuel storage for transport (NGV);
- storage for operations within the natural gas industry;
- transport of natural gas in cylinders for small-scale use, as an alternative to acetylene, or other bottled flammable gases;
- others, including transport of natural gas on a large scale from remote sources.

Of these, by far the most research has been applied to the transport application; gas storage offers a very high potential for exploitation; the third has achieved commercial status.[9] Any other uses are, as yet, conjectural. The technical problems in developing widespread use of ANG are, however, similar to all these applications.

As previously remarked, ANG aims at improving the energy density of natural gas to an extent where it may compete more directly with liquid fuels. Table 11.2 lists some typical values of the volumetric energy content of fuels currently used, or envisaged for use, in transport. We may note the high energy densities for the conventional diesel and gasoline liquids and the relatively low value for methanol, a fuel currently favoured by some groups for future US needs. By contrast, LNG has a high value but the perceived danger and handling difficulties are such as to preclude its widespread use, at least in domestic vehicles. However, the potential of LNG as a fuel shows the goals to be aimed at in developing carbons for use with ANG.

Natural gas is condensed to 1/600th of its gaseous volume [at 25 °C and 0.1 MPa (1 bar)] on liquefaction. Looked at another way, one volume of LNG is equivalent to 600 volumes of natural gas under standard conditions. This value of 600 v/v must represent an ultimate limit for ANG, as the density of molecules adsorbed in micropores is often considered as being very similar to that of the liquid state. The practical limiting value for volumetric storage in carbons is considered later, but it is worth noting that carbons currently available commercially give the equivalent storage at 3.5 MPa (35 bar) to that of CNG at 8 MPa (80 bar), and it is believed that the equivalent CNG figure could be raised to 13–15 MPa (130–150 bar) by suitable developments in adsorbent manufacture.

The other major application which is, however, in a far less advanced stage is that of storage of natural gas from the distribution system. The gas industry throughout the world is based on a transmission system developed in the USA using steel pipes operating at up to 7 MPa (1000 psig) with pressures falling to half or a third of this value at the extremities of the system. Storage is necessary to even out the varying demands, both between seasons and between night and day (diurnal variations). It is this latter application where ANG may have a role to play. Current storage is by gasholders operating at a pressure only slightly above atmospheric, and by empty cylinders ('bullets'), up to 3.5 MPa (500 psig). Enhancement of the volumetric capacity of these 'bullets', by filling them with a carbon adsorbent, could potentially lead to considerable cost savings.

Table 11.2 Energy densities[a] ($MJ\,l^{-1}$) for fuels for transport use

Liquid fuels				Natural gas						Storage battery[d]
				LNG	CNG			ANG[c]		
Diesel	Gasoline	Methanol	LPG[b]	LNG	2 MPa	10 MPa	20 MPa	Current	Future	
37	32	16	29	23	0.8	5.6[e]	9.7[e]	3.8	6	0.2–0.3

[a] Lower heating values of combustion used.
[b] 50:50 molar mixture propane–butane assumed.
[c] ANG storage assumed to be at 3.5 MPa (35 bar).
[d] Values for lead/acid and Ni/Cd.
[e] Compressibility factors (Z), assumed 0.78 at 20 MPa, 0.68 at 10 MPa, for gas of similar composition to that of Leman Bank (Table 11.1).

It is worth pointing out that the techniques for storage of natural gas are equally applicable to other permanent gases, such as oxygen or nitrogen, and possibly hydrogen, and may hence find applications in different areas.

11.2 Use of carbons for ANG

Adsorption of natural gas in micropores

Active carbon is one of the oldest adsorbents known to man and the history of scientific investigations into its mode of action is correspondingly venerable and extensive. Modern theories of adsorption of vapours on carbons and their corresponding derived isotherms are dealt with by Sing in Chapter 2, so it is not proposed to discuss them at length here.

Most investigations concern adsorption of vapours, that is, gases below their critical temperature in a region where the value of the saturation vapour pressure, p_s, has a real meaning and cannot be exceeded in the adsorption process. In the case of adsorption of natural gas on carbons at ambient temperatures, where $T/T_c > 1.5$, the principal component, methane, is clearly in the gaseous phase, the pressure of which can, in principle, be increased without limit. Adsorption by capillary condensation does not occur and we would expect that adsorbents with large meso- and macropore volumes would be less effective than those whose porosity is largely in the microporous range.

Findenegg[10] has demonstrated for homogeneous surfaces that the shape of an isotherm at higher pressures is related to the T/T_c ratio for the adsorbate and that these isotherms can, typically, be categorized in three ranges of temperature relative to T_c:

- the low-temperature region ($T < T_c$);
- the supercritical region ($T_c < T < 1.2T_c$);
- the high-temperature region ($T > 1.5T_c$).

Methane, at ambient temperatures, falls into the third category.

Investigations into the adsorption of methane at high pressures have been carried out since the early years of this century (see, for example, Brunauer[11]) and although the isotherm is approximately of Type I shape, variations have been reported. At low pressures, a gradually increasing uptake is observed[12]; as the pressure increases, the uptake may level off and a plateau is found.[13,14] At still higher pressures, the isotherm becomes complex and may go through a maximum. Menon[15] has pointed out that the maximum, followed by a decrease in apparent adsorption on raising the pressure further, will occur when the rate of increase in the density of the gas phase with pressure exceeds that of the adsorbed phase. He suggested that the pressure at which the maximum occurs may be calculated empirically for any adsorptive, independent of the adsorbent, by the expression

$$P_{max} = (T/T_c)P_c \qquad (11.1)$$

Practical isotherms for supercritical gas adsorption

One of the earliest isotherms to be derived *ab initio* was that due to Langmuir, which, in linear form, is given by

$$\frac{p}{V} = \frac{1}{bV_m} + \frac{p}{V_m} \tag{11.2}$$

where V is the volume adsorbed and V_m the volume corresponding to monolayer coverage. Plots of p/V against p, for methane on various carbons, have been found to yield straight lines from which V_m may be readily calculated.[8,13,16]

Koble and Corrigan[14a] found that the data of Ray and Box[14b] for methane on an activated carbon fitted a slightly different form of the Langmuir equation very well. Findenegg[10,42] has also shown that a type of Langmuir equation may be used to describe the surface excess of adsorbed gas for high pressures at $T/T_c > 1.5$.

It must not be supposed, however, that V_m, calculated according to equation 11.2, represents monolayer coverage in the way that Langmuir envisaged it. Gregg and Sing[17] commented that b and V_m are effectively disposable constants, which accounts for the linearity of equation 11.2. Furthermore, values for V_m, when converted into corresponding surface areas, are often far in excess of values that are physically possible.[18]

Another very early treatment of isotherm data was that due to Polanyi (as discussed by Brunauer[11]). It was taken up by Dubinin and co-workers[19] and developed into the theory of volume filling of micropores. Recently, the use[20] of the Dubinin–Astakhov (DA) equation (equation 11.3) has been suggested as being useful for high-pressure isotherms for methane and other gases where the Polanyi adsorption potential, ε, is expressed in terms of fugacity

$$a = a_0 \exp\left[-\left(\frac{\varepsilon}{E}\right)^n\right]. \tag{11.3}$$

Dubinin and co workers[21] have claimed that the characteristic energy, E, in the DA equation is effectively independent of temperature and applies equally to adsorption of methane on carbon at both room temperature and 120 K. The exponent, n, on the other hand, does need to be adjusted for use at higher temperatures. These assumptions were verified by extrapolation of $\ln p$ for a given coverage plotted against $1/T$. Schrøder Pedersen and Larsen[22] also found good linearity between $\ln p$ and $1/T$ from 254 to 297 K for methane on carbon NC 35.

Agarwal and Schwartz[20] analysed high-pressure adsorption isotherms for methane on a North American carbon, G 210. They found the pore volume, w_0, as described by Ozawa and co-workers,[23] to be reasonably constant over a limited temperature range (298–308 K). On the other hand, Barton *et al.*[16] found w_0 for similar systems to vary with temperature from 243 to 298 K. Agarwal and Schwartz[20] concluded that the DA equation represented their data for several more gases than did Ozawa and co-workers'. Findenegg's[10] corresponding state method was not analysed.

It would seem, therefore, that although it is clearly desirable to have an isotherm over a range of temperatures, no standard equation has yet emerged, and further investigation of this point would be desirable.

Comparison of methane adsorption of carbons with other adsorbents

Although microporous carbons have emerged as the best option for ANG, it is useful to give a brief discussion of the reason for this. Any material of high enough surface area, or perhaps more strictly of micropore volume, has potential for use as an adsorbent for ANG. In addition to carbons, silica, silica–alumina and pillared clays have been investigated, but it is the crystallite zeolite molecular sieves where the available volume within the framework may be as high as 50% of the total volume that would seem to have the most attractive properties. Organic polymers, of relatively high surface area, such as Amberlite XAD-4 have also been investigated[24] but show a poor uptake of methane.

Before any ANG project enters a near-commercial phase, cost has to be evaluated as accurately as possible, but it would seem that there is not a great deal of difference in the primary costs of any of these materials. Where zeolites have to be synthesized, such as for the ZSM series, then this is not so, and the extra cost would have to be offset by some specific properties that offered particular advantages.

Table 11.3 summarizes some of the data that are available on carbonaceous and inorganic oxide adsorbents. It is clear that in terms of adsorption per unit weight, the carbons are unassailable. In terms of adsorption per unit volume, which is the ultimate arbiter for ANG, the much higher packing densities of zeolite powders and pellets do bring them back more into contention. Even so, the best of the carbon results are still significantly above the best that the zeolites can offer. The results of Hayhurst and co-workers[25] on CaX zeolite are clearly very promising but the volumetric uptake is based on an assumed value of packing density that might not be realizable in practice.

In general, there is a very clear correlation between surface area (micropore volume) and gravimetric uptake. Inasmuch as the micropore volumes of zeolites are clearly defined by their crystal structure, it is unlikely that higher values for methane storage will be easily obtained. With carbons, on the other hand, careful control of burn-off will give extremely high micropore volumes (see Chapter 9 by Thwaites, Jagtoyen and Derbyshire) leading, at least in principle, to increased methane uptake.

On the other hand, increasing the microporosity decreases the density, which should otherwise be as high as possible. There is, therefore, an inherent conflict, which will be discussed more fully later, between the requirement for increased surface area and higher density. However, because carbons are more amenable to densification than zeolites, the advantage remains firmly with them.

Table 11.3 Comparison of adsorbents for ANG

Adsorbent	N_2 surface area $m^2\ g^{-1}$	Micropore volume $ml\ g^{-1}$ N_2	Packing density $g\ ml^{-1}$	CH_4 uptake at 3.5 MPa and 25 °C g per 100 g	CH_4 uptake at 3.5 MPa and 25 °C $g\ l^{-1}$ (calc.)	Source of data (Ref.)
Pillared ZCH-bentonite	225	0.16	1.09	1.75	19.1	8
Delaminated Laponite	380	—[a]	0.50	2.35	11.8	8
Zeolites:						
13X	588[b]	—	0.66	5.0	33	8,46
5A		—	0.74	4.9	36	8,46
3A		—	0.77	2.5	19	8,46
CaX			~0.7[c]	8.2	~57	25
Carbons:						
Calgon BPL	997	0.52	0.48	8.0	38.4	8
Linde UC 9LXC	1280	—	0.32	12	38	46
Anderson AX21	2342	—	0.27–0.32	20	54–64	57
Catalysts and Chemicals Europe AK-AX	3000	—	0.45	18	81	49

[a] Dashes indicate not quoted.
[b] With CO_2 at 273 K.
[c] Assumed value.

11.3 Properties of carbons for ANG applications

All the experimental and theoretical results discussed so far have stressed the essential similarity of adsorption of vapours at sub-atmospheric pressures near their normal boiling point to that at high pressures for the corresponding gas above its critical temperature. As Table 11.3 indicates, the higher the surface area or micropore volume, as determined by the adsorption of N_2 at 77 K, the higher is the adsorption of CH_4 at 25°C (298 K). The forces that lead to enhanced uptake of adsorptives in micropores are cooperative over relatively short distances and have little influence in mesopores ($d < 50$ nm) and certainly not in macropores ($d > 50$ nm). Considerable attention, therefore, has to be given to making adsorbents whose surface area is contained as much in micropores as possible. This is where zeolitic structures have a marked advantage, subject to the reservations noted above.

In addition to minimizing meso- and macropores, larger inter-particle voids should be avoided. It is possible to make carbons in monolithic shapes, such as discs, which are fully space filling; this is an area where some effort has been spent and will, no doubt, continue in the future.

Correlation of high-pressure methane adsorption with micropore volume

The vital importance of micropore volume in determining the uptake of methane-rich natural gases under pressure has already been stressed, so it is worth reviewing how good the correlation is in practice. In principle, a purely microporous adsorbent should give a purely Type I isotherm with, say, N_2 at 77 K, and whatever method is used to determine limiting uptake, whether a Langmuir plot, Dubinin–Radushkevitch (DR)[19] plot or even BET plot,[17] it should give the same value. Given that the whole process is one of filling micropores, we have a preference in this chapter for referring to micropore volume rather than assigning a somewhat fictitious surface area to the same measurement. However, in the literature, values of surface area tend to be more commonly quoted than micropore volume.

Problems arise when, as is almost invariably the case, the carbon adsorbent has a distribution of pore sizes, not only within the micropore range but merging into the smaller mesopores. Under these circumstances, the adsorption isotherm will have a more or less Type II character and will not level off in the expected Type I manner. Different methods will then tend to give different values of micropore volume/surface area which may not correlate too well with high-pressure methane uptake. One method that has recently been applied to carbon adsorbents, to differentiate the micropore volumes from the surface area due to mesopores, is the α_s plot due to Carrott and Sing,[26] described fairly fully in Chapter 2.

However, experimental correlations between high-pressure methane uptake and surface area (micropore volume) have been demonstrated.

Barton *et al.*[15] investigated a number of carbons for methane storage. They were characterized by the adsorption both of N_2 and CH_4 at 77 K. The BET areas calculated for both adsorptives were in extremely good agreement and were very similar to areas derived from applications of the Dubinin equations. The similarity in measured areas is perhaps not surprising, given the similar physical properties and the closely similar cross-sectional areas of the two molecules (0.162 nm^2 for N_2 and 0.157 nm^2 for CH_4).

The correlation between areas (micropore volumes) derived from methane adsorption at 77 K did hold qualitatively for high-pressure isotherms (up to 16 MPa) at temperatures between −30 and 25 °C, in that those carbons of highest micropore volume adsorbed most methane. A more quantitative correlation was difficult to establish because of the complex shape of the isotherms at these very high pressures.

If adsorption is limited to lower pressures, say <5 MPa, the isotherms are much simpler, showing a monotonic increase of adsorption with increasing pressure and, in some cases, an approach to a quasi-asymptotic value.

Otto[13] showed that a variety of adsorbents up to pressures of 4 MPa gave very good Langmuir plots within the temperature range 273–373 K. The correlation between the limiting uptake of methane, as derived from the Langmuir plots, and amounts of CO_2 required to saturate the surface at 273 K, as determined by BET plots, was also very good, whether the adsorbent was zeolite- or carbon-derived. Only for a 4A zeolite did the correlation fail, possibly because of molecular sizing effects.

Table 11.4 gathers together a compendium of data on methane adsorption for various commercially available carbons; results for samples specially prepared in the laboratory have, in general, been excluded. Table 11.4 relates the BET surface area as conventionally determined by nitrogen at 77 K with the gravimetric adsorption of methane at 25 °C and 3.5 MPa (35 bar). Where they have also been explicitly quoted in the literature, micropore volumes are given, as determined by extrapolation of the DR plot at low p/p_s values. The packing densities are also quoted and predicted volumetric uptake of methane, according to equations described later.

The BET surface area–uptake data are plotted in Figure 11.1 and show that a tolerably linear correlation results, to which a regression equation can be fitted,

$$G = 0.000049A + 0.025 \tag{11.4}$$

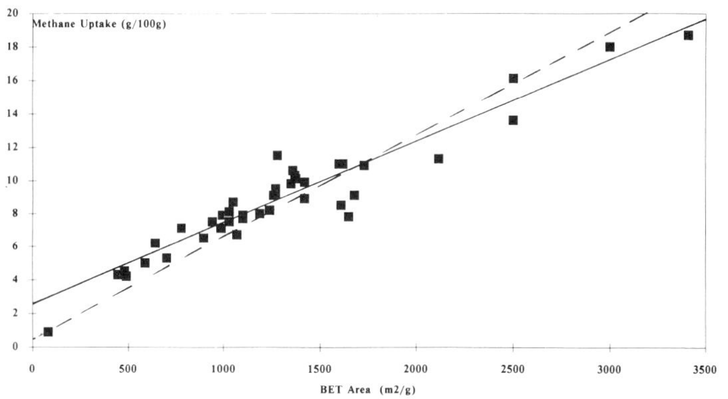

Fig. 11.1 Correlation of BET surface area with adsorption of methane at 25 °C and 3.5 Mpa for some commercial carbons. Solid line, best fit for all points; dashed line, best fit for lower values

Table 11.4 Correlation of surface area/micropore volume and methane uptake for some commercially available adsorbents

Adsorbent	Packing density g ml^{-1}	77 K nitrogen adsorption		Methane uptake at 3.5 MPa, 25°C		Source[b]
		BET surface area $m^2\ g^{-1}$	DR micropore volume ml g^1	Experimental g per 100 g	Calculated v/v[a]	
Graphon		81		0.9		a
MSC-5A		445	0.22	4.3		b
Baker, acid washed	0.37	480		4.5	54	c
Linde MS 5A		489	0.20	4.2		d
Linde MS 13X		588		5.0		e
C.L. cloth		640		6.2		f[c]
Darco SG	0.45	700		5.3	64	c
Carbotech CT 55/2	0.56	780	0.34	7.1	74	f[c]
Calgon SGL	0.49	900	0.40	6.5		f[c]
S.S. 208A		945	0.42	7.5		f[c]
Calgon BPL		988		7.1		g
Calgon BPL	0.48	997		7.9	75	h
Calgon BPL	0.44	1030	0.47	7.6	73	f[c]
ICI DXL 0-8334	0.44	1030		8.1	73	c
Ceca GAC 50G	0.56	1030		7.5	82	c
Witco JXC	0.45	1050		8.7	74	c
S.S. 208A (H)		1070	0.45	6.7		f[c]
Calgon CAL	0.49	1100	0.50	7.7	79	f[c]
Calgon BPL	0.47	1100		7.9	77	c
Carbotech CT 45/2	0.48	1190	0.51	8.0	80	f[c]
A.C. CNS 196	0.45	1190	0.49	8.0	78	f[c]
Norit R1	0.39	1240	0.63	8.2	73	f[c]
Nuchar WV-G	0.45	1260		9.1	80	f[c]
Calgon PCB	0.44	1270		9.5	79	c
Norit R3	0.42	1270	0.58	9.2	77	f[c]
Union Carbide 9LXC	0.32	1280		11.5	67	c
Kureha BAC	0.59	1350	0.60	9.8	96	f[c]
Witco 9JXC	0.46	1359		10.6	83	h
North American G 216	0.50	1370		10.3	88	c
Witco 9LCK	0.46	1375		10.1	84	h
North American G 210	0.50	1420	0.60	8.9	89	f[c]

North American G 210	0.50	1420		9.9	89	c
Cabot CSX-179-B	0.13	1600		11.0	51	c
Nuchar WV-B	0.30	1610		8.5	71	c
Barnabey Cheney 580-26	0.45	1620	0.68	11.0	89	f[c]
North American G 104	0.30	1650		7.8	72	c
North American G 101	0.30	1680		9.1	73	c
Barnabey Cheney MI	0.46	1730		10.9	94	f[c]
AD'ALL A-20		2115		11.3		f[c]
Anderson AX-21 powder	0.28	2500	1.06	16.1	84	f[c]
Anderson AX-21 pellet	0.03	2500		13.6	87	f[c]
AK-AX	0.45	3000		18.0	127	i
Osaka MCMB	0.23	3410	1.42	18.7	88	f[c]

[a] Calculated on basis of equations 11.15 and 11.19.
[b] Source of data: (a) Specovius and Findenegg[42]; (b) Ozawa *et al.*[23a]; (c) Remick and Tiller[46]; (d) Wakasugi *et al.*[23b]; (e) Otto[13]; (f) Quinn[24]; (g) Reich *et al.*[56]; (h) Innes *et al.*[8]; (i) Chaudron.[49]
[c] Some of these data are published by kind permission of Atlanta Gas Light Research Adsorbent Group (AGLARG).

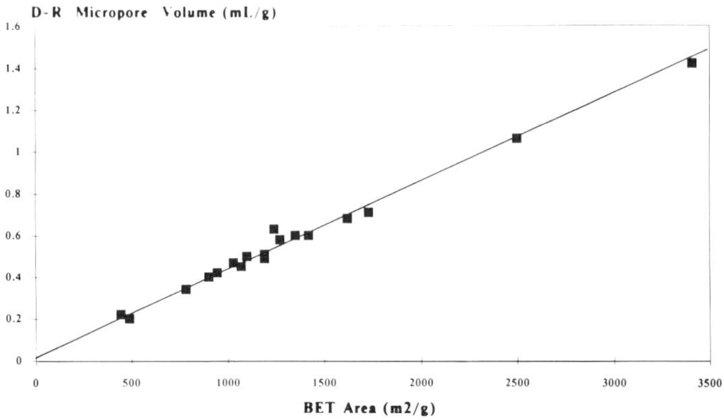

Fig. 11.2 Correlation of BET surface area with micropore volume derived from N_2 adsorption at 77 K for carbons listed in Table 11.4

where G is the fractional uptake of methane (g g^{-1}) and A is the BET surface area (m^2 g^{-1}). A plot of BET area against micropore volume is shown in Figure 11.2. The good linearity and the absence of a significant intercept at the origin indicate clearly the essential linearity of the two parameters. Again, a regression line through the points gives the relationship

$$V_{\mu} = 0.000408A + 0.034 \tag{11.5}$$

where V_{μ} is the micropore volume. Combining equations 11.4 and 11.5 gives

$$G = 0.120V_{\mu} + 0.021 \tag{11.6}$$

The relationship between uptake and BET surface area is only approximately linear. At the very highest values of surface area, the uptake may be seen to be lower than predicted, especially if bias is given to the expectation that the line should pass through the origin. The large intercept implicit in equation 11.5 is a consequence of experimental uncertainties. Moreover, there are reasons, discussed below, to suppose that carbons of high micropore volume are less able to adsorb methane at high pressures. Taking this in view, a more subjective straight line, ignoring the highest values of surface area, gives a modified relationship

$$G = 0.155V_{\mu} + 0.0006. \tag{11.7}$$

Some interest attaches to those carbons which deviate from this general relationship. In the first case, some carbons are found that adsorb more methane than would be predicted from their micropore volume. It is possible to explain this either on the basis of experimental inaccuracies in measuring the micropore volume or because the distribution of micropore size may be as important as the total micropore volume.

The first case is not unknown for carbons with very narrow entrances to micropores where diffusion at 77 K may be very slow, but is relatively fast at ambient temperatures under a high mass concentration gradient. In cases where this effect is suspected, the micropore volume should be measured with a suitable adsorptive at a higher temperature, e.g. CO_2 at 273 K.[27]

The second explanation is not yet susceptible to such a simple resolution. The sizing and distribution of sizes in micropores are topics currently of great activity. Not enough is yet known to give a definitive answer, but it may be that micropores in the 'ultramicropore' or 'primary micropore' range[26] are more effective in adsorbing methane than those of larger diameters.

The other deviation that may be noted from the correlation in Figure 11.1 is a downward one, where the methane uptake is less than would be predicted. This is more likely to happen at higher micropore volumes and may be attributed to the influence of larger pores, formally outside the micropore range ($d > 2$ nm), but which contribute increasingly to the 'surface area' which is, accordingly, a composite of contributions from micro- and mesopores. These small mesopores do not exhibit the cooperative adsorption necessary for uptake of methane at high pressures which is, accordingly, less than would be expected. In this case, care has to be taken in determining the 'true' micropore volume, for which the use of isotherm analysis, such as the *t*-plot and α-plot, may be helpful,[28] as already remarked.

In conclusion, we can say that it is clear that the correlation between methane uptake at ambient temperatures, at atmospheric pressure and above, with data from adsorption experiments at low temperature is reasonably good. As the latter are experimentally convenient and the relevant data well understood, it is clear that conventional adsorption measurements represent the first sorting test for any candidate material for ANG. This is not to say that adsorption measurements of methane at high pressures are unnecessary, but merely that they represent a second stage in the evaluation of adsorbents.

Effect of sample density on storage capacity

The previous discussion has emphasized the vital role played by microporosity in determining the uptake of methane at ambient temperatures. All other things being equal, it is clearly vital to maximize the microporosity of any adsorbent for ANG. However, we have emphasized the fact that ANG applications require maximum uptake by unit *volume* of adsorbent and thus the packing density becomes equally important. Table 11.3 illustrates the controlling effect that packing density has on uptake per unit volume.

It is therefore necessary to consider how the different contributions to the volume of the adsorbent are made up, as these control the packing density. In a vessel filled with carbon, the total volume can be considered as being apportioned in four ways, to the space occupied by the carbon atoms (V_c), to the micropores (V_μ), to the macropores (and mesopores) (V_m) and to the interparticle voids (V_{void}). Mathematically, this may be written as

$$V = \frac{1}{\rho} = V_c + V_\mu + V_m + V_{void} \tag{11.8}$$

where V is the volume associated with 1 g of material and ρ is the packing density (g ml^{-1}).

V_{void} and V_m may be lumped together as they do not contribute to the methane in the adsorbed phase but do, of course, store it at the gas-phase density.

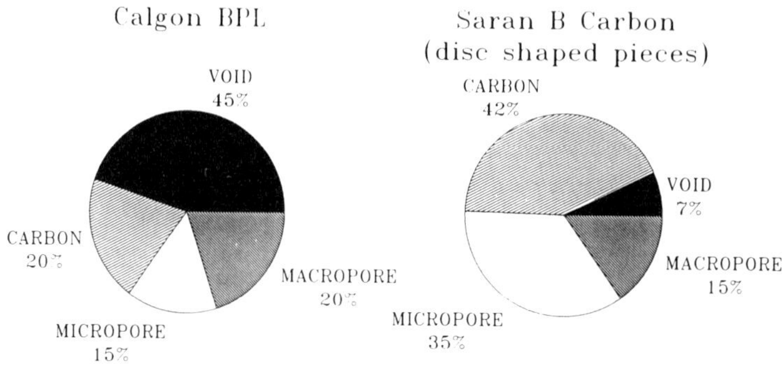

Fig. 11.3 Distribution of relative pore volumes for granules of a Calgon BPL carbon and for shaped pieces of a char derived from Saran B. Data from Quinn[24]

V_c will not vary much, provided that the microstructure of the carbon is approximately graphitic. The density of single-crystal graphite is 2.25 g cm^{-3}, so 2.2 is a reasonable working value, although lower values have been assumed by others. V_c, the volume associated with the carbon itself, is, therefore, ~0.45 cm^3 and this value may be substituted directly into equation 11.7, which reduces to

$$V = V_\mu + V_m + 0.45 \tag{11.9}$$

It is convenient to illustrate volume utilization of a storage vessel by a pie chart, as in Figure 11.3, for a typical commercial carbon (Calgon BPL) and for a Saran B carbon made as disc-shaped pieces to fill a cylindrical vessel with no apparent voidage. (Saran B is the designation used by Dow Chemical for copolymers of PVDC with 10–20% PVC.) The values for fractional volume are calculated from different density measurements. The packing density is simply the weight of the carbon in the storage vessel divided by the vessel volume. The piece density of the carbon is that obtained when the carbon is immersed in mercury at 1 atm (10 130 Pa) pressure, which gives the density of the carbon, including all pores of diameter less than 12 μm. When the pressure of mercury is increased to 414 MPa (60 000 psig), the mercury is considered to penetrate into all pores of diameter greater than 3 nm.[29] Thus, the high-pressure density (ρ_{Hg}) gives the density of the carbon, including pores of <3 nm diameter.

The various volumes (ml g^{-1} of carbon) are therefore:

Carbon = 0.45

Micropore = $(1/\rho_{Hg}) - 0.45$

Macropore = $(1/\text{piece density}) - (1/\rho_{Hg})$

Void = $(1/\text{packing density}) - (1/\text{piece density})$

This definition of a micropore does not strictly follow that of IUPAC[30] ($d < 2$ nm), but is a convenient one which follows the divisions imposed by the mercury porosimetry method.

The density of adsorbed methane in a micropore at 25 °C and infinite pressure is estimated to approach that of liquid methane at its normal boiling point.[31]

Accordingly, the excess storage capacity of a vessel filled with carbon over that of the gas phase is reflected in the fractional volume that is in micropores. A carbon with a high packing density, a large micropore volume and low macropore volume would, therefore, be ideal to provide a high storage capacity. However, since interparticle voids can be a large percentage of the volume of a vessel packed with irregularly shaped granules, carbons which are made to a specific shape to fill the vessel efficiently are more desirable. Figure 11.3 illustrates this point. The granular carbon-filled vessel has only 15% micropore, with 65% combined void and macropore volume, whereas the vessel filled with disc-shaped Saran B carbon has 35% micropore and only 22% combined void and macropore volume.

Work by Buczek and Czepirski[32] illustrates well the compromise which, sometimes, may have to be made between bulk density and gravimetric uptake. They found that abrading a commercial carbon (Norit R2) in granulated form gave markedly different uptake, according to the degree of abrasion. If 40% of the outer carbon was removed, the specific surface area of the remainder increased slightly, while the bulk density increased from 0.437 to 0.450 g cm^{-3}, partly because of better packing of the smaller particles, but also because the other material had been over-activated in manufacture and contained a greater proportion of larger meso- and macropores. The uptake of methane per unit volume was increased by about 12% over the original material. On the other hand, removing 80% of the original carbon gave material of still higher density but much reduced specific surface area, an indication of the effect of diffusion restriction during activating. This highly-abraded carbon had a poorer volumetric uptake than the original.

Mechanical properties

The form of a carbon adsorbent is also an important practical consideration. When it is in the form of dust, or abrades to one in use, problems can arise when filling and emptying a storage vessel. Adsorbent can be lost by transport in the gas stream and cause damage to engine components by abrasion. If filters are used to prevent this, they can become clogged, creating difficulties in use. The adsorbent carbon, therefore, needs to be resistant to attrition caused by abrasion between the particles. If the storage vessel is to be used in a vehicle where it will be subjected to constant vibration, then a non-friable carbon is required. Unfortunately, many of the carbons which have good adsorptive capacity and are accordingly highly porous, are also soft and friable. This is particularly true of carbons which have been chemically activated by potassium hydroxide or like materials. Coconut and other shell-derived carbons are generally much more resistant to attrition.

Another advantage of producing carbons in an efficient space-filling form is that they are also strong and resist attrition. Monolithic carbon blocks which fit the storage vessel would be ideal, but would be very difficult to make. However, it may be possible to make shaped pieces, which nest or fit together, and these should minimize attrition.

Devising test methods for attrition of carbons that represent what actually occurs during use is not easy. Results can be influenced in an unpredictable manner by the size, shape and density of the carbon particles. The need for a test which is independent of these variables has increased with the increasing use of adsorbent carbons in both gas–solid and liquid–solid applications. The US Naval Laboratory, in particular, has extensively investigated the attrition of activated carbon to dust, and has developed a test method.[33] This measures a rate of dust attrition for the carbon by subjecting it to an acceleration of 4 *g* in a vibrator. This type of test would seem to resemble more closely the conditions under which a storage tank is likely to be used, as opposed to other attrition-testing methods, such as the stir bar[34] or ball-pan.[35]

Limitations of natural gas adsorption in microporous carbons

The question now arises of whether we can predict the uptake of natural gas (methane) that we are likely to achieve in practice. A simplified model, which considers the distribution of porosity in the adsorbent and which estimates the density of methane adsorbed in micropores, does enable us to give some estimate.

For the purpose of discussion, we can define a loading factor, L, as follows

$$L = \frac{\text{volume of methane (std) adsorbed at temperature } T \text{ and pressure } P}{\text{volume of carbon}}$$

[Note: standard volumes of gases are normally referred to 0°C and 1 atm (10 130 Pa) pressure. There are some local variations (for example, the British gas industry specifies 15 °C), but they are of minor account in our discussion.] We shall specify the temperature of adsorption as 25 °C and the pressure as 3.4 MPa (500 psia), although any other desired conditions could be chosen, but earlier arguments support our present choice and data taken under these conditions are plentiful. From the foregoing discussion, we assume that L is some function of micropore volume, that is,

$$L = f(V_{\mu}) \tag{11.10}$$

Practical isotherms are normally measured in terms of fractional weight increase, which we have called G. Again,

$$G = f'(V_{\mu}) \tag{11.11}$$

where G is the fractional weight increase under our standard conditions of adsorption (3.4 MPa and 25 °C).

The form of $f'(V_{\mu})$ has been established from experimental data, as shown in Figure 11.1, for which linear regression lines have been drawn, as in equations 11.4 and 11.6.

The coefficient of V_{μ}, in equation 11.7, 0.155, may be regarded as the density of the adsorbed methane in units of g ml^{-1}. It is interesting to compare it with the density of liquid methane (0.42 g ml^{-1}) at its boiling point (190 K) and the value of 0.16 g ml^{-1} at its critical temperature (112 K). On purely theoretical grounds, Dignam[3] calculated a limiting value of 0.27 g ml^{-1} for the adsorbed state. Barton *et al.*[31] applied the Dubinin–Raduschkevitch plot to their results for methane adsorbed on carbons at 25 °C to derive values of limiting uptake at infinite pressure. Division of these values by those of micropore volume measured from the high-pressure mercury densities (as described in the next section, on microporosity) gave a value of the limiting density of adsorbed methane of 0.36 g ml^{-1}. At 3.4 MPa, about 75% of this density was reached, giving a density for adsorbed methane at this pressure of 0.27 g ml^{-1}, higher than that indicated by our present empirical correlation but in good agreement with Dignam's prediction. Had Barton *et al.* used the DR micropore volumes, rather than those derived from density measurements, the value for the limiting methane density in the micropores would have been 0.28 g ml^{-1}.

In equations 11.6 and 11.7, G has units of g g^{-1}. It is straightforward to convert G to G', the equivalent standard volume of methane at 298 K and 0.1 MPa. Using equation 11.7, for example,

$$G' = 237\,V_{\mu} + 9.2 \tag{11.12}$$

Dividing this by the volume of carbon (from equation 11.8), we can express the methane stored volumetrically (our previously defined L) as

$$\frac{G'}{V} = L = \frac{237\,V_{\mu} + 9.2}{V_{\mu} + V_{\mathrm{m}} + 0.45} \tag{11.13}$$

We may note that the denominator of equation 11.13, $V_{\mu} + V_{\mathrm{m}} + 0.45$, is the reciprocal of the packing density, ρ (equation 11.8), and that, accordingly, we may write

$$L = (237\,V_{\mu} + 9.2)\,\rho \tag{11.14}$$

or, using the relationship in equation 11.6,

$$L = (183\,V_{\mu} + 32)\rho \tag{11.15}$$

These are extremely convenient equations to use because they can be used to predict the loading, L, in terms of volume of gas stored by adsorption per unit volume of adsorbent in terms of two easily-determined parameters, V_{μ} and ρ. They are of general applicability whatever the form of the carbon adsorbent, whether powder, granules or formed shapes. It is important to stress, however, that the micropore volume and the packing density should be measured on samples from the same batch, especially where formed shapes are used.

If f is considered to be the fractional volume of the carbon which is micropore, i.e. $f = V_{\mu}/V$ then equation 11.15 becomes

$$L = 237f + 9.2\rho \tag{11.16}$$

This equation directly relates L with the relative distribution of porosity in a practical adsorbent,

as $f \rightarrow 1$, $\rho \rightarrow 0$ and so $L \rightarrow 237$

Since the volume of carbon is irreducible at 0.45 ml g^{-1} and so, in practical terms, f can never reach unity, the maximum value for L must be less than this limiting value. It may be possible to envisage a totally microporous carbon which is two-thirds, or even three-quarters, micropore. Such a carbon, with 75% microporosity, would have a density of $(0.25 \times 2.2) = 0.55\ \text{g ml}^{-1}$, with virtually every carbon atom being a surface atom. The limiting value of L then reduces to ~183 v/v.

Bearing in mind that V_μ may also be determined, in principle, by high-pressure mercury porosimetry, we may write

$$L = \left[237\left(\frac{1}{\rho_{Hg}} - 0.45\right) + 9.2\right]\rho \quad (11.17)$$

L may now be predicted solely in terms of density measurements, although the cautionary remarks given later in the section on mercury porosimetry should be borne in mind.

Although the combined macropore volume and voidage, V_m, do not store natural gas by adsorption, they do so at the density of the gas phase, and some allowance has to be made for this.

At the standard conditions of 3.4 MPa and 298 K, there are $23.6\ \text{g l}^{-1}$ of methane in the gas phase, or $36\ \text{ml ml}^{-1}$, so the amount stored in the voids and macropores is clearly $36V_m\rho$. V_m may be expressed in terms of the other volume and density parameters as

$$V_m = \frac{1}{\rho} - V_\mu - 0.45 \quad (11.18)$$

The amount of gas storage, L_G, for a unit volume of adsorbent is therefore

$$L_G = 36\rho\left(\frac{1}{\rho} - V_\mu - 0.45\right) \quad (11.19)$$

L_G is by no means negligible; for carbons with micropore volumes between 0.4 and $0.8\ \text{ml g}^{-1}$ and supplied in the form of granules, it is ~20. It tends not to vary much from sample to sample because V_μ and ρ tend to compensate for each other, in the sense that samples of high microporosity have low packing (and piece) densities.

Adding L and L_G together gives the total amount of gas stored, both in the adsorbed phase and in the void space and macropores. Comparison of the predicted values given by these equations and results obtained for a number of carbons using an apparatus described in figures 11.5 and 11.6 gives a very satisfactory agreement, although it tends to underpredict performance by about 10%.

Whether to use equation 11.14 or 11.15 to predict L is a matter of circumstances. Equation 11.14 better represents the adsorbed loading for adsorbents of lower micropore volume, say $V_\mu < 1$. For $V_\mu > 1$, equation 11.15 represents known experimental data more accurately.

The important feature to be borne in mind is the equal importance that packing density has with micropore volume in determining uptake of methane. Although equation 11.16 predicts a theoretical maximum of 237 at 3.4 MPa and 25°C, values approaching this (170) have been attained.[36] The high uptake was obtained not by striving for excessive microporosity, but by making shaped sizes of high packing density.

11.4 Characterization of carbons for adsorbed natural gas

Microporosity

As described in the section on the effect of sample density on storage capacity, we can calculate a value of the micropore volume from density measurements, using the relationship

$$V_{\text{micro}} = \frac{1}{\rho_{\text{Hg}}} - \frac{1}{\rho_{\text{c}}} \tag{11.20}$$

V_{micro} has units of ml g^{-1} and ρ_{Hg} is the measured density at a mercury intrusion pressure of 414 MPa (60000 psig), as this density includes the volume of pores <3 nm, in addition to that of the carbon substrate. In the section referred to above, we assumed a value for ρ_c, the density of the substrate, of 2.2 g ml^{-1} but, in principle, it should be possible to determine this by helium pycnometric measurements. In practice, the helium density may well be in error in highly microporous samples because of adsorption of helium; moreover, no great error is introduced by assuming V_c fixed at 0.45 ml.

Micropore volume is conventionally measured by the adsorption of gases at low temperatures and applying various analyses to the resulting isotherms.[19,37,38] Table 11.5 shows the values obtained by applying the different methods to the isotherms of N_2 at 77 K for a number of different carbons,[24] from which it is clear that the three techniques of isotherm analysis can give similar values, provided that the micropore volumes are not too large (<0.6 ml g^{-1}).

On the other hand, values obtained by the density method described above are consistently low. A similar discrepancy has already been commented on by Harrison *et al.*[39] for highly porous powdery carbons. The discrepancy may perhaps be accounted for by contributions by pores of >3 nm to the adsorption of N_2 at higher values of p/p_s. Use of the α_s plot to analyse the isotherms may give some clearer indication that this is occurring.

It has already been suggested that the distribution of micropore size, as

Table 11.5 Values of micropore volumes (ml g^{-1}) for some selected carbons using different methods of determination

	Method			
Carbon	Density	Gurvitsch	Dubinin	Pierce
Calgon BPL	0.35	0.43	0.47	0.42
Calgon SGL	0.30	0.36	0.40	0.31
Norit R1	0.42	0.57	0.63	0.54
Sutcliffe Speakman 208A	0.33	0.38	0.42	0.35

opposed to the total micropore volume, is a key factor in the uptake of methane at ambient temperatures. Evidence has already been adduced to support this suggestion but a specific example may make the point clearer. Two carbons, Calgon BPL and Saran A (made by carbonizing 100% PVDC homopolymer) were found to have essentially the same micropore volumes (0.47 and 0.46 ml g^{-1}, equivalent to ~1000 $m^2 g^{-1}$) as determined from the DR isotherm. However, the methane uptakes at 25°C and 3.4 MPa were 0.075 and 0.098 g^{-1}.

Saran A carbon has been shown[40] to have very uniform slit-shaped pores of width 1–2 nm and these pores would seem to be more suitable for adsorption of methane than the, possibly, wider pores of BPL. Accordingly, the relationship shown in Figure 11.1 may represent an over-simplification and a more consistently predictive relationship would have to weight the micropore volume with some sort of distribution function.

An even more important influence of micropore size may be on the reversible deliverability of methane under cyclic conditions. In all cases, this is lower (and sometimes much lower) than the initial uptake, but this aspect is discussed later.

Mercury porosimetry

The vital importance of density in determining the performance of carbons for storing ANG has already been pointed out in a previous discussion. The densities, as determined by intrusion of mercury, are particularly useful predictive parameters.

The topic of mercury porosimetry does not need to be discussed at length here. Any porosimeter is likely to be usable, but one which has an intrusion pressure up to 414 MPa (60000 psig) can force mercury into pores down almost to the micropore range. Intrusion and extrusion curves are, invariably, found to be non-coincident. The pressures used are such as, possibly, to break down the pore walls, so this needs to be borne in mind in detailed interpretation of the results.

Determination of micropore volume by difference, using the density of the carbon at 414 MPa (60000 psig), may perhaps have to be regarded with some caution with highly microporous carbons. The intrusion volume may still be increasing steeply at the highest intrusion pressure and, hence, be critically dependent on it. This is not unusual for microporous materials and may be attributable, in part, to an overlap between micro- and mesopores. It does mean that the values for micropore volume derived from density measurements could have some uncertainty attached to them. In such cases, it may be best to use the value obtained from N_2 adsorption.

High-pressure isotherms

Adsorption isotherms are normally obtained either gravimetrically, by direct measurement of the weight increase of a suspended adsorbent as the partial pressure of the adsorbate is gradually varied or, volumetrically, by adding a dose of gas from a known volume. Both methods have been successfully used for high-pressure isotherms of methane/natural gas, although the greater convenience of gravimetry and the availability of commercial instruments (for example, the Sartorius range) have tended to favour this, although both techniques have their drawbacks.

For the gravimetric method, a buoyancy correction must be made, although most adsorption isotherms are measured at low pressure where correction for buoyancy or deviation from ideality is small. This requires a knowledge of the value for the density of the carbon and of the equation of state, to correct for the deviation of methane from ideality, for example, the virial equation of Douslin *et al.*[41] The technique of Specovius and Findenegg,[42] where a non-porous counterbalance of similar mass and density to the adsorbent sample is used, helps to keep the buoyancy correction to a minimum.

A high-pressure volumetric apparatus, similar to that described by Masters and Gesser,[43] can be used to determine the isotherm, providing that appropriate corrections are made. The dead space (void) volume is usually calibrated by using helium, but carbons of high surface area adsorb it to some extent, although this can be reduced by using a higher temperature for calibration. Correction for the non-ideality of methane has to be made just as for the gravimetric method. Perhaps the greatest source of error in the volumetric method is the cumulative error as each point on the isotherm is determined.

Nonetheless, when appropriate corrections are made, using either technique, a true adsorption isotherm is measured independently of which method is used. This is illustrated by Figure 11.4, which shows the 25 °C methane isotherm on a highly microporous carbon. The isotherm was measured both gravimetrically, using the Sartorius 4436 high-pressure microbalance, and volumetrically with a sample 50 times larger on the apparatus described by Barton *et al.*[16] The coincidence of the points on the isotherm is excellent and shows no significant deviation for either method; this has been shown to be the case for other carbons.

For the purpose of measuring the amount of gas stored in a vessel, the volumetric method is much more direct. In this case, it is not necessary to try to differentiate between the adsorbed gas and that remaining in the void and macropore volume, but merely to measure the total volume stored in the vessel. Errors in determining void volume, due to helium adsorption, are therefore

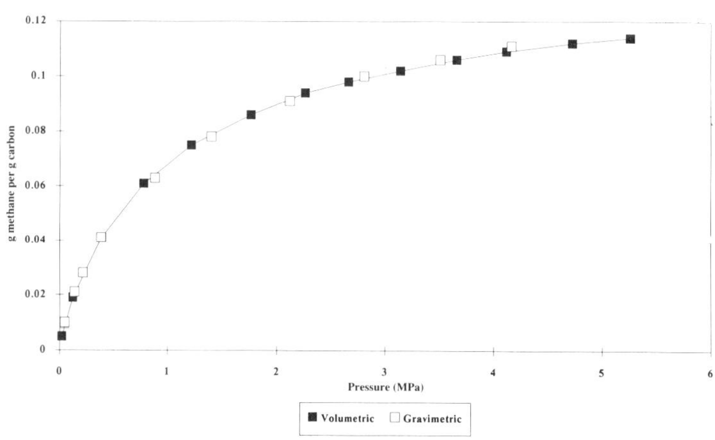

Fig. 11.4 Comparison of data points for a methane isotherm determined at 25 °C (■) volumetrically and (□) gravimetrically. Data from Quinn[24]

obviated. The only uncertainty in extrapolating such results to a larger scale lies in the assumption that the packing density of the carbon in the small experimental vessel is the same as would be obtained in filling a larger storage vessel with it.

Cyclic testing of carbons for ANG

The characterization methods described so far have used conventional techniques which are well developed and which use only a few grams of carbon, at most. They are therefore particularly suited to laboratory studies in the early stages of development of an adsorbent. When candidate carbons are identified for further study, they need to be tested on a larger scale under conditions which, at least partially, replicate those likely to be encountered in the field.

One apparatus which has proved particularly useful allows cyclic testing between 0.1 and 3.5 MPa (0 and 500 psig). With such an apparatus, the effects of various parameters on the deliverable, as opposed to the stored, volume of natural gas over a large number of cycles of operation, may easily be investigated. Not only methane but also a real natural gas may be used to determine the effect of repeated cycling on performance.

Such apparatus is in use at the laboratories of both authors. One of them is shown in Figures 11.5 and 11.6. It consists of a high-pressure stainless-steel vessel with an internal volume of either 0.5 or 1.0 l, filled with the carbon under test. This seems to be the minimum size that relates directly with that of storage vessels to be used, for example, in the transport field. The container is attached to a manifold through which natural gas may be admitted or vented. The amount

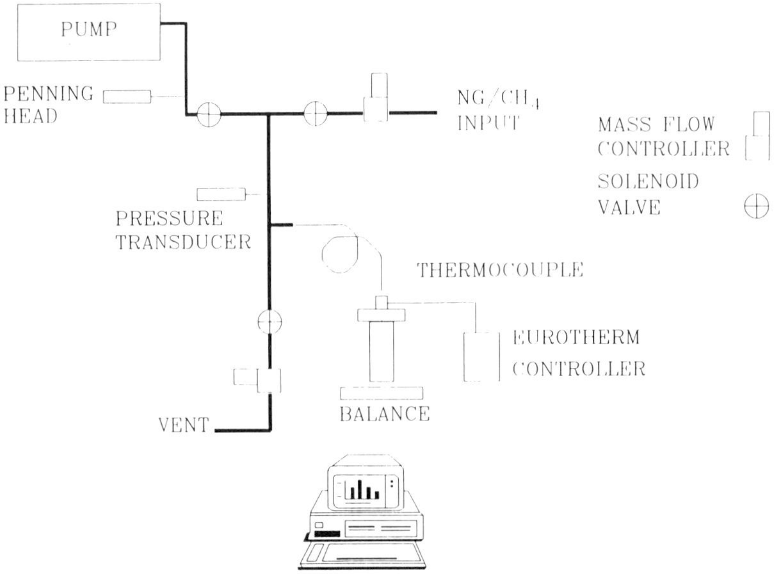

Fig. 11.5 Diagram of components of an apparatus for cyclic charging and discharging of a carbon with natural gas

Fig. 11.6 Photograph of cycling test apparatus. Components as labelled in Figure 11.5

of gas in the storage vessel, both in the adsorbed state and in the gaseous state in inter-particle voids and macropores, is measured directly by placing the vessel on the pan of an electronic gravimetric balance.

This latter is not a critical item; any suitable commercial balance with an adequate load capacity and sufficient resolution of weight changes will suffice. In this particular instance, a Salter Model EP/12KA balance of capacity 12 kg and resolution 0.1 g has worked satisfactorily.

Pressure is conveniently measured by transducers and the gas flow, in and out of the vessel, by mass flow controllers. It is useful to be able to measure heat changes during adsorption and desorption; a thermocouple with the active junction in the middle of the bed is used for this.

All the data may be collected at the interface of a suitable microcomputer which can, in turn, control the admission and venting of natural gas so that cyclic testing is fully automatic. The system can be used to measure the effect of packing density on the relative reversible storage capacity of different carbons, temperature excursions during cycling and the long-term stability of the adsorbent system.

Figure 11.7 shows the cyclic charge and discharge of methane on a typical carbon. The regularity and reversibility of each cycle are particularly notable. The other important point to note is that the first uptake is significantly larger than any thereafter. The pressure was cycled between 3.5 and 0.1 MPa (35 and 1 bar) and this residual gas represents a sort of 'cushion gas' that is not normally accessible to cyclic operation. It illustrates well the difference between adsorbed storage volume and deliverable volume.

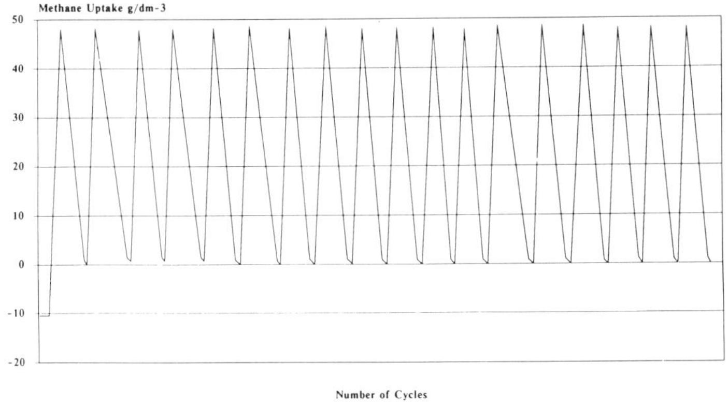

Fig. 11.7 Typical cyclic results for alternate charge and discharge of methane between 0.1 and 3.5 MPa (1 and 35 bar) obtained using the apparatus shown in Figures 11.5 and 11.6[48]

11.5 Practical application of carbons for ANG

Choice of suitable adsorbents

The properties of carbon for potential use with ANG have been very fully described in the previous discussion. It is clear that material with the highest space-filling potential, strong and non-friable, and with a high uptake of methane is required. These first two parameters are still in the process of development by carbon manufacturers and by end-users in the transport field but, as already mentioned, carbons having a methane uptake of 170 v/v have already been described,[36] based on selective activation of a Saran char. Process studies suggest that carbons with such high capacity would be of great interest in a number of applications including vehicle fuel storage, although, currently, the deliverable capacity of materials generally available is only just over half of this value.

Maximizing the delivered volume

It has already been stressed that there is a shortfall, often considerable, between the amount of methane that is stored on a carbon and the amount that can be recovered by desorption down to atmospheric pressure. This shortfall exists as a result of the shape of the methane isotherm at ambient temperatures.

Figure 11.8 shows some typical experimental isotherms for carbons that have been proposed as adsorbents for natural gas. To clarify the argument, they have all been normalized by taking the uptake of methane at 3.4 MPa (500 psia) as having unit value in every case. Ideally, an adsorbent should show complete reversibility between zero pressure [0.1 MPa (1 bar) absolute] and the working pressure of 3.4 MPa (34 bar); above 3.4 MPa, further adsorption is of no practical benefit and the isotherm should, ideally, level off. CNG is, of course,

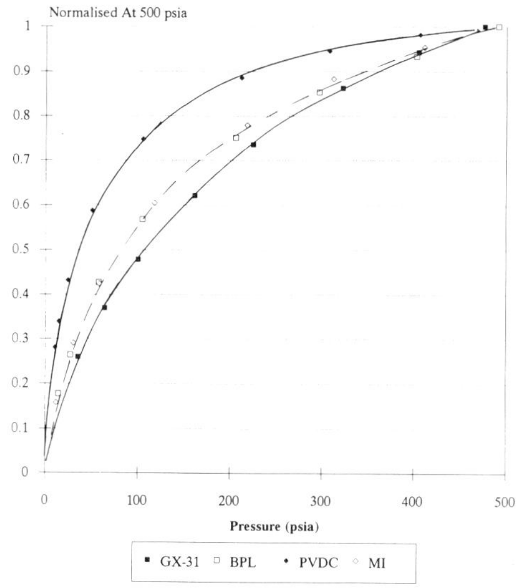

Fig. 11.8 Comparison of the shape of natural gas isotherms for some different carbons. The values for adsorption have been normalized by taking the value of 3.4 MPa as 1.0. (◆) PVDC char; (◇) MI (coconut shell); (□) BPL (coal-derived); (■) GX-31 (Amoco)

fully reversible from 0.1 to 3.4 MPa (1 to 34 bar) but the continued increase of stored gas above this pressure would, significantly, be of no practical use.

No adsorbent has completely ideal characteristics. Considering the important region near atmospheric first, we can see that some carbons, notably that derived from PVDC, have isotherms that increase very steeply with pressure and at the origin $dG/dp \rightarrow \infty$, where G is uptake. Bearing in mind the fact that p does not go below 0.1 MPa (1 bar) for a freely vented system, it is clear that there is a greater amount of 'cushion gas' that cannot be removed, except by evacuation to $<$0.1 MPa (1 bar) for such a carbon, compared with, say, GX 31, where dG/dp varies much less steeply.

For pressures $>$3.4 MPa, exactly the converse is true. Adsorption that increases above this pressure is of no practical use and, as a result, carbons with a flatter G/p characteristic, such as the PVDC carbon, are closer to the ideal. There is therefore a conflict between the requirement for high uptake at medium pressures, coupled with easy release on the one hand, and minimum adsorption at near 0.1 MPa (1 bar) pressure on the other.

Yet again, the distribution of micropore sizes controls this behaviour by determining the shape of the adsorption isotherm. A practical adsorbent may be chosen on the basis of a compromise, to optimize the ratio of stored-to-deliverable gas. Alternatively, consideration may need to be given to ways in which the 'cushion gas' can be evolved, either by evacuation or by heating.

Enhancement of storage by ANG

It is now appropriate to examine those circumstances under which ANG may be used, such that it has a large enough enhancement of storage of methane over

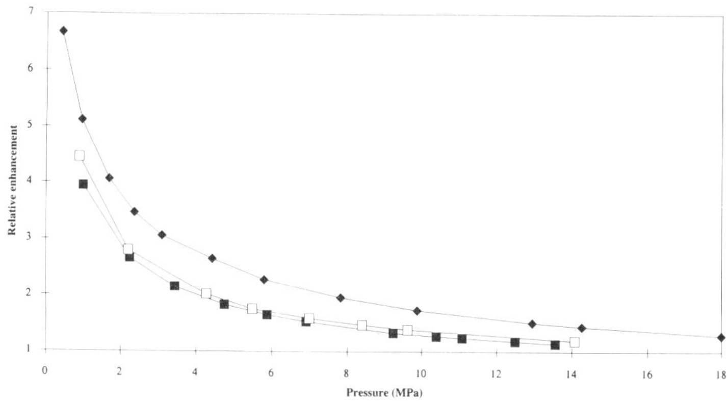

Fig. 11.9 Enhancement of natural gas storage relative to an equivalent empty vessel for three carbons: (◆) Amoco GX 32 (packing density 0.29 g ml^{-1}); (□) Norit R1 (packing density 0.40 g ml^{-1}); (■) Calgon BPL (packing density 0.44 g ml^{-1})

CNG to make it economically feasible. An operating pressure of 3.5 MPa has been taken as typical for the purposes of discussion so far, but practical advantages over CNG may well be obtained[44] at even higher pressures, such as 5.2 MPa (750 psig). Figure 11.9 plots the storage relative to that of an empty vessel of the same volumetric size, for three different carbons, as a function of increasing pressure. The data include methane stored in the gas phase, in voids and in macropores, and also in the adsorbed phase because this represents what would happen in a commercial installation.

The relative storage decreases sharply as pressure increases, representing the inverse of the adsorption isotherm. Thus, if one were constrained to using low pressures, say 0.7 MPa (7 bar), the enhancement over the storage in an empty vessel would be >5 for GX-32, although it must be borne in mind that much of that would be effectively non-recoverable for cyclic operation. At 3.5 MPa, the enhancement is ~3 for GX-32 and 2.2–2.3 for BPL and Norit R1. This represents storage of >100 and 77–80 v/v, respectively, and such values may be taken as typical of current practice.

Higher values are possible if the packing density can be increased. For example, at 3.4 MPa and 25 °C, shaped pieces of PVDC carbon stored 100 g CH_4 per litre of storage vessel. This corresponds to >153 v/v and an enhancement of 4.5 over the density in the gas phase at this pressure. These values indicate what is possible but, as yet, only through careful laboratory preparation of individual pieces. Current research is being directed at realizing such a performance by carbons made on a commercial scale at an economic price.

Effects of adsorption enthalpy

Methane has a heat of adsorption on to carbon, where there is a non-specific interaction, of 15–18 kJ mol^{-1}. This is normally of little consequence in adsorption experiments, as the sample quickly equilibrates thermally with its surroundings. On the scale of application envisaged commercially for ANG, the enthalpy

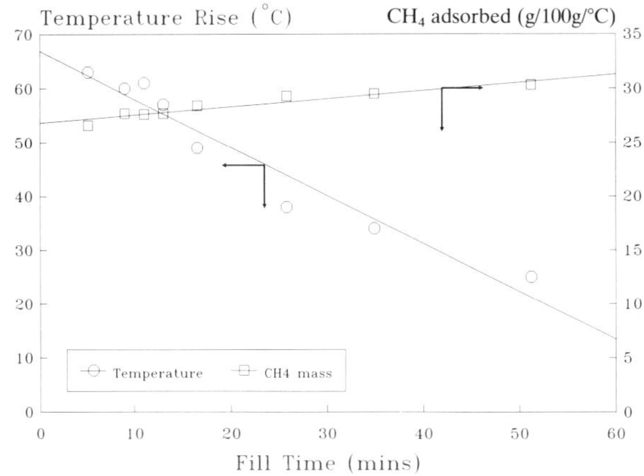

Fig. 11.10 Effect of filling time on (○) temperature of bed and (□) gravimetric uptake of methane at 3.5 MPa (35 bar). Ambient temperature 21 °C

of adsorption and desorption causes temperature excursions that may cause considerable operational problems.

Practical measurements show that the exotherm on charging an ANG vessel is around 70 °C.[44–48] The actual figure will depend on the adsorption capacity of the carbon and its packing density, in addition to the heat transfer properties of the carbon-containing vessel–ambient surroundings. Calculations[48] using the heat of adsorption and the specific heats of carbon and methane suggest that, under adiabatic conditions, a carbon adsorbing 80 v/v of methane would show a temperature rise of 72 °C. It should be noted that the temperature rise increases more or less pro rata with the volumetric adsorption. A carbon of similar density with a capacity 150 v/v would show a rise of ~140 °C.

A major factor influencing the maximum exotherm is the rate of fill, because the heat release is instantaneous on adsorption but the removal of heat is limited by the heat transfer properties already mentioned. Figure 11.10 represents the effect of fill time both of the magnitude of the exotherm and on the uptake of methane in a 0.5 l vessel filled to 3.5 MPa. The maximum temperature is measured in the centre of the carbon bed which increases in a more or less linear fashion with decreasing fill time. Extrapolation of the line to zero fill time gives an adiabatic temperature rise of 70 °C, in excellent agreement with the calculated value of 72 °C given above. The decrease in capacity with decreasing fill time arises because the uptake of methane is lower at higher temperatures. Nonetheless, the decrease in performance is only about 10% over the range of fill times investigated, rather than the nearly 50% predicted, from examination of the respective isotherms at 25 and 95 °C from other work.[46]

This relatively small decrease in capacity has been remarked upon by other workers[36] and is, presumably, related to the relatively small scale of the experiments. The storage vessel has by far the greatest thermal mass of any of the system components and has a relatively large surface area exposed to the ambient air.

On desorption, a strong endotherm develops. From an ambient temperature of 25 °C, a minimum of nearly −40 °C has been measured[47] for a rapid discharge. This endotherm will be, to an extent, self-regulating as the rate of desorption of gas will fall to zero if the temperature becomes low enough.

Consequence of temperature changes on operations for ANG

The temperature changes described above have to be taken into account for any practical ANG system. For example, for transport applications, some consideration should be given to the thermal properties of any storage cylinder used. A high surface area-to-volume ratio should be aimed at, with possibly some finning to help heat transfer to the surrounding air. In the absence of published data on the rate of heat transfer in model systems, it is difficult to say much more, but it is likely that the limiting step may be the gas film diffusion on the inside of the vessel, in which case it may be difficult to overcome it. A reliable, predictive model for heat transfer across an ANG vessel is, therefore, seen as a highly desirable research objective.

For some rates of fill, such as overnight in a garage attached to domestic dwellings, the heat evolution is easily dissipated and represents little problem and the maximum capacity for storage is obtained. For fast fills, of 15 mins or less, there are problems as an appreciable loss of maximum capacity occurs. This may have to be accepted but may be of little consequence for 'topping-up' a partly full ANG tank.

An interesting approach to overcome the problem has been explored by workers at the Institute of Gas Technology (IGT), Chicago.[47] They included within the ANG vessel some material (disodium hydrogen phosphate was typical) that underwent a phase change at near-ambient temperature, such that heat was absorbed or evolved isothermally. This thermal energy storage (TES) system was claimed to give near-isothermal operation on both the fill and empty cycles. The commercial feasibility of the system will depend on the cost of the TES material and whether the displacement of active carbon adsorbent by it is acceptable. Work is continuing on this concept.

For large-scale applications of ANG to, say, gas storage for diurnal load variations, the problem of the heat management becomes more critical. On the basis that a slow overnight fill is usual, it may be possible to dissipate the heat of adsorption in existing vessels. If these are enclosed, then some heat-transfer plant must be installed to cool the bed by gas recirculation. Process studies on a conceptual system have suggested that the heat transfer facilities could form an appreciable proportion of the installed costs.[50]

Use of real natural gases for ANG

So far, we have assumed natural gas and methane to be synonymous but, as Table 11.1 shows, there are appreciable quantities of minor compounds in natural gas that may affect the performance of an ANG system. Hydrocarbons above C_4 may cause problems by irreversible adsorption on extended operation. Golovoy and Blais[45b] found that after 100 cycles of operation on natural gas there was a 20–30% loss of adsorption capacity, although the loss depended on the logarithm of the number of cycles and was therefore still acceptable after 1000 cycles. For pure methane, on the other hand, Barton *et al.*[51] found no fall-off in capacity at all over 100 cycles. Repeating the experiments with a number of different carbons on Canadian natural gas[36] replicated the results of Golovoy and Blais, although a levelling off of storage capacity was found after 40 cycles, with no further deterioration detected.

These results suggested that the higher hydrocarbons may be adsorbed effectively irreversibly and reduce the adsorption of methane. In fact, the near-irreversibility of adsorption of nonane is the basis of a method introduced by Gregg and Langford[52] for analysing the influence of microporosity. Golovoy and Blais[45b] produced strong evidence for the non-reversibility of adsorption of hydrocarbons by thermally desorbing and analysing the residual products from carbons subjected to extended use. Alkanes from C_4 to C_9, alkyl-substituted C_5–C_6 cycloalkanes, benzene and toluene were all detected. On the other hand, C_2 and C_3 products were not, suggesting that all the major hydrocarbon components, comprising >99.7% of the total, are reversibly stored, an encouraging result.

Some consideration should possibly be given to the influence of CO_2 and H_2O by preferential adsorption on the adsorbent. Table 11.1 gives a range of concentrations of CO_2 in natural gas from 0.04 to 1.00% molar and of water from 75–180 ppm (v/v). At 3.5 MPa (35 bar) total pressure, these correspond to partial pressures of 1.4–3.5 kPa (0.014–0.035 bar) and 0.26–0.63 kPa (2.6–6.3 mbar), respectively; at 25 °C, the corresponding p/p_s values are 0.0002–0.0005 and 0.08–0.20.

At these p/p_s values, CO_2 is unlikely to interfere with the adsorption of methane. Further, the miscibility of liquid CO_2 and CH_4 suggests that this will carry over to the adsorbed phase and that adsorbed CO_2 will desorb with CH_4. Only for natural gases with CO_2 content substantially above 1% will much account have to be taken of this aspect. Golovoy and Blais[45b] did not comment specifically on CO_2 being desorbed from the adsorbed residue, although it is not clear whether the natural gas used in their study contained CO_2 or not. The evidence currently available suggests that CO_2 presents no serious problem.

The water content of natural gas may, however, be on the threshold of being a nuisance. Kenny and Sing[53] investigated the adsorption of water vapour on carbons and found very variable behaviour. For hydrophobic carbons, which include those derived from Saran or from coconut shell without added activators, there is little uptake at low pressures, but a sudden burst of adsorption for p/p_s >0.3 occurs. Evans[54] reported similar effects.

Again, it seems that the appropriate choice of carbon may avoid problems. Some carbons which contain cationic residues from their manufacturing process may show much greater sensitivity towards water adsorption.

For completeness, we may refer briefly to the fate of sulphur-containing compounds, present naturally or added as stenching agents at the ppm level. These will be irreversibly adsorbed by almost all adsorbents, but their low concentration should not lead to significant loss of adsorption capacity.

To avoid the loss of adsorption capacity that does occur with repeated use of natural gas, as opposed to methane, it is entirely feasible to use a 'guard' adsorption bed which selectively adsorbs all the components that are held irreversibly at ambient temperatures. The Atlanta Gas Light Adsorbent Research Group[55] has used a small guard bed on a cyclic test unit, capable of storing 500 ft^3 (14.15 m^3) through adsorption at 2.1 MPa (300 psi). This bed was shown to be effective in adsorbing all other components of natural gas, with the exception of ethane. The guard bed could be heated during the desorption part of the cycle. After 1500 fill–empty cycles using natural gas, a 3% deterioration in storage capacity was observed. Such a system with a guard bed has been used on Chevy S10 trucks powered by ANG, for some years, with no operational problems. The guard bed is regenerated, when required, by heating it during running and the desorbed components are fed into the inlet manifold. Thus, the adsorbent in the main storage vessel is protected from adsorbates which are difficult to desorb and which gradually reduce the effective storage capacity of the adsorbent. Without such a guard bed, reductions of up to 29% in storage capacity were observed over 1500 cycles.

Such a system is clearly easy to apply to any ANG system where the maximum capacity for reversible storage must always be maintained. It has the merit that none of the adsorbed material need be lost for use, small though this may be. A carbon of wider pore size than in the primary vessel may be used, so that the higher hydrocarbons are adsorbed, but not methane.

Kinetic effects in the adsorption process

The adsorption and desorption processes in physical adsorption are, normally, fairly rapid and would not be expected to be a serious problem in ANG applications. On the other hand, Koresh *et al.*[58] have recently reported the failure of a carbon cloth (TCM 128) to equilibrate methane uptake at 308 K after many months. This behaviour seems to be a peculiarity of this particular material. Although surface areas were not explicitly stated, the micropore volume is ~0.12 ml g^{-1} (data taken from Figure 12 of earlier work.)[59] On the other hand, the uptake of methane at 1.5 MPa (15 bar) (probably not equilibrated) was ~1.4 g of carbon, about half the amount that our correlation in Figures 11.1 and 11.2 would suggest.

This points to a remarkable diffusion restriction in this carbon because of its extremely narrow ultramicropores, a point elaborated further by Koresh *et al.*[58] It also underlines the vital importance of having not only a high micropore volume but also one where extremely small pores are avoided.

In practice, using the much more highly-activated carbons that are necessary to obtain high micropore volumes does not, apparently, lead to any kinetic problems, either on adsorption or desorption. Our own work[24,48] showed no significant failure to reach equilibrium rapidly and other workers have reported the same experience.[8,13] The thermal effects during sorption are a far more serious practical problem.

11.6 Concluding remarks

This review has concentrated on the basic physical chemistry of evaluating carbons for use with ANG. The barriers to further progress have been identified but they are, gradually, being overcome. Best current practice gives a reversible storage on a kilogram scale of natural gas of 80–100 v/v at 3.5 MPa and 25°C. Further improvement would seem to depend more on producing carbons of high packing density, especially by making shaped pieces, rather than making carbons of greater gravimetric adsorption capacity than at present.

Nearly all the studies published so far on ANG applications have focused on the transport aspect, which is clearly seen as vitally important to future provision of vehicles of low environmental pollution. CNG holds the current lead as a means of using natural gas as a fuel, and most cost studies have concentrated on identifying a break-even point where ANG proves cheaper than CNG. For example, Golovoy and Blais (Ford Motor Company)[45b] suggested that 7 MPa represented a compromise pressure where increased adsorption gave a conceptual road vehicle a driving range with the target of 160 km. Above this pressure, the advantage of ANG over CNG became too small to be worth pursuing, although it has to be said that the carbon they used had a storage capacity of only 80 v/v at 3.5 MPa.

A feasible target to aim at for a commercial development of space-filling shapes of high adsorbent capacity carbon is ~150 v/v at 3.5 MPa, equivalent to storage at a CNG pressure of ~15 MPa. This is coming close to the current CNG pressures of 20 MPa and, if achieved, would give a natural gas vehicle (NGV) a range of only 25% less than that for CNG cylinders but at a markedly lower compression cost.

11.7 Acknowledgements

The authors thank their various colleagues who provided experimental data or other help with this review. N.D.P. expresses thanks to British Gas plc and D.F.Q. to the Atlanta Gas Light Adsorbent Research Group (AGLARG) for permission to publish this chapter.

11.8 References

1. *BP Review of World Gas*, 1987: BP Gas International, London.
2. PARKYNS, N.D. 1988: in *Energy Production Processes*. Rugby: Institution of Chemical Engineers, p. 35.
3. DIGNAM, M.J. 1982: *Study of the feasibility of developing new adsorbent materials for compressed natural gas*, Report 33259 to Ontario Ministry of Transportation. Toronto.
4. SAGE, B.H., REAMER, H.H., OLDS, R.H. and LACEY, W.N. 1942: *Industrial and Engineering Chemistry* **34**, 1108.
5. OLDS, R.H. 1953: in Farkas, A. (ed.) *Critical Behaviour of Hydrocarbons*. New York: Academic Press, Vol. 2.
6. MELVIN, A. 1988: *Natural Gas, Basic Science and Technology*. Bristol: Hilger.
7. HAGAN, M. 1962: *Clathrate Inclusion Compounds*. New York: Reinhold.
8. INNES, R.A., LUTINSKI, F.E., OCCELLI, M.L. and KENNEDY, J.V. 1984: *Report AC 01-84CE50071*. Washington, DC: US Department of Energy.
9. TALU, O. 1990: *Adsorption News* **1**, 3. Cincinnatti, OH: G-Tec Co.
10. FINDENEGG, G.H. 1983: in Belfort, G. and Myers, A. *Fundamentals of Adsorption*. Proceedings of Engineering Foundation Conference, New York, Engineering Foundation, p. 207.
11. BRUNAUER, S. 1944: *Physical Adsorption of Gases and Vapours*. Oxford: Oxford University Press, 1st edn.
12. McBAIN, J.W. and BRITTON, G.T. 1930: *Journal of the American Chemical Society* **52**, 2198.
13. OTTO, K. 1982: *Alternative Energy Sources IV*. Ann Arbor, MI: Ann Arbor Science, Vol. 6, p. 241.

14. (a) KOBLE, R.A. and CORRIGAN, T.E. 1952: *Industrial Engineering Chemistry* **42**, 383; (b) RAY, G.C. and BOX, E.O. 1950: *Industrial Engineering Chemistry* **42**, 1315.
15. MENON, P.G. *Chemical Reviews* **68** (1968) 277; *Journal of Physical Chemistry* **72** (1968) 2695.
16. BARTON, S.S., DACEY, J.R. and QUINN, D.F. 1983: in Belfort, G. and Myers, A. *Fundamentals of Adsorption, 1st Engineering Foundations·Conference*, New York, Engineering Foundation, p. 65.
17. GREGG, S.J. and SING, K.S.W. 1982: *Adsorption, Surface Area and Porosity*. London: Academic Press, 2nd edn.
18. EVERETT, D.H. and POWL, J.C. 1976: *Journal of the Chemical Society, Faraday Transactions 1* **72**, 619.
19. DUBININ, M.M. 1955: *Quarterly Review* **9**, 101, and 1975: *Progress in Surface and Membrane Science* **9** 1.
20. AGARWAL, R.K. and SCHWARTZ, J.A. 1988: *Carbon* **26**, 873.
21. FOMKIN, A.A., CHKHAIDZE, E.V., SERPINSKII, V.V., TSITSISHVILI, G.V. and DUBININ, M.M. 1989: *Izvestiya Akademii Nauk SSSR, Seriya Khimicheskaya* 858 (English translation).
22. SCHRØDER PEDERSEN, A. and LARSEN, B. 1989: *Report Risø-M-2781*. Denmark: Risø National Laboratory.
23. (a) OZAWA, S., KUSUMI, S. and OGINO, Y. 1976: *Journal of Colloid Interface Science* **56**, 83; (b) WAKASUGI, Y., OSAWA, S. and OGINO, Y. 1981: *Journal of Colloid Interface Science* **79**, 399.
24. QUINN, D.F., unpublished results.
25. ZHANG, S.-Y., HAYHURST, D.T. and TALU, O. 1989: Paper presented to AIChemE, San Francisco.
26. CARROTT, P.J.M. and SING, K.S.W. 1989: in Unger, K.K., Rouquerol, J., Sing, K.S.W. and Kral, H. (eds) *Characterisation of Porous Solids*. Amsterdam: Elsevier, p. 77.
27. MARSH, H. and SIEMIENIEWSKA, T. 1965: *Fuel* **44**, 355.
28. CARROTT, P.J.M., ROBERTS, R.A. and SING, K.S.W. 1989: in Unger, K.K., Rouquerol, J., Sing, K.S.W. and Kral, H. (eds) *Characterisation of Porous Solids*. Amsterdam: Elsevier, p. 89.
29. ROOTARE, H.M. and PRENSLOW, C.F. 1967: *Journal of Physical Chemistry* **71**, 2733.
30. SING, K.S.W., EVERETT, D.H., HAUL, R.A.W., PIEROTTI, R.A., ROUQUEROL, J. and SIEMIENEWSKA, T. 1985: *Pure and Applied Chemistry* **57**, 603.
31. BARTON, S.S., HOLLAND, J.A. and QUINN, D.F. 1986: in Liapis, A.I. (ed.) *Fundamentals of Adsorption, 2nd Engineering Foundations Conference*. New York: Engineering Foundation, p. 99.
32. BUCZEK, B. and CZEPIRSKI, L. 1987: *Adsorption Science and Technology* **4**, 217.
33. DIETZ, V.R., LAKCHMANAN, J.L., SELWYN, S. and RESING, J. 1983: *Report 5074*. Washington, DC: US Naval Research Laboratory.
34. DIETZ, V.R. 1978: *Publications of Technical Papers and Proceedings of the Annual Meeting – Sugar Industry Technologists* **37**, 473.
35. *Method ASTM D 3802:79* 1979: Philadelphia: American Society for Testing and Materials.
36. QUINN, D.F. 1988: Unpublished report to AGLARG.
37. GURVITSCH, L. 1915: *Zhurnal – Russkoe Fiziko-Khimicheskoe Obshchestvo* **47**, 805.
38. PIERCE, C. 1968: *Journal of Physical Chemistry* **72**, 3763.
39. HARRISON, B.H., BARTON, S.S., DACEY, J.R. and SELLORS, J. 1979: *Journal of Colloid Interface Science* **71**, 367.
40. DACEY, J.R. and THOMAS, D.G. 1954: *Transactions of the Faraday Society* **50**, 740.

41. DOUSLIN, D.R., HARRISON, R.H., MOORE, R.T. and McCULLOUGH, J.P. 1964: *Journal of Chemical Engineering Data* **9**, 358.
42. SPECOVIUS, J. and FINDENEGG, G.H. 1978: *Berichte Bunsengesellschaft für Physikalische Chemie* **82**, 174.
43. MASTERS, K.J. and GESSER, H.D. 1981: *Journal of Physics E: Scientific Instruments* **14**, 1043.
44. GOLOVOY, A. 1983: in *SAE Conference Proceedings.* Warrendale, PA15096, p. 129.
45. (a) BRASLAW, J., NASEA, J. and GOLOVOY, A. 1982: *Alternative Energy Sources IV.* Ann Arbor, MI: Ann Arbor Science, Vol. 6, p. 261; (b) GOLOVOY, A. and BLAIS, E.J. 1983: in *Alternate Fuels for Special Ignition Engines, SP-559 SAE Conference Proceedings.* Warrendale, PA 15096, p. 47.
46. (a) REMICK, R.J. and TILLER, A.J. 1986: in Smith, M. (ed.) *Proceedings of Gaseous Fuels for Transportation Conference, Vancouver, B.C.* p. 479; (b) REMICK, R.J., ELKINS, R.H., CAMARA, R.H. and BULICZ, R. 1984: *Report 0327-1.* Washington, DC: US Department of Energy and NASA.
47. JASIONOWSKI, W.J., TILLER, A.J., FATA, J.A., ARNOLD, J.M., GAUTHIER, S.W. and SHIKARI, Y.A. 1989: Paper presented at International Gas Research Conference, Tokyo.
48. PEARSON, S., British Gas, unpublished results.
49. CHAUDRON, G. 1989: Report to International Gas Union, Natural Gas Vehicle Task Force.
50. British Gas, Midlands Research Station, unpublished results.
51. BARTON, S.S., HOLLAND, J.A. and QUINN, D.F. 1985: *The Development of Adsorbent Carbon for Storage of Compressed Natural Gas, Report AF-85-01.* Toronto: Ontario Ministry of Transportation.
52. GREGG, S.J. and LANGFORD, J.F. 1969: *Transactions of the Faraday Society* **65**, 1394.
53. KENNY, M.B. and SING, K.S.W. 1990: *Abstracts of 'Carbone 90', Paris.*
54. EVANS, M.J.B. 1987: *Carbon* **25**, 81.
55. Atlanta Gas Light Co., Atlanta, GA: unpublished results.
56. REICH, R., SIEGLER, W.T. and ROGERS, K.A. 1980: *Industrial Engineering Chemistry. Process Design Development* **19**, 336.
57. STRISNA, F., COLEMAN, R.E. and NOTARO, F. 1985: *Enhancement and Automotive CNG Storage via Adsorbents. NYSERDA Report 85-11.* Albany, NY.
58. KORESH, J.E., KIM, T.-H., WALKER, D.R.B. and KOROS, W.J. 1990: *Journal of the Chemical Society, Faraday Transactions 1* **86**, 2267.
59. KORESH, J.E. and SOFFER, A. 1980: *Journal of the Chemical Society, Faraday Transactions 1* **76**, 2457.

Index